Electric Vehicle Integration in a Smart Microgrid Environment

Electric Vehicle Integration in a Smart Microgrid Environment

Edited by
Mohammad Saad Alam
Mahesh Krishnamurthy

CRC Press
Taylor & Francis Group
Boca Raton London New York

CRC Press is an imprint of the
Taylor & Francis Group, an **informa** business

First edition published 2021
by CRC Press
6000 Broken Sound Parkway NW, Suite 300, Boca Raton, FL 33487-2742

and by CRC Press
2 Park Square, Milton Park, Abingdon, Oxon, OX14 4RN

© 2021 selection and editorial matter, Mohammad Saad Alam and Mahesh Krishnamurthy individual chapters, the contributors

CRC Press is an imprint of Taylor & Francis Group, LLC

MATLAB® and Simulink® are trademarks of The MathWorks, Inc. and are used with permission. The MathWorks does not warrant the accuracy of the text or exercises in this book. This book's use or discussion of MATLAB® and Simulink® software or related products does not constitute endorsement or sponsorship by The MathWorks of a particular pedagogical approach or particular use of the MATLAB® and Simulink® software.

The right of Mohammad Saad Alam and Mahesh Krishnamurthy to be identified as the author[/s] of the editorial material, and of the authors for their individual chapters, has been asserted in accordance with sections 77 and 78 of the Copyright, Designs and Patents Act 1988.

Reasonable efforts have been made to publish reliable data and information, but the author and publisher cannot assume responsibility for the validity of all materials or the consequences of their use. The authors and publishers have attempted to trace the copyright holders of all materials reproduced in this publication and apologize to copyright holders if permission to publish in this form has not been obtained. If any copyright material has not been acknowledged, please write and let us know so we may rectify in any future reprint.

Except as permitted under U.S. Copyright Law, no part of this book may be reprinted, reproduced, transmitted, or utilized in any form by any electronic, mechanical, or other means, now known or hereafter invented, including photocopying, microfilming, and recording, or in any information storage or retrieval system, without written permission from the publishers.

For permission to photocopy or use material electronically from this work, access www.copyright.com or contact the Copyright Clearance Center, Inc. (CCC), 222 Rosewood Drive, Danvers, MA 01923, 978-750-8400. For works that are not available on CCC, please contact mpkbookspermissions@tandf.co.uk

Trademark notice: Product or corporate names may be trademarks or registered trademarks and are used only for identification and explanation without intent to infringe.

ISBN: 978-0-367-42391-9 (hbk)
ISBN: 978-1-032-01050-2 (pbk)
ISBN: 978-0-367-42392-6 (ebk)

Typeset in Times
by codeMantra

Contents

Foreword vii
Preface... ix
Editors xiii
Contributors xv

Chapter 1 Trends in Electric Vehicles, Distribution Systems, EV Charging Infrastructure, and Microgrids 1

April Bolduc

Chapter 2 Fog Computing for Smart Grids: Challenges and Solutions 7

Linna Ruan, Shaoyong Guo, Xuesong Qiu, and Rajkumar Buyya

Chapter 3 Opportunities and Challenges in Electric Vehicle Fleet Charging Management 33

Chu Sun, Syed Qaseem Ali, and Geza Joos

Chapter 4 Challenges to Build a EV Friendly Ecosystem: Brazilian Benchmark 73

Ana Carolina Rodrigues Teixeira

Chapter 5 Coordinated Operation of Electric Vehicle Charging and Renewable Power Generation Integrated in a Microgrid 97

Alberto Borghetti, Fabio Napolitano, Camilo Orozco Corredor, and Fabio Tossani

Chapter 6 Energy Storage Sizing for Plug-in Electric Vehicle Charging Stations 119

I. Safak Bayram, Ryan Sims, Edward Corr, Stuart Galloway, and Graeme Burt

Chapter 7 Innovative Methods for State of the Charge Estimation for EV Battery Management Systems 143

Zeeshan Ahmad Khan and Franz Kreupl

Chapter 8 High-Voltage Battery Life Cycle Analysis with Repurposing in Energy Storage Systems (ESS) for Electric Vehicles....................... 181

Mamdouh Ahmed Ezzeldin, Ahmed Alaa-eldin Hafez, Mohamed Adel Kohif, Marim Salah Faroun, and Hossam Hassan Ammar

Chapter 9 Charging Infrastructure for Electric Taxi Fleets 213

Chandana Sasidharan, Anirudh Ray, and Shyamasis Das

Chapter 10 Machine Learning-Based Day-Ahead Market Energy Usage Bidding for Smart Microgrids...249

Mohd Saqib, Sanjeev Anand Sahu, Mohd Sakib, and Esaam A. Al Ammar

Chapter 11 Smart Microgrid-Integrated EV Wireless Charging Station 267

Aqueel Ahmad, Yasser Rafat, Samir M. Shariff, and Rakan Chabaan

Chapter 12 Shielding Techniques of IPT System for Electric Vehicles' Stationary Charging 279

Ahmed A. S. Mohamed and Ahmed A. Shaier

Chapter 13 Economic Placement of EV Charging Stations within Urban Areas 295

Ahmed Ibrahim AbdelAzim

Chapter 14 Environmental Impact of the Recycling and Disposal of EV Batteries 313

Zeeshan Ahmad Arfeen, Rabia Hassan, Mehreen Kausar Azam, and Md Pauzi Abdullah

Chapter 15 Design and Operation of a Low-Cost Microgrid-Integrated EV for Developing Countries: A Case Study 335

Syed Muhammad Amrr, Mahdi Shafaati Shemami, Hanan K. M. Irfan, and M. S. Jamil Asghar

Index... 359

Foreword

It is an honor for me to add my comments to a very important book by Dr. Mohammad Saad Alam and Dr. Mahesh Krishnamurthy, *Electric Vehicle Integration in Smart Microgrid Environment*.

Transport electrification, as a response to the global decarbonization goal, has advanced significantly over the decade. The IEA, in its Global Electric Vehicle (EV) Outlook 2020, projects an increase in the global EV stock (excluding two/three wheelers) from around 8 million in 2019 to 50 million by 2025 and close to 140 million vehicles by 2030, constituting about 7% of the global vehicle fleet by 2030. The smooth and successful adoption of projected EVs will require an ecosystem that facilitates the efficient operation of EVs. However, this large-scale integration of EVs with the grid comes with its set of technical and economic challenges. The high penetration of EVs can affect the power quality of the distribution network. Challenges facing the utilities such as power quality issues, proper planning of charging stations, interoperability of charging infrastructure, peak power demand, battery management, etc. are important issues to be addressed.

In this edited book, the authors have examined the current energy scenario for microgrids and have discussed the challenges and opportunities due to the increasing penetration of distributed power generation systems and EVs into the microgrids. Through the course of 15 chapters, the various aspects of EV integration with the grid are presented – current trends in EV and their charging stations, the opportunities and challenges, EV-connected microgrid planning, power market operation and planning, innovative methods for EV battery management, and economic and environmental impacts. This book is a timely endeavor as it has captured the state-of-the-art technologies and trends in smart microgrid management with EV integration, EV charging infrastructure technologies, smart charging, and deployment.

I believe, the contents of this book will expose the readers to subjects that could potentially alter the paradigm for integration of transportation electrification to microgrids and emerging EV charging infrastructure technologies and would serve as a valuable tool and guide to a wide spectrum of stakeholders such as power system architects, practitioners, developers, new researchers, and graduate-level students, especially for developing countries of the Middle East and South East Asia, to produce an even more capable and diverse insight in various domains from smart home, smart cities, industry, business, and consumer applications. I would like to congratulate Dr. Mohammad Saad Alam and Dr. Mahesh Krishnamurthy for their keen interest in the emerging area of integration of EVs with the grid of the future and thank them for their efforts on introducing state-of-the-art topics to all of us.

Reji Kumar Pillai
President, India Smart Grid Forum (ISGF)
Chairman, Global Smart Energy Federation (GSEF)

Preface

Electric vehicles (EVs) offer numerous benefits in comparison with traditional gasoline-powered vehicles, including lower operating costs, zero tail-pipe emissions and the potential to run on locally generated renewable energy. Wider adoption of EVs could take us closer to reducing the environmental impacts of the transportation infrastructure and achieving energy independence. Although these are very important goals, several challenges still exist in the wider adoption of EVs that need careful consideration. This leaves considerable scope for research and innovation. Among the current issues being discussed, one of the biggest challenges for automotive manufacturers, utilities, and customers is the issue of charging battery for EVs to recreate the "filling a tank" experience without overwhelming an ageing grid. A solution to this problem has been presented in the form of a microgrid concept. The ability to generate, control, and expend energy locally can reduce computational complexity, shorten response times, and create solutions that are cost-effective, while being easier to deploy and maintain.

A cursory review of existing literature in the area of grid integration of EVs shows that there are several important aspects that need to be carefully studied. This book provides a technological insight into several of these topics and includes specific case studies that will help the reader identify the context of the present application.

With growing acceptance of EVs, it is very important to consider the overall ecosystem in context. Chapter 1 deals with the fundamentals of the EVs, microgrid, and charging infrastructure, starting with energy consumption in charging stations and laying groundwork for integration of renewable energy sources. Chapter 2 explores numerical methods in smart grids. Chapter 2 evaluates fog computing-enabled smart grid applications and identifies key challenges and the possible approaches.

In order to evaluate the impact of electrification, it is important to consider fleet operation of EVs. This topic is tackled in Chapter 3. Starting with fundamentals of fleet charging management for EVs, it introduces charging technologies and aggregated chargers towards providing V2G services and identifies challenges associated with their deployment. A specific application example is considered in Chapter 9, which presents a holistic overview of electric taxi operations with case studies focusing on charging infrastructure. The chapter also provides multi-criterion decision-making tools for the selection of charging technology and sites for fleet charging. A case study for an EV ecosystem in the Brazilian system is studied in Chapter 4, where specific challenges are presented in adoption of EVs in the country.

Chapter 5 focuses on the coordinated operation of EVs and renewable energy in a microgrid by studying parking lots equipped with bidirectional charging stations and renewable generation such as photovoltaic (PV) panels and stationary battery storage units. In order to accommodate peak power demand in a microgrid system, stationary energy storage systems are being seen as very strong candidates, which is discussed in Chapters 6–8. Chapter 6 takes a practical outlook and presents criteria for sizing the energy storage system in an EV charging station. It presents case studies to show relationships between energy storage size, grid power, and PEV demand and also explores methods to reduce peak electricity consumption and the station's monthly electricity bill. Chapter 7 provides an innovative approach for reducing the computational effort for SoC estimation by adaptively resampling the current measurements running at a fixed sample rate. The algorithm is not limited to new developments but can also be implemented in existing commercial systems. For validating the algorithm, a worldwide harmonized light vehicle test procedure is applied with charging and resting phases. To estimate SoC, a modified 2-RC model in combination with an extended Kalman filter is implemented. Chapter 8 and 14 takes a deep dive into the environmental impacts of batteries and explores the environmental impacts and challenges in repurposing and disposal of high-voltage batteries.

In order to evaluate the impact of electrification, it is important to consider fleet operation of EVs. This topic is tackled in Chapters 3 and 9. Starting with fundamentals of fleet charging management for EVs, it introduces charging technologies and aggregated chargers towards providing V2G services and identifies challenges associated with their deployment. A specific application example is considered in Chapter 9, which presents a holistic overview of electric taxi operations with case studies focusing on charging infrastructure. The chapter also provides multi-criterion decision-making tools for the selection of charging technology and sites for fleet charging.

Chapter 10 deals with various aspects of grid integration of EVs. Chapter 10 presents an expert system of the bidding process using cloud-based artificial intelligence. The next level of the charging infrastructure deals with emerging charging technologies. Chapter 11 takes a wireless approach to the charging concept and demonstrates a PV-based charging approach. Chapter 12 explores shielding types, impact, and design in inductive power transfer by exploring three main shielding techniques, passive, active, and reactive.

To address practical challenges in implementing EV charging in different environments, a case study for an EV ecosystem in the Brazilian system is studied in Chapter 4, where specific challenges are presented in adoption of EVs in the country. Chapter 13 discusses technical and economic challenges and practical constraints of choosing the locations of EV charging stations within a large metropolitan area highlighting their merits and shortcomings of various approaches citing examples from the city of Cairo, Egypt. For the Indian market, Chapter 15 explores the design and realization of a low-cost solar PV-based microgrid system for providing alternate mode of EV charging while mitigating the load-shedding scenarios.

It is expected that this book will serve as a reference for a larger audience such as power system architects, practitioners, developers, researchers, and graduate-level students, especially for developing countries of the Middle East and Southeast Asia.

Mohammad Saad Alam
Mahesh Krishnamurthy

MATLAB® is a registered trademark of The MathWorks, Inc. For product information, please contact:

The MathWorks, Inc.
3 Apple Hill Drive
Natick, MA 01760-2098 USA
Tel: 508-647-7000
Fax: 508-647-7001
E-mail: info@mathworks.com
Web: www.mathworks.com

Editors

Mohammad Saad Alam is a professor at the Department of Electrical Engineering, Aligarh Muslim University and Founding Director of the Center of Advanced Research in Electrified Transportation (CARET). Before joining Aligarh Muslim University, he has worked in the North American Automotive Industry in the Electric Vehicle research and product development. His current research interests include electric mobility and connectivity, xEV charging infrastructure, smart Microgrid n, large-scale new and renewable energy integration, high-voltage electric energy storage systems, transactive energy, block chain application to Transportation electrification and big data analytics in the sustainable energy industry. He has coauthored over 150 publications. He is an associate editor of the *IEEE Transactions on Transportation Electrification* and the *Journal of Modern Power System and Clean Energy*.

Mahesh Krishnamurthy is a professor of Electrical Engineering and the director of the Electric Drives and Energy Conversion Lab and Grainger Power Electronics and Motor Drives Laboratory at the Illinois Institute of Technology. Before joining Illinois Tech, he worked as a design engineer at EF technologies in Arlington, Texas. His research primarily focuses on design, analysis, and control of power electronics, electric machines, motor drives and energy storage for electrified transportation, renewable energy, and industrial applications.

Dr. Krishnamurthy was the recipient of the 2006–2007 IEEE VTS-Transportation Electronics Fellowship Award for his contributions. Since 2015, he has been serving as a Distinguished Speaker with the IEEE-Vehicular Technology Society after serving as a Distinguished Lecturer from 2011–2013 to 2013–2015. He has co-authored over 125 scientific articles, book chapters, and technical reports. He is currently the advisor for the Formula Electric racecar team at Illinois Tech, which won the prestigious Fiat Chrysler Innovation award at the SAE Formula Hybrid Competition and the NASA Robotic Mining Competition team. He has received several teaching and research awards, including the 2019 Thomas Jacobius Excellence in Inter-professional Education Award (single award), 2017 Bauer Family Teaching Excellence Award (single award), and the 2017 Armour Excellence in Education Award (single award) at Illinois Tech.

Dr. Krishnamurthy was the General Chair for the 2014 IEEE Transportation Electrification Conference and Exposition. In the past, he has served as the Technical Program Chair for the 2011 Vehicle Power and Propulsion Conference and 2013 IEEE- Transportation Electrification Conference. Dr. Krishnamurthy has served as the guest Editor or associate Editor for several IEEE journals including the Special Section of IEEE Transactions on Vehicular Technology on Sustainable Transportation Systems, Special Issue of IEEE Transactions on Power Electronics on Transportation Electrification and Vehicle Systems, and Special Issue of IEEE Journal of Emerging and Selected Topics in Power Electronics on Transportation Electrification. He is currently serving as an editor for IEEE Transportation Electrification Magazine, Chair for the IEEE Power Electronics Society Technical Committee on Vehicle and Transportation Systems, and editor-in-chief for IEEE Transactions on Transportation Electrification.

Contributors

Ahmed Ibrahim AbdelAzim
Ethos Esco Consultancy
Cairo, Egypt

Md Pauzi Abdullah
School of Electrical Engineering
University Technology Malaysia
Johor Bahru, Malaysia

Aqueel Ahmad
Center of Advanced Research in
  Electrified Transportation
Aligarh Muslim University
Aligarh, India

Hossam Hassan Ammar
School of Engineering and Applied
  Science
Nile University
Giza, Egypt
and
Smart Engineering Systems Research
  Center (SESC)
Nile University
Giza, Egypt

Essam A. Al-Ammar
King Saud University
Riyadh, Saudi Arabia

Syed Muhammad Amrr
Indian Institute of Technology Delhi
New Delhi, India

Zeeshan Ahmad Arfeen
School of Electrical Engineering
University Technology Malaysia
Johor Bahru, Malaysia
and
Electrical Engineering Department
The Islamia University of Bahawalpur
  (IUB)
Bahawalpur, Pakistan

M. S. Jamil Asghar
Center of Advanced Research in
  Electrified Transporation
Aligarh Muslim University
Aligarh, India

Mehreen Kausar Azam
College of Engineering and Sciences
Institute of Business Management Sindh
Karachi, Pakistan
and
N.E.D University of Engineering &
  Technology
Karachi, Pakistan

I. Safak Bayram
Department of Electronic and Electrical
  Engineering, Faculty of Engineering
University of Strathclyde
Glasgow, United Kingdom

April Bolduc
S Curve Strategies
San Diego, USA

Alberto Borghetti
Department of Electrical Engineering
University of Bologna
Bologna, Italy

Graeme Burt
Department of Electronic and Electrical
  Engineering, Faculty of Engineering
University of Strathclyde
Glasgow, United Kingdom

Rajkumar Buyya
School of Computing and Information
  Systems
The University of Melbourne
Melbourne, Australia

xv

Rakan Chabaan
Hyundai Kia America Technical
 Center Inc
Superior Township,
Michigan, USA

Edward Corr
Power Networks Demonstration Centre
University of Strathclyde
Glasgow, United Kingdom

Camilo Orozco Corredor
Department of Electrical Engineering
University of Bologna
Bologna, Italy

Shyamasis Das
Alliance for an Energy Efficient
 Economy (AEEE)
India

Mamdouh Ahmed Ezzeldin
School of Engineering and Applied
 Science
Nile University
Giza, Egypt
and
Smart Engineering Systems Research
 Center (SESC)
Nile University
Giza, Egypt

Marim Salah Faroun
School of Engineering and Applied
 Science
Nile University
Giza, Egypt

Stuart Galloway
Department of Electronic and Electrical
 Engineering
Faculty of Engineering
University of Strathclyde
Glasgow, United Kingdom

Shaoyong Guo
State Key Laboratory of Networking &
 Switching Technology
Beijing University of Posts and
 Telecommunications
Beijing, China

Ahmed Alaa-eldin Hafez
School of Engineering and Applied
 Science
Nile University
Giza, Egypt

Rabia Hassan
College of Engineering and Sciences
Institute of Business Management Sindh
Karachi, Pakistan

Hanan K. M. Irfan
Abul Kalam Azad University of
 Technology
Kolkata, India

Geza Joos
McGill University
Montreal, Canada

Zeeshan Ahmad Khan
Development Engineer
TKI Automotive GmbH
Germany

Mohamed Adel Kohif
School of Engineering and Applied
 Science
Nile University
Giza, Egypt

Franz Kreupl
Chair of Hybrid Electronic Systems
Technical University Munich
Munich, Germany

Ahmed A. S. Mohamed
Center for Integrated Mobility Sciences
National Renewable Energy Laboratory
 (NREL)
Golden, Colorado

Contributors

Fabio Napolitano
Department of Electrical Engineering
University of Bologna
Bologna, Italy

Syed Qaseem Ali
Opal-RT Technologies
Canada

Xuesong Qiu
State Key Laboratory of Networking & Switching Technology
Beijing University of Posts and Telecommunications
Beijing, China

Yasser Rafat
Center of Advanced Research in Electrified Transportation
Aligarh Muslim University
Aligarh, India

Anirudh Ray
School of Planning and Architecture (SPA)
India

Linna Ruan
School of Computing and Information Systems
The University of Melbourne
Melbourne, Australia
and
State Key Laboratory of Networking & Switching Technology
Beijing University of Posts and Telecommunications
Beijing, China

Sanjeev Anand Sahu
Indian Institute of Technology (Indian School of Mines)
Dhanbad, India

Mohd Sakib
Aligarh Muslim University
Aligarh, India

Mohd Saqib
Indian Institute of Technology (Indian School of Mines)
Dhanbad, India

Chandana Sasidharan
Alliance for an Energy Efficient Economy (AEEE)
India

Ahmed A. Shaier
Zagazig University
Zagazig, Egypt

Samir M. Shariff
Department of Electrical Engineering
Taibah University
Medina, Saudi Arabia

Mahdi Shafaati Shemami
Center of Advanced Research in Electrified Transporation
Aligarh Muslim University
Aligarh, India

Ryan Sims
Power Networks Demonstration Centre
University of Strathclyde
Glasgow, United Kingdom

Chu Sun
McGill University
Montreal, Canada

Ana Carolina Rodrigues Teixeira
University of São Paulo
São Paulo, Brazil

Fabio Tossani
Department of Electrical Engineering
University of Bologna
Bologna, Italy

1 Trends in Electric Vehicles, Distribution Systems, EV Charging Infrastructure, and Microgrids

April Bolduc
S Curve Strategies

CONTENTS

1.1 Introduction: Transportation Electrification Trends 1
1.2 Distribution System Trends 2
1.3 Charging Technology Trends 4

1.1 INTRODUCTION: TRANSPORTATION ELECTRIFICATION TRENDS

With the rapid growth of transportation electrification, efficient electric vehicle (EV) integration with the grid is becoming exponentially more important. Geographically, China is leading the EV and electric bus market, followed by Europe and then the United States. Automakers continue to accelerate their EV manufacturing efforts to comply with increasingly stringent regulations in these countries. While pandemics like COVID-19 can demonstrate initial delays in manufacturing, the overall impact of such world events is low. By 2022, more than 500 models of EVs will be available globally due to competitive pricing and consumer choice, making EVs attractive to new buyers in the market.[1]

Passenger EV sales has grown from 450,000 in 2015 to 2.1 million in 2019 as battery prices decrease, battery capacity improves for a longer driving range, the installation of charging infrastructure continues, and EV sales move into new markets. Globally, sales will increase to 8.5 million by 2025, 26 million by 2030, and 54 million by 2040 when over half of all passenger vehicles sold are electric.[2]

As for the electricity consumption required by this growing technology grows, the rise in EV sales increases the demand for more fast charging stations. If the U.S.

---

[1] BNEF https://about.bnef.com/electric-vehicle-outlook/.
[2] BNEF https://about.bnef.com/electric-vehicle-outlook/.

reaches its forecasted growth of more than 20 million EVs by 2030, the vehicles could require annual energy consumption of 93 terawatt-hours (TWh).[3] If these vehicles demonstrate larger battery capacities and rates of charge as current automakers are demonstrating, the collective electricity consumption could reach between 58 and 336 TWh annually.[4] By 2040, passenger electric cars could consume 1,290 TWh, while commercial EVs consume 389 TWh and electric buses consume 216 TWh.[5]

To prepare the electric grid to support such a need, there is much being done across the globe. Key drivers for such support include policy requirements for regions to reduce pollution and meet air quality goals and recognition of EV electricity consumption as an opportunity by the electric power industry to sustain electric load growth reduced by energy efficiency. Additionally, the demonstrated ability of grid-integrated technologies such as smart microgrids and managed charging is needed to smooth the grid transition to accommodate this load – even for the most congested grids with intermittent power supply across the globe.

The grid must be able to integrate this technology while meeting both the capacity needs of transportation electrification and the need for increased renewable energy to reduce greenhouse gas emissions.

1.2 DISTRIBUTION SYSTEM TRENDS

Many utilities are taking a leading role in facilitating transportation electrification. Trends in increased infrastructure investment, collaboration across utilities, and grid modernization are apparent. Atlas EV Hub tracks the number of U.S. investor-owned utility transportation electrification programs being implemented. By April 2020, almost $3 billion in utility investments were approved or pending approval to support this growth.[6] Increasingly, programs have moved from a focus on light-duty EVs to medium- and heavy-duty transportation electrification due to the benefits these vehicles can provide the grid, while at the same time, heavy-duty vehicle charging could require 1 megawatt per charge.

California utilities have made the majority of this investment and are now creating a collaborative 10-year plan across the state's different utilities that looks to minimize transportation electrification grid impacts and accelerate EV adoption. The state's climate, air quality, and economic development goals require broad electrification of both passenger and fleet vehicles and require support for the widespread adoption of transportation electrification.[7] Over the past decade, numerous utility transportation electrification programs have been filed with their regulating body, the California Public Utilities Commission, in number and scale. During this time, the regulator assessed the utility programs that did not contain transportation infrastructure deployment planning strategies or projections on how to include incremental transportation electrification load into their distribution and transmission systems.

---

[3] EEI/IEI, November 2018, EV Sales Forecast and the Charging Infrastructure Required through 2030.
[4] National Renewable Energy Laboratory, 2018, Electrification Futures Study: Scenarios of Electric Technology Adoption and Power Consumption for the United States.
[5] BNEF https://about.bnef.com/electric-vehicle-outlook/.
[6] Atlas EV Hub, 2020. Utility Filings Dashboard. www.atlastevhub.com.
[7] California Senate Bill 350, DeLeon, 2015.

Therefore, they proposed a "transportation electrification framework" requiring the utilities to develop an overarching 10-year plan that details investments in transportation electrification infrastructure.[8]

The goal of this framework is to create a process that best harnesses lessons learned from past regulator proceedings, research, and transportation electrification efforts taking place in the state, as well as create a competitive market. Such a 10-year plan can provide guidance and standardize the key components of transportation electrification programs, such as charging vendor criteria, open access, cybersecurity, safety, and the length of time a utility should take to interconnect EV charging infrastructure. Most importantly, a plan like this can encourage utilities to collaborate across their distribution planning departments to assess the research from EV charging pilots from within their territories and across the globe to more fully understand the possible impacts of increasing the load from EVs and how to best use technology to integrate these efforts with the grid.

An example of such a collaboration is the West Coast Clean Transit Corridor Initiative in the U.S. made up of **nine electric utilities** and **two agencies** representing more than two dozen municipal utilities that worked together to develop a study to electrify 1,300 miles of interstate from the Mexican to the Canadian border for freight haulers and delivery trucks.[9] The study proposes a phased approach that could lead to significant reductions of pollution from freight transportation along the Pacific Coast providing a roadmap for electric utilities to electrify transportation in a coordinated fashion. The first phase would involve installing 27 charging sites along Interstate-5 at 50-mile intervals for medium-duty EVs, such as delivery vans, by 2025. A second phase would expand 14 of the 27 charging sites to also accommodate charging for electric big rigs by 2030 when it is estimated that 8% of all trucks on the road in California could be electric. Of the 27 proposed sites, 16 are in California, 5 in Oregon, and 6 in Washington. The study also demonstrated that an additional 41 sites on highways connecting to Interstate-5 should be considered for electrification.

Near- and long-term distribution planning such as this can help determine the number of shovel-ready charging infrastructure locations vs. those that will trigger expensive distribution upgrades. For example, a transit agency converting its fleets to electric buses over time could trigger the need for a new substation upgrade.[10] For a majority of grids, improving the modeling and transparency into a distribution system's hosting capacity can provide visibility of gaps in grid infrastructure when aligned with possible charging site locations. This visibility supports charging infrastructure deployment in regions where the incremental load would not trigger distribution system upgrades, and where load management technology could defer otherwise necessary upgrades.

While these gaps are identified and modeled by grid modernization planning departments, parallel efforts can be performed to design charging infrastructure programs in distribution system locations where the grid currently has the capacity and

---

[8] https://docs.cpuc.ca.gov/PublishedDocs/Efile/G000/M326/K281/326281940.PDF.
[9] West Coast Clean Transit Corridor Initiative: Interstate 5 Corridor California, Oregon, Washington, June 2020, www.westcoastcleantransit.com.
[10] https://ww2.arb.ca.gov/rulemaking/2018/innovative-clean-transit-2018.

where costly upgrades can be avoided. The advancement of smart charging technology and the implementation of these efforts in EV charging infrastructure is one of the best ways to reduce distribution impacts.

1.3 CHARGING TECHNOLOGY TRENDS

The global EV charging infrastructure market is projected to reach $140 billion by 2030 and grow at an estimated annual rate of 31%.[11] Germany, home to major automakers such as Volkswagen, BMW, and Mercedes, significantly propelled their demand for EV charging infrastructure by passing a policy to ban internal combustion engines by 2030. Such a rapid pace of adoption will be assisted by charging innovation and the ability to both manage charging loads and reduce on-peak charging by incentivizing drivers to shift their charging time when there is the most capacity on the grid. For drivers to participate in any such advancement technology, the ease of use for the driver or commercial fleet operator must not be hindered.

Standardization of charging technology accessibility and interoperability is a growing trend. The way the first EV charging technologies in China and the U.S. evolved are broadly similar, but fast charging in China has one standard, known as China GB/T, while the U.S. has three EV fast charging standards: CHAdeMO, SAE Combo, and Tesla.[12] Considerations around charging options for EV owners include the ease in accessibility at the place and time it is needed and that it is competitively priced.

EVs are often compared to the phenomenon of rooftop solar installations and frequently cluster in a particular neighborhood as awareness grows about the benefits of the technology, leading to increased adoption. Utility EV time-of-use rates are made available only to those with an electric car. Until recently, they have been the only mechanism to encourage off-peak charging. Incentivizing drivers with lower rates to charge at times of the day when there is more capacity on the grid has proven effective. Utilities should consider being mindful that unintended consequences can arise if large numbers of vehicles start to shift to the same time causing new distribution load spikes. Managed charging is a solution that helps to intelligently stagger vehicle charging and avoid grid spikes. Transportation electrification programs across the globe are increasingly including managed charging in their efforts. Managed charging can be implemented by one-directional load control of the vehicle telematics or charging station. The goal of managed charging is to avoid costly grid upgrades and effectively integrate EVs into the grid to help accelerate adoption and advance a clean, smart, and affordable energy system. Wind and solar energies are now the cheapest sources of electricity across more than two-thirds of the world, and by 2030, they undercut commissioned coal and gas almost everywhere, further incentivizing the transition to transportation electrification.[13]

---

[11] Electric Vehicle Charging Infrastructure Market: Global Opportunity and Trend Analysis, 2019–2030, Research and Markets.
[12] https://energypolicy.columbia.edu/sites/default/files/file-uploads/EV_ChargingChina-CGEP_Report_Final.pdf.
[13] https://about.bnef.com/new-energy-outlook/.

Managed charging intelligence can be found within the charger or the vehicle itself and ideally respond to a signal sent from the utility or entity requesting the load shift and can effectively manage charging efforts to benefit grid needs. Implementing such charging across fleets of commercial EVs is another way to maximize grid benefits. EV charging station technology is also evolving with new standards to improve open access across different EV charging vendors and networks to simplify the charging experience for drivers. A key to this success is interoperability or the capability of drivers to use other vendor's charging networks without having to sign up for each one separately as has been the requirement in the past. This is made possible through software that provides the exchange of driver payment data across platforms.

In most cases, a grid modernized for transportation electrification must meet the capacity needs of EVs as well as the need for clean renewable energy to fuel cars for consumers and fleets, especially when it includes grid integrated technologies such as managed charging, open source, and interoperability to smooth grid peaks. Developing partnerships among grid planning departments to share smart charging pilot results and leverage EV program data from around the world will improve grid impacts with seamless managed charging programs that are invisible to the customer and ensure both drivers and electric fleet owners have a positive experience and continue to grow the adoption of EVs.

https://www.researchandmarkets.com/reports/5023828/electric-vehicle-charging-infrastructure-market?utm_source=dynamic&utm_medium=BW&utm_code=v8g9wg&utm_campaign=1389168+-+Global+Electric+Vehicle+Charging+Infrastructure+Market+(2019+to+2030)+-+Opportunity+and+Trend+Analysis&utm_exec=jamu273bwd.

2 Fog Computing for Smart Grids
Challenges and Solutions

Linna Ruan
Beijing University of Posts and Telecommunications

Shaoyong Guo and Xuesong Qiu
Beijing University of Posts and Telecommunications

Rajkumar Buyya
The University of Melbourne

CONTENTS

2.1	Introduction	8
2.2	SGs	9
	2.2.1 Architecture	9
	2.2.2 Current and Upcoming Problems	12
2.3	Fog Computing-Driven SG Architecture	15
	2.3.1 Features	15
	2.3.2 Fog Computing Complements the Cloud	16
	2.3.3 Fog Computing Helps Address SG Problems	17
2.4	Current Solutions for Applying Fog Computing to SGs	17
	2.4.1 Fog-based SG Architecture	17
	2.4.2 Mainly Discussed Applications	17
	2.4.3 Key Problems Focused in Strategy Design	20
	2.4.4 Fog+	23
2.5	Research Challenges and Future Directions	26
	2.5.1 Security and Privacy	26
	2.5.2 Huge Amounts of Data Processing	26
	2.5.3 Fog and Cloud Combination	28
	2.5.4 Fog Device Deployment	28
2.6	Summary and Conclusions	28
References		29

2.1 INTRODUCTION

In recent years, significant climate change, such as global warming and air quality deterioration, threatens all the lives on Earth and attracts worldwide concern about harmful gas emission as well as energy issues. Given this background, the traditional power grids are transformed to smart grids (SGs) to enhance energy efficiency and system reliability, providing a promising solution to address environmental problems. Following this trend, microgrids, as small-scale local power systems, are also proposed to optimize energy management individually or through collaboration with main grid. The two types of grids are implemented at the utility level and facility level respectively, while both contribute to energy system and environment. This chapter mainly focuses on SGs, the large-scale conception. Due to many commonalities, most of the discussion also fits for microgrids.

SGs enable two-way communication and integrate renewable resources for power generation, being used to support smart cities and other energy required scenarios. Apart from these benefits, it is subject to some problems during implementation, mainly reflected in four aspects. (1) A huge amount of data generated by SG devices requires robust processing capability. (2) The emerging delay-sensitive applications propose instant response requirements. (3) Transmission of all data over the uplink increases the burden on the communication channels. (4) Uploading data to cloud through the open Internet increases the risk of privacy violations.

The traditional mode of processing data in Cloud shows its limitations in this background, mainly due to the limited transmission resources and long response delay and, in particular, to the data privacy risk. Moving the processing of emergency data to the edge side is regarded as an efficient way to address these problems, which is also the main intention of fog computing. Therefore, as one of the advanced technologies and the vertical downward extension of cloud computing, fog computing is discussed to be employed in SGs to enhance edge-side processing capability, reduce response time, relieve the burden of core network, and protect user privacy. During the application of fog computing, two problems are deputed most. First, how to deal with the relationship between the two computing modes; should fog computing replace or complement cloud computing. Second, fog computing benefits SGs on multiple aspects, while some new problems also emerged, how to cope with that.

In this chapter, we aim to conduct a comprehensive analysis of fog computing in SGs. We begin with a brief introduction of SGs and fog computing, focusing on their features, components, advantages, and challenges. Then, we discuss the application of fog computing in SGs by analyzing the existing research and the current solutions, so as to illustrate the application scenarios and summarize the frequently used methods. Further, we outline the challenges of fog-enabled SGs and the future research directions. Finally, we conclude the chapter.

In general, we hope to clarify three problems through this chapter: Why is fog computing suitable for SGs? What are the current solutions? And what challenges may exist for future applications?

2.2 SGs

SGs are defined in various ways by different organizations. In the United States, SGs are viewed as a large-scale solution to realize energy transformation from global network to the localized. While in China, SGs are defined as an approach that ensures energy supply based on physical network. For Europe, SGs mean a broader RE (renewable energy)-based system with society participation and countries' integration [1]. Although there are differences in the definition of SGs, consensus has been reached on three aspects. (1) SGs are envisioned as the next-generation electrical energy distribution network and an important part of smart cities. With the reliable communication system, SGs can manage energy more intelligently and effectively; (2) SGs allow two-way both electrical flow and information flow interaction between demand side and supply side, which makes energy consumption and pricing strategy easier to be monitored. In addition, supply-demand match, efficient energy utilization, and energy cost reduction can be realized; (3) SGs allow devices to interact information and are suitable for Internet of Things scenarios. In a nutshell, SGs integrate advanced information and communication technologies into the physical power system to

- enhance the level of system automation and hence contribute to operation efficiency;
- improve system security and reliability;
- fit the requirements of sustainable development better by using cleaner electricity resources and storage devices;
- enhance energy efficiency by facilitating two-way information interaction;
- allow customers to monitor their energy consumption and schedule electricity usage plan, which would benefit both themselves and the systems.

2.2.1 ARCHITECTURE

A SG can be described from three perspectives. First, from the perspective of functions, SGs can be viewed as the combination of physical infrastructures and information technologies. Second, considering the core processes and participants, a SG system consists of seven domains. Third, in view of the coverage scale, small-scale microgrid as an important component of SGs has been hotly discussed in recent years. The details of these three ways are introduced as below.

1. Perspective 1: Functions (physical and information domain integration)

    The SG is a product of cross-domain integration, which distinguishes it from the traditional power grid. The physical domain refers to the electrical power systems, including the energy generation, transmission, distribution, and consumption. While the information domain refers to information technologies, which are used to automatically transmit and retrieve data when necessary [2]. This integration is also reflected in interaction flows, relatively corresponding to electrical flows and information flows.

2. Perspective 2: Core processes and participants

The National Institute of Standards and Technology (NIST) illustrates the functions of a SG with a conceptual model as shown in Figure 2.1, which defines seven important domains. The concept of each domain is explained as follows.

- Bulk generation domain

  Generate electricity for later transmission and distribution and finally for residential, commercial, or industrial use. The generation sources include traditional sources (such as fossil fuel and coal) and distributed energy sources (such as solar and wind power).

- Transmission domain

  It is commonly defined as the carrier for long distance power transmission. In some specific scenarios, it also has the capability of electricity storage and generation.

- Distribution domain

  Distribute electricity to or from (when the surplus power generated by distributed resources needs to be sent back to the market) customers. Similar to power transmission, the distribution domain has the capability of electricity storage and generation in some cases.

- Customers

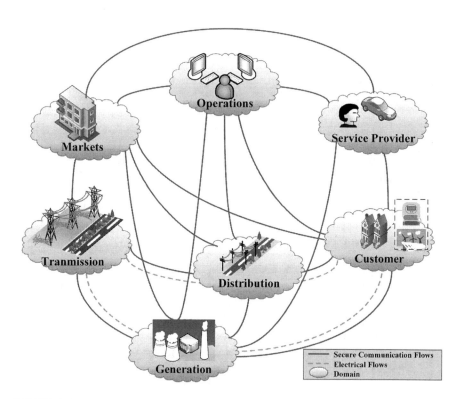

FIGURE 2.1 Conceptual model of SGs defined by NIST.

Fog Computing for Smart Grids

They are the end users of electricity. According to the consumption habits and levels, they are divided into three types: residential, commercial, and industrial. Besides consuming electricity, customers may also generate and store electricity by embedding distributed resource infrastructures and batteries. In addition, demand-side management allows them to monitor and manage their energy usage.
- Service providers
   Organizations that provide services for electricity users and utilities.
- Operations
   The managers of the electricity movement.
- Markets
   A trading place for operators and customers. In power grid systems, the markets are divided into wholesale markets and retail markets, depending on the transaction mode.

3. Perspective 3: Coverage scale
   By integrating distributed resources, a new form of SGs has been formulated, called microgrids. Extending the literature [3], its architecture is depicted in Figure 2.2.

Similar to SGs, microgrids are defined by different organizations. For example, the Microgrid Exchange Group has defined microgrids as "a group of interconnected loads and distributed energy resources within clearly defined electrical boundaries that act as a single controllable entity with respect to the grid", while the consensus on microgrids lies in four aspects. (1) They are an important component of SGs [4]. (2) Compared with SGs, microgrids refer to a smaller distributed local power system [4–5]. (3) Distributed generators and power-storage units are included. (4) They can operate either in conjunction with the main grid (excessive power that cannot be totally consumed locally can be sold to utilities through electricity market) or in an isolated mode (only provide services for end customers, which differentiates it from the centralized power generation form). Such obvious benefits listed below make microgrids a hot topic for SG researchers.

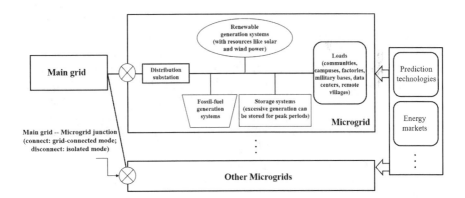

FIGURE 2.2 A microgrid architecture.

- Localized mode eases the integration of distributed and renewable energy sources (such as solar and wind power), relieving the burden of generators in peak load periods and therefore reducing harmful gas emission.
- Locating close to demand side makes microgrids easier to get the knowledge of users' needs, resulting in efficiency increase and transmission cost reduction [6]. It provides better service by ensuring the energy supply of critical loads (loads generated by devices or organizations, which require continuous power supply, such as military equipment, hospitals, and data centers [2]), system quality, resilience, and reliability and allows customers to join demand-side management in an easier way.
- Provide strong support to the main grid by handling local urgent issues, such as the variability of renewables and the sensitive loads, which require local generation, and provide auxiliary services to the bulk power system.
- Local-global controlling can be realized. With both regional requirements and overall performance taken into account, microgrids offer better performance insurance and more opportunities for multi-technology integration (electric vehicle, residential energy storage, rooftop photovoltaic systems and smart flexible appliances [2]), which is beneficial for SG development.

Many researchers have investigated the relationship between SGs and microgrids. They are widely regarded as the implementation of new era grids at utility level and facility level, respectively [4–5], [7]. Though with some differences in construction, both contribute significantly to the energy system and environment.

2.2.2 Current and Upcoming Problems

1. Latency requirements

   Emerging mission-critical and delay-sensitive SG applications, such as demand response, emergency restoration process, and substation monitoring, require low round trip latency [8]. Based on characteristics, these applications can be categorized into two types, the flexible real-time ones and the fault-tolerant but continuity-required ones. A typical application of the first type is demand-side management, which allows customers to monitor their electricity consumption in almost real time. However, cloud computing cannot meet the requirements, due to long distance data transmission, possible channel congestions, and server failures. The second type cannot be satisfied by cloud computing either, because possible connection failures and processing delay can cause interruption, making it difficult for cloud to support a continuity-required service.

2. New security challenges

   While integrating various information technologies, SG systems are facing new security challenges. We categorize the challenges on device level, communication, and system level.

- Device level—security measure upgrade

    Resource-constrained devices. SGs have many resource-constrained devices that hold limit capability to upgrade the security corresponding hardware and software in their lifespans. Moreover, a SG environment includes more participants, more technologies, and frequently information interaction compared to traditional power grid, improving energy efficiency while being more vulnerable to security attacks. Hence, there is an urgent need to figure out an effective protection method for resource-constrained devices [9].

    Large number of devices. As mentioned above, resource constraints pose a challenge to the security protection of devices. In some cases, connecting to cloud seems a proper way to upgrade the security credentials and software. But with the exponentially growing number of devices, it is impractical to allow all devices to do that, which is a resource-, energy-, and time-consuming process. Therefore, in the face of a large number of devices, how to ensure system security is still a big challenge.

- Communication and system level

    Cope with security problems while ensuring operation. When faced with security issues, shut down-then-fix is the common way currently. However, shutdown also means interruption, which is intolerable to mission-critical or delay-sensitive services. Like the example proposed in Ref. [9], if an electric power generator chooses to shut down when met malware attack, severe disruption will be caused, leading to power outages. Therefore, fix-while-operate is a prospect mode and still a challenge for SGs.

    Keep private data private. In SGs, the collected data are usually stored in cloud-based data centers for further processing or future use. Since the private information contained in the dataset is valuable for making energy strategy and then making benefits, it is preferred by intruders, and even service providers or cloud operators. Therefore, it is important to ensure transmission security during the way from smart meters to the central cloud to prevent data leakage. However, the data are transmitted through open Internet, and the number of connected users continues to increase, making it harder to figure out a strong privacy protection solution [10].

    Robustness. Communication network is one of the adding components that differentiate SGs from traditional power grids. Hence, its robustness directly impacts how a SG system would be judged. A robustness communication network means that it can keep normal or recover quickly even in such terrible situations, like natural disasters or human intervention. In this case, we care about whether there are advanced technologies that can be included to deal with emergencies intelligently in addition to existing resource-consuming solutions (providing redundant links or power backup facilities).

    Reliability. Reliability has always been viewed as a challenge for one system especially for SGs due to the high outage cost. According to a

Sun Microsystems analysis, blackouts cost approximately US$1 million every minute to electric companies [8]. The main reasons leading to outages lie in three aspects. (1) Lack of accurate knowledge of system status in real time; (2) Lack of prediction and analysis capabilities; (3) Lack of timely and effective response measures. SGs offer better communication, autonomous control, and management methods to relieve these problems. However, how to extend the current framework to handle diverse and sophisticated issues in the future still seems a problem.

3. Distributed control

As mentioned before, the basic components of SGs are geo-distributed, which is inefficient to be processed with remote-centralized cloud. Therefore, a distributed computing platform is preferred to provide location-based services and analytics, location-free billing and charging, and many more [11].

4. Prediction responsiveness

Prediction is the basis of making predecisions, and its accuracy directly impacts whether a strategy proposed is appropriate. In a SG environment, demand and generation prediction are studied most. Demand prediction is divided into long-term prediction and short-term prediction according to the time interval. It is important for demand-side management, while generation prediction is usually used for renewable energy resources, such as solar panels and wind turbines. Its accuracy mainly depends on the weather prediction, and it is an important component of microgrids. In the past several years, cloud platform is the carrier for the two types of predictions. However, instant decision-making is required for some specific applications recently and nearly real-time prediction is expected, which cannot be satisfied by cloud computing. In this case, how to enhance prediction responsiveness while meeting the requirements of computing capability is a challenge.

5. Supply-demand match

Communication network enables information interaction between power providers and consumers. It enhances energy efficiency by supporting system balancing and provides incentives to customers to optimize their electricity usage by cutting or shifting peak period demand. Supply-demand match intends to realize less energy waste and higher energy efficiency, beneficial for both the system and environment. However, the demand and renewable resource generation is always changing dynamically, which requires real-time information flow to support customer's immediate participation. The requirement cannot be satisfied by centralized cloud processing due to large response delay. Given this background, how to implement real-time information interaction between providers and consumers is a problem.

6. Complexity complicates the system management

The SG is an increasingly complex system since the quantity and rate sharply increased data and various technologies applied. Besides, there are some new services that should be supported, such as two-way communications, real-time information interaction, and demand-side management. Therefore, managing the complex system to realize all these functions, guaranteeing their requirements, and balancing the interests of all participants are really a challenge.

2.3　FOG COMPUTING-DRIVEN SG ARCHITECTURE

Fog computing was first proposed by Cisco as the vertical downward extension of cloud computing. By providing computing, communication, controlling, and network storage capability at the proximity of data source, fog computing contributes to response time reduction and complements edge-side processing capability. It is mainly used to handle mission-critical and delay-sensitive applications. From the tech giants to manufactures, fog computing is discussed to be used in many scenarios, especially the SGs, which have such challenges mentioned above and view fog computing as a proper technology to break the barriers. Before delving into the application of fog computing in SGs, we give a brief introduction about the main features of fog and discuss how to deal with the relationship between fog computing and cloud computing.

2.3.1　Features

The features of fog can be simplified as AESR (Awareness, Efficiency, Scalability, Responsiveness) illustrated as follows. They also reflect the advantages of fog computing on different aspects.

- Awareness

    The awareness refers to two aspects: objective awareness and location awareness. In a SG, users' preferences are various, such as profit, quality of experience (QoE), and energy efficiency. Since fog nodes are geographically distributed, each masters a relatively small area, making the nodes easier to get users' expectations and preferences and providing suitable and even customized strategies. It is like the general saying "the right is the best", awareness makes fog computing a good service provider.

- Efficiency

    In a broader perspective, fog computing is regarded as the added computing nodes between the end devices and the cloud. Moreover, the capabilities of fog computing are not limited to computing, communication, and storage, and these basic functions make it also a good resource manager and task scheduler. It integrates all the edge-side resources, such as smart appliances and computers, and finds the best place for task processing with the combination of resource scheduling. In this way, fog computing attains high efficiency in terms of both resource and system operation.

- Scalability

    Fog platform locates close to users and is small in size, making it easier to adjust according to environment requirements, supporting infrastructures update and scaling with less cost. In addition, fog permits even small groups to access public programming interfaces and copes with new emerging services well with good scalability.

- Responsiveness

    Quick response is one of the main advantages of fog computing and also the motivation for its proposal. Fog platforms implement data processing close to users, significantly shortening the transmission link, which makes actuators obtain data analysis results and operation suggestions almost in real time, meeting the requirements of mission-critical and delay-sensitive applications. This is essential for not only the SG stable operation but also for enabling millisecond reaction times of embedded AI to support emerging artificial applications, which is mentioned as one solution for applying fog computing to SGs in the next section.

2.3.2 Fog Computing Complements the Cloud

After introducing the features of fog computing, we should explain why it is still proposed in the context of 'cloud computing everything' and then clarify what kind of relationship exists between them. We will analyze from the following aspects.

- Latency

    As mentioned before, geo-distributed fog computing enables quick response due to locate proximity to end users. It offers users an opportunity to obtain analysis results timely and then go on operation or cope with urgent issues. As a comparison, centralized cloud computing is a time-consuming process, caused by long distance data transmission both uplink and downlink. Moreover, huge amounts of data transmitted to cloud put great pressure on network channels, which may lead to congestion or even interruption. In a word, fog computing can complement cloud on real-time performance and reduce the possibility of channel congestions.
- Accessibility

    Fog platforms locate at the edge side, enhancing the possibility for end devices to get served, especially for resource-constrained ones. Besides, fog computing costs less either on time or energy compared to cloud computing and processes data locally, protecting the system from channel congestions. In a conclusion, fog complements cloud on high accessibility, providing a more general and affordable solution for devices and services.
- Privacy

    Data privacy protection, which means the protection of sharing confidential data with the third parties, is important for the reliability of a SG system. Similarly, we analyze the data processing mode of fog and cloud computing, so as to show their performance differences on data privacy. Fog computing enables data to be processed separately, indicating that private data can only be accessed by the fog while public data also can be transmitted to the cloud. While cloud works on shared background [12] and data is transmitted through Internet, each link has a risk of data leakage, which endangers the safe and stable operation of SG systems. Therefore, fog complements cloud by providing another location for data, protecting data privacy while ensuring efficiency.

After the above analysis, it is easy to get a conclusion: cloud has powerful computing capability, while fog outperforms on latency, availability, and privacy. Now we are able to answer the question "what relationship exists between cloud computing and fog computing?". An appropriate answer is that fog computing is a complement of cloud computing. They have their own advantages, and no one can replace the other. In a specific scenario, which paradigm to apply depends on the requirements of services and pursuing of users. Therefore, combining the centralized cloud and the distributed fog nodes to create a hybrid fog-cloud platform is the best way currently to address SG problems.

2.3.3 Fog Computing Helps Address SG Problems

After clarifying the relationship between fog and cloud, let us discuss one of the main application scenarios—SGs. The specific solutions will be introduced in the next section, while before that, we want to illustrate why fog computing is suitable for SGs by comparing the earlier mentioned challenges faced by SGs (under traditional cloud computing form) with the performance supplement that fog can provide. The analysis can be summarized in Table 2.1.

2.4 CURRENT SOLUTIONS FOR APPLYING FOG COMPUTING TO SGs

A lot of research has discussed how to strengthen SGs with fog computing. Since edge computing is interchangeably defined as fog computing in most of the cases, both of the two computing paradigms in SG environments are discussed in this chapter and are uniformly called fog computing. In this section, we depict a generic architecture for fog-enabled SGs (fog-SGs) and review the mainly discussed services, key problems in strategy design, and other technologies that may provide further performance improvement.

2.4.1 Fog-based SG Architecture

A fog-based SG architecture is proposed in Figure 2.3. It contains three layers, which are the infrastructure layer, constructed with residential, commercial, and industrial buildings, acting as power demand side; the access layer, with fog and cloud computing, providing computing, communication, and storage capabilities. Fog servers are deployed with base stations and the supply layer, which is mainly responsible for power generation, transmission, and distribution. The main services supported by fog computing are also listed in the architecture, and the details are shown as below.

2.4.2 Mainly Discussed Applications

From the perspective of core links of SGs, we introduce how can fog computing benefit the applications.

TABLE 2.1
Fog Provides Effective Ways to Address Smart Grid Problems

Classifications	Smart Grid Challenges	How Fog Can Help
Latency constraints	The flexible real-time applications and the fault-tolerant but continuity-required applications have real-time response or continuous operation requirements, which cannot be satisfied by cloud.	Fog locates at the proximity of end devices, strengthens edge computation, communication, controlling, and storage capabilities, provides delay-reduced services, avoids the risk of channel congestions, and ensures consistent operation.
New security challenges	On device level, continuous upgrading of security measures is hard to realize.	Edge resource is empowered with fog and able to support security infrastructure upgrade.
	On communication and system level, service continuity cannot be guaranteed. Private data face the risk of leakage. Robustness and reliability are also not that satisfied.	Fog computing provides service with reduced delay, which can ensure the continuous operation. Private data are processed at the edge; only public data can be further transmitted to the cloud.
Distributed control	The basic components of SGs are geo-distributed. Centralized cloud is high-cost and not that suitable. A distributed paradigm is preferred.	Fog computing follows distributed form and is able to provide location-based services and analytics and location-free billing and charging.
Prediction	Demand and generation prediction are preferred in future SGs. Cloud can meet the requirement of computation while fail to update information in real time.	Fog nodes have the computing capability to do basic prediction and can send back the results with short delay. It can catch the dynamic changes of information and update within latency limit.
Supply-demand match	Demand and renewable resource generation is always changing, while cloud cannot process this frequently changing status.	Fog computing interacts the demand and pricing strategies between customers and providers timely, which facilitates the demand response process.
Complexity	With the rapid increase of data volume and rate, and the need to support various technologies and services, the SG system becomes more and more complex. Complexity complicates the control of SGs.	Fog computing can undertake data analysis, support delay-sensitive services, and relieve the burden of end devices, network, and cloud. Distributed mode means tasks can be split for processing, which decreases the complexity of SGs.

- Power supply

  For power supply, fog computing is mainly used to set price strategy, balance supply-demand, and identify abnormal fluctuations. Demand-side management is one of the important applications in SGs, which has a potential of cutting off or shifting electric demand in peak periods. In this application, real-time pricing is the main motivation for customers, and supply-demand balancing is one of the aims. Due to the suitable

Fog Computing for Smart Grids

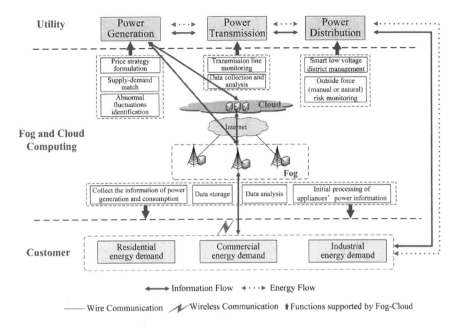

FIGURE 2.3 An architecture of fog-cloud-enabled SGs.

computing and communication capabilities and providing delay-reduced services, fog computing is envisioned as a well-suited technology to facilitate demand-side management. Besides, in Ref. [13], fog nodes are considered to be embedded in charging points and detect the abnormal. It mentions that fog computing can identify the abnormal status by analyzing sudden power fluctuations and then report to cloud for further management.

- Power transmission

For power transmission, line status monitoring is a widely mentioned service that applies fog computing [13–15]. Transmission line monitoring is important for obtaining full knowledge of the equipment condition (especially during bad weathers, such as high temperature, heavy rain, strong wind, or snowstorm), supporting safe and stable operation of power systems. In this application, graph, video, and data information is collected by unmanned aerial vehicles [13] or video sensors [15], which are controlled by edge network, and then the information is sent to fog nodes to filter and process. The analysis results reflect whether there is a possibility of failure and effectively alleviate the bad effects as a result. In this process, quick response, data privacy, and system reliability are really important, which can be better satisfied by fog computing. Of course, if there is a large amount of data waiting for processing or the status is too complicated for fog, hybrid fog-cloud will be adopted, in which preprocessing and further processing are considered to be executed, respectively.

- Power distribution

   For power distribution, fog computing is considered to be used in smart low voltage (LV) district management and outside force (manual or natural) risk monitoring [16]. For LV district management, the transformer terminal unit is enhanced with fog computing, which is used as low voltage side agent. LV topology identification, distribution fault diagnosis, and line loss analysis therefore can be realized with less delay and have light pressure on storage and processing. For monitoring, the role of fog is similar to when it is used for power transmission. In traditional mode, the collected data are stored in a local recorder and then sent to cloud for processing. By facilitating fog computing, lightweight data can be processed locally, and warning information can be sent almost in real time.

- Substation

   For substation, the function of fog computing is to monitor the operation status and equipment environment and analyze data, similar to the power transmission process, while the difference is the source of data. In this scenario, the data mainly come from a large number of various sensors. With fog computing, most of the information and data processing can be implemented locally and respond timely to ensure the warning before events, the suppression during events, and the reviews after events [13].

- Power consumption

   For microgrid systems, fog computing is used to collect the information of power generation and users' electricity consumption in real time and then abstract their behavior mode based on power information in the time dimension. During system operation, the behavior mode is used to judge the power balance level and identify abnormal status [13].

For advanced metering systems, fog devices are embedded in power meter concentrator to support the storage and analysis of data as well as services. The enhanced real-time interaction and price prediction capabilities empower the demand-side management application.

For smart home, most smart appliances require initial processing of power information. With time and cost-saving characteristics, fog computing would be preferred to obtain better performance. In addition, fog nodes can act as user agents, interacting information between users and cloud, providing local data collection, operation status monitoring, and small-scale controlling functions, and contributing to emerging applications, such as demand response and fault diagnosis [16]. Table 2.2 summarizes the mentioned SG applications supported by fog computing.

2.4.3 Key Problems Focused in Strategy Design

Specific problems that are often discussed in fog-SGs include resource management, task scheduling, security and privacy protection, and comprehensive ones. The solutions provided are illustrated as follows.

TABLE 2.2
SG Applications Supported by Fog Computing

Applications	Functions Realized	Benefits Provided	References
Power supply	Price strategy formulation (demand prediction)	Supports demand-side management.	[13]
	Supply-demand match	Enhances energy efficiency.	
	Abnormal fluctuation identification	Identifies abnormal situations and sends warning information timely. Enhances system reliability.	
Power transmission	Transmission line monitoring	Conceives the status of the equipment timely.	[13–15]
	Data collection and analysis	Ensures the warning before events, the suppression during events, and the reviews after events. Supports safe and stable operation of the power system.	
Power distribution	Smart low voltage (LV) district management	Fast topology identification, distribution fault diagnosis, and line loss analysis.	[16]
	Outside force (manual or natural) risk monitoring	Lightweight data localized processing. Finds abnormal situations and sends warning information timely. Enhances system reliability.	
Power substation	Operation status and equipment environment monitoring	Gets the condition of the operation status and equipment environment timely.	[13]
	Data collection and analysis	Ensures the warning before events, the suppression during events, and the reviews after events. Supports safe and stable operation of the power system.	
Power consumption	Power generation and users' electricity consumption information collection.	Supports demand-side management.	[13,16]
	Price prediction	Helps users optimize electricity usage plan.	
	Abstracts the behavior mode	Judges the power balance level and identifies abnormal status.	

- Resource management

  A residential scenario is considered in Ref. [17]. It proposes a cloud-fog-based SG architecture, in which multiple buildings are formulated as end user layer, and each building corresponds to one fog device. By applying the cloud-fog mode, quick response can be realized. In addition, particle swarm optimization (PSO), round robin, and throttled algorithms are used to implement electric load balancing. It is verified that PSO outperforms the other two algorithms in terms of response time and total cost.

The work [18] focuses on a residential scenario and aims to optimize energy consumption, which is important for demand-side management application. A cloud-fog architecture is proposed, and each fog node manages energy demand scheduling of several buildings. It aims to reduce total energy cost and formulates the problem as a distributed cooperative demand scheduling game.

- Task scheduling

  A fog computing system is considered to provide strong storage and computing resources in SG communication network. The work [19] aims to minimize the total cost for the system running with subject to the tasks' requirements. A green greedy algorithm is designed to provide a solution for the optimization problem.

  Another work [20] takes SG communication network into consideration and designs a service caching and task offloading mechanism, to realize network load balancing and accelerate response. A computing migration model is also proposed to support task offloading from central cloud to the edge network.

- Security and privacy protection

  By facilitating information communication, benefits such as energy saving and customers' satisfaction improvement can be realized. However, it also means SG is more vulnerable to attacks, which can be summarized as device attack, data attack, privacy attack, and network attack [21]. Therefore, confidentiality, integrity, and availability (CIA), as the key judgments of a security policy, should be highly guaranteed.

  In Ref. [21], a fog computing-based strategy is proposed to detect cybersecurity incidents in SGs. It models the anomaly detection process on the basis of generally used Open-Fog Reference Architecture. Sensors, actuators, smart meters, and a central monitoring unit are included in the model, and fog is embedded in phasor measurement units to do device measurement. The main process is data acquisition, data preparation, feature extraction (using k means algorithm) and behavior modeling, anomaly detection, and score provision. Hence, unusual power usage can be detected, which is probable for an attack on a SG system.

  For failure recovery, a fog computing-based dispatching model is proposed in Ref. [22]. In that paper, each regional power grid is in charge of a fog node, which is responsible for fault information storage, repairing resource allocation, and making dispatching plan with short delay. A dispatching algorithm based on genetic algorithm is carried out in fog nodes, realizing cost reduction and satisfaction increase during the repair process.

  The work [23] presents a security fog-SG model, in which access authentication, data security, and real time protection are described as expected functions. Then, for physical layer authentication, it adopts k-nearest

neighbors as a solution. A differential privacy data distortion technique consisting of Laplace and Gaussian mechanisms is provided in Ref. [13].
- Comprehensive problems

    Due to the complex environment of SGs, some research also tried to figure out solutions for comprehensive problems to balance multiple performance metrics, such as security, efficiency, and functionality. Reference [24] builds a fog computing-based SG model and proposes a concrete solution for both aggregation communication and data availability. By encrypting under a double trapdoor cryptosystem, security data aggregation is implemented. The solution is designed for service providers to realize dynamic control and electricity distribution. Similarly, a privacy-preserving authentication and data aggregation scheme for fog-based SGs is proposed in Ref. [25]. In that scheme, short randomizable signature and blind signature are used for anonymous authentication, and then fog nodes are used to solve billing problems. Table 2.3 summarizes the mentioned problems and solutions.

2.4.4 Fog+

Fog complements cloud in multiple aspects, while for a fog-SGs system, it is not enough. By integrating other advanced technologies, the performances are possible to be further improved. We select three hot discussed technologies, which are blockchain, AI, and SDN. They are expected to empower energy security and privacy protection, prediction, and resource flexibility for Fog-SGs, respectively.

- Fog + Blockchain

    As mentioned before, fog computing assists SG security on transmission and service continuous operation. Nonetheless, the security and privacy protection of fog-SG systems are still a challenge, since fog server itself is less secure than cloud, and there exists interaction and service migration between heterogeneous fog nodes [26]. Hence, this problem has been stressed in recent years, and some security solutions are proposed.

    However, since SGs have many resource-constrained devices, most of the conventional methods (certificate authority-based, ring signature, blind signature, and group signature) are not suitable, due to bad traceability and participation flexibility, high computation, and communication overhead. Given this background, blockchain is envisioned as a new chance for SGs. Blockchain allows network participants to record the system in a distributed shared ledger. In a blockchain enabled system, fog servers need to join, while end uses are not required to, which prevents users' identity leakage. At the same time, the change in the participation status of one user will not influence others, as the registration is identity-based. The smart contract included also supports traceability and revocability [27]. Therefore, blockchain is chosen as fitting all the security requirements of SGs, and its scalability requirements can be complemented by fog computing. Moreover, both blockchain and fog computing follow decentralized network form, which makes them possible to integrate.

TABLE 2.3
Key Problems and Solutions

Classification	Problems	Objectives	Solutions	References
Resource management	Electric load balancing	Reduces response time and total cost	Builds a cloud-fog-based SG architecture. Proposes an algorithm based on PSO.	[17]
	Energy demand scheduling	Reduces total energy cost	Builds a cloud-fog-based SG architecture. Formulates a distributed cooperative demand scheduling game.	[18]
Task scheduling	Task scheduling	Minimizes the total cost for system running	Proposes a green greedy algorithm.	[19]
	Service caching and task offloading	Balances network load and reduces communication delay	Proposes a computing migration model. A load-balancing algorithm based on popularity and centrality.	[20]
Security and privacy protection	Cybersecurity incident detection	Detects anomalies and reduces communication delay	Proposes an anomaly detection process model and a detection method based on fog computing.	[21]
	Failure recovery	Reduces cost and increases satisfaction during the repair process	Proposes a dispatching model based on fog computing and a dispatching algorithm based on genetic algorithm.	[22]
	Accesses authentication, data security, and real-time protection	Realizes physical security and improves energy efficiency	Proposes a secure fog-SG model, a physical security approach, and an electricity forecasting method.	[23]
Comprehensive problems	Aggregation communication and data availability	Realizes dynamic control and electricity distribution	Proposes a fog computing-based SG model. A concrete solution for both aggregation communication and data availability.	[24]
	Security and privacy issues in fog-based SG communication	Anonymous authentication and billing	Proposes a privacy-preserving authentication and data aggregation scheme for fog-based SGs.	[25]

In Ref. [27], a blockchain-based mutual authentication and key agreement protocol is proposed for fog-SG systems. In addition to providing basic security properties, it also offers an efficient method for key update and revocation as well as conditional identity anonymity with less costs. From another perspective, [28] proposed a permissioned blockchain edge model for SG network. It focuses on privacy protections and energy security issues and formulates an optimal security-aware strategy by smart contracts.

- Fog + AI

SGs require accurate prediction of demand, pricing, and generation capability to support resource preparing or policy setting and contribute to SG applications, such as demand-side response and monitoring. The basic elements of prediction are huge amounts of data, computation capability, and intelligent algorithms. Evaluation metrics are speed, accuracy, memory, and energy [29]. The amount of data generated by the SG infrastructures increases sharply both in terms of quantity and rate, which could be the first element of prediction. Fog enhances edge-side computation capability and can provide delay-reduced services and match the second element. Then, for the third element, artificial intelligence algorithms seem to be the best choice currently. The following two cases illustrate how AI is supported by fog computing and how it is applied in SGs.

The work [16] focuses on distribution outside force damage monitoring application. For the processing platform, the fast and accurate vehicles' identification with AI is the most critical segment. Since the AI algorithm is a resource-consuming process, the hybrid fog-cloud is envisioned as the most suitable technology to enhance its operation performance and save system resources.

Demand and dynamic pricing predictions are considered in Ref. [13]. Demand prediction is important to let providers understand the needs of users, guide the generation plan, and contribute to electric resource balancing. Dynamic pricing acts as the main incentive for customers to respond to demand shifting or cutting off in peak period and therefore directly relates to users' benefits. The suitable AI approaches are listed as Auto Regressive Integrated Moving Average models, Auto Regressive models, ANN, fuzzy logic, and (long short-term memory) LSTM in that paper.

- Fog + SDN

Fog computing offers SGs benefits on multiple aspects. However, for the increasing scale of SG networks, how to transmit data to the fog servers or sometimes to the cloud could bring significant effect on resource efficiency. To face the challenge of data routing, involvement of software-defined networking (SDN) is viewed as a potential solution of fog-SGs, mainly due to driving a more manageable and flexible network with a global view brought by decoupling the network control plane and data plane. Reference [30] proposed an SDN-based data forwarding scheme for fog-enabled SGs. With the global information provided by SDN, the shortest path is calculated, and a path recovery mechanism is designed to avoid link failures. SDN is mainly used to facilitate multicasting and routing schemes for SGs [30].

2.5 RESEARCH CHALLENGES AND FUTURE DIRECTIONS

2.5.1 Security and Privacy

As mentioned before, the security of fog computing is still a challenge, also for fog-SGs. The reasons are threefold: (1) heterogeneous. As the emerging of Internet and SGs, fog computing will be combined with various technologies and access to other systems. Interacting information and migrating services between heterogeneous devices in large or small scales face the danger of malicious attacks. (2) Dynamism and openness. Fog computing is a small edge-side computing platform and runs in an open and dynamic environment, which is vulnerable to attacks. (3) Data dispersion. Since each fog node has limited capability, data are usually transmitted to geo-distributed nodes to store or process. It brings a risk of data leakage, packet loss or incorrect organization, and breaking data integrity.

In conclusion, security issues of fog-SGs, such as edge control, data dispose, computation, and communication still need new ideas to match the distributed, heterogeneous, and complex environment. Blockchain is a probable method to address identity risks, while the research is still at the initial stage. Scalable enhancement and layered mechanisms are the research directions, and how to deal with outsourcing services as well as off-chain status need further discussion.

2.5.2 Huge Amounts of Data Processing

With the emerging of SG infrastructures, various system data increase both in rate and amount, which brings high complexity and aggravates the burden for data processing. The challenges brought by huge amounts of data are mainly reflected in the following aspects.

- Data control, security, and privacy

    The quantity, diversity, and rate of data in SGs increase dramatically. Based on broad consensus, fog computing strengthens the computation, communication, controlling, and storage capability of edge side and is envisioned as a good choice to cope with the data explosion of SGs. However, a more accurate statement is that fog can relieve the problems, not solve. Due to the characteristics, fog computing platform means a set of capacity limited nodes. Indeed, the nodes can interact to support powerful data processing, but either for tech giants or manufacturers, it is really challenging. We will analyze the reasons from different stages of data processing. First, data split. When the resource of one fog node is not sufficient to hold the data collected by sensors, advanced metering infrastructure, or other devices, data will be split into several parts to process. However, what split granularity should be selected. Large granularity is beneficial for organization and data integrity, while small granularity can use resources with high efficiency and is more likely to be processed quickly. In this case, the choice seems dependent on what performance we pursue. Is it easy? Don't get the conclusion too early, let's continue. Then, data distribution. This stage contains three problems, which node to send, how much to send, and which path to select. For the first problem, the node with the richest available resources, or closest distance, or with the best comprehensive performance in combination with the degree of credibility is always selected as the destination. For the second problem, it is usually decided along with the first one and depends on what we are pursuing for, the system balance, the short delay, the energy efficiency, or anything else. Now for the third problem, two choices usually are mentioned, the shortest one or the most secure one. Finally, data storage. Where to store and how much to store. Should the data be sent back to end device, retained at the fog node, or sent to the cloud. We also have to take a decision.

Now, it is easier to reach a conclusion. During data processing, there are several problems to be considered in each stage, and each problem has several choices. It seems that these choices all depend on our goals, but the goals of different stages may not be the same. To achieve the best comprehensive performance, we must make various considerations and trade-offs between different performances with subject to the requirements of specific services. In the existing research, the solutions proposed are usually based on a specific application with such assumptions, which has a limited value for a real industrial scenario. Therefore, for the data processing of fog-SGs, how to deal with more and more complicated situations is very challenging and needs more discussion.

- Artificial intelligence integration

    Prediction is one of the important aspects to show the smartness of SGs. As we discussed, the large amount of data collected from SG devices is the basis of prediction, and the artificial intelligence (AI) algorithm is an essential tool to guarantee the accuracy. However, since the AI algorithms are resource-consuming, in fog-enabled SG environments, some aspects are still challenging: (1) limited capability. Considering the complexity of prediction, it is usually carried out in cloud traditionally. However, due to the emerging applications and higher requirements proposed, real-time prediction is expected in SGs. Fog computing can realize quick response for most of the services, but it is not sure for prediction, depending on the complexity of the algorithm. Therefore, the edge side still needs other technologies to facilitate capability of fog-SGs. (2) Follow the dynamics of algorithms. AI algorithms are not that stable, which means that they are more like an art, designed for the specific problem, and innovation always comes out, including more parameters, more layers, or new structures. Enabling fog-SG systems keeping up with new AI designs is obviously another challenge.

2.5.3 Fog and Cloud Combination

The hybrid fog-cloud computing is envisioned as the most promising mode to be integrated in SGs. But how to collaborate the two kinds of paradigms to show their own best and also realize the complement is really a challenge. Resource management problems are discussed most, which contain resource aggregation, task offloading, caching, and storage. Due to virtualization technologies, the resource can be customized as containers or virtual machines, which enhances the resource flexibility but makes it more complex to formulate resource management strategies. Besides, the participation process of data and applications complicates the problem further. In the future, how to make generic strategies and customized strategies for specific SG scenarios with given optimization objectives still needs more discussion.

2.5.4 Fog Device Deployment

How to integrate fog computing into SGs has been discussed in recent years. However, most of the solutions have such assumptions and not suitable for the real status. Consider the deployment of fog computing; there are usually two choices. (1) Embedding fog computing functions in the existing SG equipment. (2) Designing new fog computing equipment. The first option costs less but requires manufacturers' permission for the equipment transforming. It is necessary to ensure the performance of the embedded part, especially the safety performance, while not bringing negative impact on the existing performances. The second option costs higher, but it does not need to negotiate with the device manufacturers but just needs to consider the access with other devices. But both of the options require the support of the government and power suppliers, which control device accessing and ensure the stable operation of SGs. As discussed above, the deployment and realization of fog functions in a real SG are still a challenge.

2.6 SUMMARY AND CONCLUSIONS

In this chapter, we investigated the application of fog computing in SGs from three perspectives: characteristics, solutions, and challenges. Through the brief introduction to the components and features of SGs and fog computing, two problems are analyzed. (1) Why fog computing is still proposed in the context of 'cloud computing everything' and how to deal with the relationship between fog and cloud computing? A proper answer is: fog computing is proposed as a complement of cloud computing. They have their own advantages, and no one can replace the other. Under a specific scenario, which paradigm to apply depends on the constraints of devices, requirements of services, and users. The hybrid fog-cloud computing is a generic and the most promising mode to be integrated in future SGs. (2) Why fog computing is suitable for SG systems? The answer is: the challenges faced by SGs (under traditional cloud computing form), such as latency constraints, new security issues, distributed control, prediction, demand-supply match, and complexity problems, all can be relieved with the performance supplement that fog provides.

Through the analysis of the existing research, we identified the state-of-the-art of fog-SGs. A generic architecture is proposed, and the main applications of fog computing are discussed from the perspective of core links of SGs. In summary, fog computing is mainly used for delay-sensitive or mission-critical applications, to support prediction, monitoring, and information interaction. These functions are the basis of microgrids, demand-side management, and communication network, which are the emerging areas in SGs.

From the perspective of theoretical research, hot issues in policy design are also introduced. They are divided into resource-related, task-related, security-related, and comprehensive issues, which usually are summed up as an optimization problem with such constraints. The optimization objectives mainly include delay, energy consumption, satisfaction, system balancing, resource efficiency, safety, credibility, etc. Since applications in SGs usually involve multiple participants and their benefits need to be balanced, game theories are commonly used to solve these problems. In addition, optimization algorithms such as genetic evolution algorithms are widely used as solutions.

In response to the evolvement of industrial applications and technologies, we also proposed research challenges and future directions for fog-enabled SGs from the aspects of security and privacy, huge amounts of data processing, fog and cloud combination, and fog device deployment.

In summary, this chapter has analyzed the advantages and potential value of applying fog computing in SGs. Considering many commonalities, many concepts presented in this chapter also apply to microgrids. Obviously, fog computing acts as a strong tool for designing optimized SG systems that can fulfill the emerging requirements of applications with its outstanding "compute, communicate, storage, and control" capability.

REFERENCES

1. X. Yu, C. Cecati, T. Dillon and M. G. Simões, "The new frontier of smart grids," *IEEE Industrial Electronics Magazine*, vol. 5, no. 3, pp. 49–63, Sept. 2011.
2. R. K. Barik et al., "FogGrid: Leveraging fog computing for enhanced smart grid network," *14th IEEE India Council International Conference (INDICON)*, Roorkee, 2017, pp. 1–6.
3. M. Krarti, "Utility Rate Structures and Grid Integration," *Optimal Design and Retrofit of Energy Efficient Buildings, Communities, and Urban Centers*, 2018, pp. 189–245.
4. Y. Yolda, A. Nen, S. M. Muyeen, A.V. Vasilakos and İ. Alan, "Enhancing smart grid with microgrids: Challenges and opportunities," *Renewable and Sustainable Energy Reviews*, vol. 72, pp. 205–214, 2017.
5. M. A. Sofla and R. King, "Control method for multi-microgrid systems in smart grid environment—Stability, optimization and smart demand participation," *IEEE PES Innovative Smart Grid Technologies (ISGT)*, Washington, DC, 2012, pp. 1–5.
6. F. Jalali, A. Vishwanath, J. de Hoog and F. Suits, "Interconnecting Fog computing and microgrids for greening IoT," *IEEE Innovative Smart Grid Technologies - Asia (ISGT-Asia)*, Melbourne, 2016, pp. 693–698.
7. S. Parhizi, H. Lotfi, A. Khodaei and S. Bahramirad, "State of the art in research on microgrids: A review," *IEEE Access*, vol. 3, pp. 890–925, 2015.
8. V. C. Gungor et al., "Smart grid and smart homes: Key players and pilot projects," *IEEE Industrial Electronics Magazine*, vol. 6, no. 4, pp. 18–34, Dec. 2012.

9. M. Chiang and T. Zhang, "Fog and IoT: An overview of research opportunities," *IEEE Internet of Things Journal*, vol. 3, no. 6, pp. 854–864, Dec. 2016.
10. F. Y. Okay and S. Ozdemir, "A fog computing based smart grid model," *International Symposium on Networks*, Computers and Communications (ISNCC), Yasmine Hammamet, 2016, pp. 1–6.
11. D. N. Palanichamy and K. I. Wong, "Fog computing for smart grid development and implementation," *IEEE International Conference on Intelligent Techniques in Control, Optimization and Signal Processing (INCOS)*, Tamilnadu, 2019, pp. 1–6.
12. A. Kumari, S. Tanwar, S. Tyagi, N. Kumar, M. S. Obaidat and J. J. P. C. Rodrigues, "Fog computing for smart grid systems in the 5G environment: Challenges and solutions," *IEEE Wireless Communications*, vol. 26, no. 3, pp. 47–53, June 2019.
13. S. Chen et al., "Internet of things based smart grids supported by intelligent edge computing," *IEEE Access*, vol. 7, pp. 74089–74102, 2019.
14. Y. Huang, Y. Lu, F. Wang, X. Fan, J. Liu and V. C. M. Leung, "An edge computing framework for real-time monitoring in smart grid," *IEEE International Conference on Industrial Internet (ICII)*, Seattle, WA, 2018, pp. 99–108.
15. Y. Zhang, K. Liang, S. Zhang and Y. He, "Applications of edge computing in PIoT," *IEEE Conference on Energy Internet and Energy System Integration (EI2)*, Beijing, 2017, pp. 1–4.
16. C. Jinming, J. Wei, J. Hao, G. Yajuan, N. Guoji and C. Wu, "Application prospect of edge computing in smart distribution," *China International Conference on Electricity Distribution (CICED)*, Tianjin, 2018, pp. 1370–1375.
17. S. Zahoor, N. Javaid, A. Khan, B. Ruqia, F. J. Muhammad and M. Zahid, "A cloud-fog-based smart grid model for efficient resource utilization," *14th International Wireless Communications & Mobile Computing Conference (IWCMC)*, Limassol, 2018, pp. 1154–1160.
18. S. Chouikhi, L. Merghem-Boulahia and M. Esseghir, "A fog computing architecture for energy demand scheduling in smart grid," *15th International Wireless Communications & Mobile Computing Conference (IWCMC)*, Tangier, 2019, pp. 1815–1821.
19. J. Yao, Z. Li, Y. Li, J. Bai, J. Wang and P. Lin, "Cost-efficient tasks scheduling for smart grid communication network with edge computing system," *15th International Wireless Communications & Mobile Computing Conference (IWCMC)*, Tangier, 2019, pp. 272–277.
20. M. Li, L. Rui, X. Qiu, S. Guo and X. Yu, "Design of a service caching and task offloading mechanism in smart grid edge network," *15th International Wireless Communications & Mobile Computing Conference (IWCMC)*, Tangier, 2019, pp. 249–254.
21. R. El-Awadi, A. Fernández-Vilas and R. P. Díaz Redondo, "Fog computing solution for distributed anomaly detection in smart grids," *International Conference on Wireless and Mobile Computing, Networking and Communications (WiMob)*, Barcelona, 2019, pp. 348–353.
22. N. Zhang, D. Liu, J. Zhao, Z. Wang, Y. Sun and Z. Li, "A repair dispatching algorithm based on fog computing in smart grid," *28th Wireless and Optical Communications Conference (WOCC)*, Beijing, , 2019, pp. 1–4.
23. A. Xu et al., "Efficiency and security for edge computing assisted smart grids," *IEEE Globecom Workshops (GC Wkshps)*, Waikoloa, HI, 2019, pp. 1–5.
24. J. Liu, J. Weng, A. Yang, Y. Chen and X. Lin, "Enabling efficient and privacy-preserving aggregation communication and function query for fog computing-based smart grid," *IEEE Transactions on Smart Grid*, vol. 11, no. 1, pp. 247–257, Jan. 2020.
25. L. Zhu et al., "Privacy-preserving authentication and data aggregation for fog-based smart grid," *IEEE Communications Magazine*, vol. 57, no. 6, pp. 80–85, Jun. 2019.

26. R. Yang, F. R. Yu, P. Si, Z. Yang and Y. Zhang, "Integrated blockchain and edge computing systems: A survey, some research issues and challenges," *IEEE Communications Surveys & Tutorials*, vol. 21, no. 2, pp. 1508–1532, Secondquarter 2019.
27. J. Wang, L. Wu, K. R. Choo and D. He, "Blockchain-based anonymous authentication with key management for smart grid edge computing infrastructure," *IEEE Transactions on Industrial Informatics*, vol. 16, no. 3, pp. 1984–1992, Mar. 2020.
28. K. Gai, Y. Wu, L. Zhu, L. Xu and Y. Zhang, "Permissioned blockchain and edge computing empowered privacy-preserving smart grid networks," *IEEE Internet of Things Journal*, vol. 6, no. 5, pp. 7992–8004, Oct. 2019.
29. J. Chen and X. Ran, "Deep learning with edge computing: A review," *Proceedings of the IEEE*, vol. 107, no. 8, pp. 1655–1674, Aug. 2019.
30. F. Y. Okay, S. Ozdemir and M. Demirci, "SDN-based data forwarding in fog-enabled smart grids," *1st Global Power, Energy and Communication Conference (GPECOM)*, Nevsehir, 2019, pp. 62–67.

3 Opportunities and Challenges in Electric Vehicle Fleet Charging Management

Chu Sun
McGill University

Syed Qaseem Ali
Opal-RT Technologies

Geza Joos
McGill University

CONTENTS

3.1	Introduction	34
3.2	EV Chargers	35
	3.2.1 Interfaces and Standards	37
	3.2.2 Features and Topologies	40
	3.2.3 Controls	40
	3.2.4 Capabilities	41
3.3	EV Aggregation	42
3.4	Available Ancillary Grid Services with Aggregated EVs	45
	3.4.1 Frequency Response and Regulation	45
	3.4.2 Power Smoothing	47
	3.4.3 Load/Generation Following	49
	3.4.4 Spinning Reserve	50
	3.4.5 Reactive Power Support	51
	3.4.6 Voltage Support	51
	3.4.7 Discussion	53
3.5	Case Studies	53
	3.5.1 Frequency Regulation	53
	3.5.2 Power Smoothing	55
	3.5.3 Load/Generation Following	55
	3.5.4 Spinning Reserve	55
	3.5.5 Voltage and Reactive Power Support	56

3.6   Challenges and Future Research Directions..56
    3.6.1   Technology Initiatives..56
    3.6.2   Economical Aspects 58
    3.6.3   Environmental Aspects...59
    3.6.4   Safety and Security...59
    3.6.5   Future Directions..60
3.7   Conclusion 61
References..61

3.1 INTRODUCTION

Sales of plugin electric vehicles (EVs) have been continuously increasing around the world over the last few years. Numerous manufacturers, new and old, are getting into the EV market [1,2]. The increased number of EV chargers is expected to impact the grid in terms of load increase, requiring infrastructure upgrade and capacity expansion [3]. Due to the schedulable nature of the EV loads, they can be leveraged to the benefit of the grid by proper coordination [4]. Since the batteries in the EVs can serve for energy storage as well, they can be used to provide vehicle-to-grid (V2G) services [5]. They can also be used in conjunction with various renewable energy resources to increase reliability and hosting capacity of the power system while reducing the effects of load increase on the grid [6]. Since individual vehicle charger interfaces and batteries are not capable of making a significant impact on providing these services, an aggregating entity is necessary to assemble enough power to be considered as a player in the market [7]. Some of the V2G services have been shown to provide economic benefits to the vehicle owners as well as the aggregating entities [8] and increased reliability and stability for the system operator [9]. The economic benefits, however, can be significantly reduced by battery degradation due to cycling [10]. The economic impact is highly dependent on the market operation in which the EVs participate and depends on the coordination and scheduling framework. The technology and infrastructure required are common to all V2G service platforms. Several aspects of the V2G technology are considered. Various battery charger topologies are reviewed in Ref. [11–13], and the impact of V2G technologies on the distribution grid are considered in Ref. [14,15]. A review of the various frameworks of the V2G and their extension to vehicle to home and vehicle to vehicle are presented in Ref. [16].

    This chapter provides a review of charger topologies and their associated control principles along with the functions that need to be implemented by the EV and charger manufacturers to provide V2G services. It also reviews the literature on V2G service frameworks and technologies proposed for the vehicles and the associated control functions that need to be implemented in the chargers by their manufacturers to enable such use. Challenges that impede a wider acceptance and implementation of V2G are also presented. This review will help the reader to understand the relations between the various power electronic interfaces of EV battery chargers and their capabilities in providing V2G services.

    The chapter is organized as follows: Section 3.2 gives a brief overview and classification of the typical topologies of EV battery chargers; Section 3.3 introduces the idea of EV aggregation, while Section 3.4 presents the ancillary services that can be provided using EV aggregation. Section 3.5 presents case studies of EV aggregation providing ancillary services from experiences around the world. Various challenges and future trends of EV fleet aggregation are presented in Section 3.6 with Section 3.7 concluding the chapter.

3.2 EV CHARGERS

Chargers are typically classified according to their interfaces and power levels as shown in Table 3.1. The table also shows the types of power interfaces for each level and their placement, i.e., on-board or off-board. The charging time of a given vehicle battery varies with the power of the charger used, e.g., high-power chargers will result in reduced charging times.

Many power electronic interfaces for EV chargers have been proposed in the literature. They can be broadly classified into chargers that transfer power to the vehicle by induction [17] or by conduction [18]. Inductive power transfer (IPT) chargers are in a development stage and require wider deployment and greater maturity to compete with the conductive charging systems. Conductive chargers, however, are widely deployed with various features and capacities.

Chargers can completely reside on-board of the EV with only the power supply outside, it can partially reside on and off-board of the EV, or it can completely reside off-board of the vehicle with a direct DC interface to the battery. The equipment that resides off-board of the vehicle is called the EV supply equipment (EVSE). The EVSE can be connected to the battery via several paths including directly, via a DC/DC converter, via a dedicated charger, or via another component or set of components on the electric power train for the EV as shown in Figure 3.1.

The batteries used in ful EVs are rated at 400–600 V. Since the battery voltage may vary by up to 150 V from fully charged to deeply discharged, an additional DC/DC converter can be used to keep the output DC voltage constant. This output voltage creates the high voltage (HV) bus. The traction motor is then connected to this bus via a DC/AC inverter. The DC/DC converter connected to the battery and the traction inverter is designed to handle bidirectional power flow. Power flows in the opposite direction, from the motor to the battery, during regenerative braking and charges the battery. Another DC/DC converter then steps down the HV bus voltage to the low voltage (LV) bus level for the auxiliary loads of the vehicle including air conditioning, power steering, and lighting. The LV bus voltage is conventionally selected to be between 12 and 48 V for safety reasons [19]. The charger output can be connected directly to the battery terminals or to the HV bus having an input that is single phase AC, three-phase AC, or DC depending on the power level and the design of the charger.

The charging system of the EV electric power train shown in Figure 3.1 requires independent power electronic converters for traction drive and charger, which increases the overall volume, weight, and cost. To address this, an integrated charging system is proposed, which uses the winding of the traction motor as filter inductor for the EV charger, while the same power electronic interface can be employed for traction [12]. The same principle can be applied to the integrated EV charger with six-phase permanent magnet

TABLE 3.1
Charger Classification

Charging Level	Charging Power	Vehicle Interface	Charger Placement	Power Interface
AC Level 1	1.4 kW (12 A) 1.92 kW (16 A)	Single phase 120 V_{AC}	On-board	Any single phase for opportunity charging – residential
AC Level 2	≤19.2 kW (80 A)	Single phase 208–240 V_{AC}	On or off-board	Dedicated EVSE – commercial and residential
AC Level 3	>19.2 kW (80 A)	Three phase 208–480 V_{AC}	On or off-board	Any three-phase outlet or dedicated EVSE – commercial and industrial
DC Level 1	50 kW (80 A)	200–500 V_{DC}	Off-board	Dedicated EVSE – commercial and industrial
DC Level 2	100 kW (200 A)	200–500 V_{DC}	Off-board	Dedicated EVSE – commercial and industrial
DC Level 3	>100 kW (200 A)	200–500 V_{DC}	Off-board	Dedicated EVSE – commercial and industrial

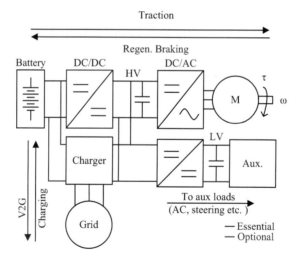

FIGURE 3.1 A typical electric power train of an EV.

synchronous motor drive. However, the torque generated during charging must be canceled to avoid motor rotation during charging [20].

The following subsections discuss further possible classifications of chargers according to the typical standards applicable to EV chargers along with their interfaces, their topologies, their controls, and finally according to their capabilities in terms of sourcing and sinking active and reactive power.

3.2.1 INTERFACES AND STANDARDS

The input voltage and the power ratings for various chargers are defined as different levels by the SAE J1772 standard (for North America area) [18]. The classifications are summarized in Table 3.1. The table also presents the operating voltage levels, the charger placement, and the conventionally used interfaces. Injected harmonics and other power quality requirements for systems connected to the grid are standardized by the IEEE and IEC [21–23] and by the SAE specifically for EV chargers in J2894-1 [24]. Vehicles operating in the V2G mode are viewed as a distributed resource (storage), for which the IEEE 1547 standard [25] could be applied. Electromagnetic emission limits are defined by IEEE in Ref. [26], and the limits for electromagnetic immunity are defined by the SAE in Ref. [27]. Galvanic isolation from the grid is a preferred option to meet the safety requirements of the EVSE handler. However, for high-power chargers using transformers to isolate from the grid increases the weight, size, and cost of the charging system and can be avoided by using residual current monitoring devices and disconnecting the system if the residual current increases beyond the limits set by various countries (5–30 mA).

The standards set a minimum charging efficiency of 90%, the charging THD to 10%, the discharging THD to 5%, and the minimum power factor while charging to 0.95. Individual limits for various harmonic groups are also defined in applicable standards. Various other standards exist for charge couplers and communications. They are outside the scope of this review.

The commonly used interfaces and connectors for the chargers are shown in Table 3.2, along with their diagram, ratings, regions where they are most commonly used, and the standards that define them. The first column of the table shows the interfaces that the chargers are plugged into. There are more variants of such interfaces, but only the most common ones are listed.

The standards related to V2G services involve grid integration requirement, EV connector, EV charging system, and communication links. The standards of EV charging connectors are summarized in Table 3.2. The integration of EVs and charging stations as distributed energy resources (DERs) with utility grid is specified in IEEE Std.1547-2018 [25]. For EV charging systems, four charging modes are defined, and their major parameters are summarized in Table 3.3 [28,29]. The relevant standards for each mode in different regions of the world are listed in Table 3.4 [30], ranging from charging station and electromagnetic compatibility (EMC) to the protection system. For the communication between EV and EVSE, generally three standards can be found [31]: (1) ISO 15118, which is based on power line communication (PLC) and applies to AC or DC charger. It lacks sufficient security measures and suffers from poor response times (60 seconds). (2) SAE J2847, also using PLC and suitable for AC or DC charging application. The communication is based on the SEP 2.0 protocol and has small response time (less than 1 seconds), but it is mainly designed for aggregation with nonvehicle smart home technologies. The use cases of SAE J2847 standards are defined in SAE J2836 series standards. (3) CHAdeMO, which is based on controller area network and applies to DC charging only. Its response time is also less than 1 s, but there is

TABLE 3.2
Charging Interfaces for EVs

Plug Type	Connector Name	Ratings	Regions	Supported Charging Type	Standards
	CEE 7/7, 7/16 NEMA 1-15, 5-15 Type 1 Type G	110–240 V_{AC} 2.5–15 A_{AC}	All over the world	1φ AC: Level 1 & 2	CEE 7; NEMA WD-6; IEC TR 60083:2015
	SAE J1772-2009	120–240 V_{AC} 16–80 A_{AC}	North America; Japan	1φ AC: Level 1 & 2	SAE J1772
	SAE J1772 CCS Type 1	120–240 V_{AC} 16–80 A_{AC} 200–600 V_{DC} 200 A_{DC}	North America; Japan	1φ AC: Level 1 & 2; DC Level 3	SAE J1772
	Mennekes Type 2	250–400 V_{AC} 63 A_{AC}	Europe	1φ & 3φ AC: Level 1, 2 & 3	IEC 62196
	EU DC CCS Combo Type 2	250–400 V_{AC} 63 A_{AC} 200–850 V_{DC} 65–200 A_{DC}	Europe	1φ & 3φ AC: Level 1, 2 & 3 DC Level 3	IEC 62196-3
	Chademo Yazaki	500 V_{DC} 120 A_{DC}	Europe, Japan, and North America	DC Level 1 & 2	IEEE 2030.1.1TM 2015
	Tesla	110–250 V_{AC} 12–80 A_{AC}	Europe and North America	1φ AC: Level 1, 2	Tesla proprietary
	GB/T 20234.3	750/1,000 V_{DC} 80–250 A_{DC}	China	DC: Level 1, 2 & 3	GB/T 20234.3-2015

no vehicle identification or guaranteed security, and it only works with CHAdeMO chargers (high-capacity DC). For V2G services to be available, communication should also be established among EV, EVSE, and the grid side. The information model of EV or EVSE can be defined based on common information model standards such as IEC 61850, IEC 61970, IEC 61968, and IEC 62325. The relevant communication protocols include Open Charge Point Protocol (OCPP), Hubject's Open Interchange Protocol (OICP), and other proprietary protocols. Some metering communication standards on the EV side and EVSE side are also defined. A summary of the communication standards is given in Table 3.5 [32–36].

TABLE 3.3
Different Charging Modes and Their Major Parameters

Mode	Maximum Current and Voltage	Description
Mode 1	16 A and 250 V (AC, 1-phase) 16 A and 480 V (AC, 3-phase)	Conductive connection between EV and a standard socket-outlet of an AC supply network without communication or additional safety features
Mode 2	32 A and 250 V (AC, 1-phase) 32 A and 480 V (AC, 3-phase)	Conductive connection between EV and a standard socket-outlet of an AC supply network with communication and additional protection features (in the cable)
Mode 3	(1): 32 A and 250 V (AC, 1-phase) (2): 70 A and 250 V (AC, 1-phase) 63 A and 480 V (AC, 3-phase) (3): 16/32 A and 250 V (AC, 1-phase) 63 A and 480 V (AC, 3-phase)	Conductive connection between EV and an AC EVSE permanently connected to an AC supply network with communication and additional protection features (permanent on the wall).
Mode 4	(AA): 200 A and 600 V (DC) (BB): 250 A and 600 V (DC) (EE): 200 A and 600 V (DC) (FF): 200 A and 1,000 V (DC)	Conductive connection between EV and an AC or DC supply network utilizing a DC EVSE, with (high-level) communication and additional safety features (permanent in the installation)

TABLE 3.4
Standards of EV Charging System in Different Regions

Mode	USA/Canada	EU	China	Japan
Mode 1	Not allowed	CE; LVD 2014/35/EU (In some countries), EN 61851-1, EN 61851-22	Not allowed	Restricted
Mode 2	NRTL; UL 2594, UL 2231-1, UL 2231-2, UL 991, UL 1998, CSA C22.2 No. 280, CSA C22.2 No. 281-1, CSA C22.2 No. 281.2, CSA C22.2 No. 0.8	CE; LVD 2014/35/EU, EMC 2014/30/EU, EN 61851-1, EN 61851-22, EN 62752	GB/T 18487.1-2015, GB 22794-2008, NB/T 42077-2016	PSE; IEC 61851, JIS C 8221, JARI A 0101
Mode 3	NRTL; UL 2594, UL 2231-1, UL 2231-2, UL 991, UL 1998, CSA C22.2 No. 280, CSA C22.2 No. 281-1, CSA C22.2 No. 281.2, CSA C22.2 No. 0.8	CE; LVD 2014/35/EU, EMC 2014/30/EU, RED 2014/53/EU; EN 61851-1, EN 61851-22	GB/T 18487.1-2015, NB/T 33002, NB/T 33008.2	PSE; IEC 61851, JIS C 8221, JARI A 0101
Mode 4	NRTL; UL 2202, UL 2231-1, UL 2231-2, UL 991, UL 1998, CSA-C22.2 No. 107.1, CSA C22.2 No. 281-1, CSA C22.2 No. 281.2, CSA C22.2 No. 0.8	CE; LVD 2014/35/EU, EMC 2014/30/EU, RED 2014/53/EU; EN 61851-1, EN 61851-23, EN 61851-24	GB/T 18487.1-2015, NB/T 33001, NB/T 33008.1	CHAdeMO; JIS D 61851-23:2014, JIS D 61851-24:2014, JIS TS D 0007

TABLE 3.5
Communication Standards for V2G Services

EV-EVSE	EVSE-Grid
Metering Communication: ANSI C12.22 and IEEE 1703	Automation of DER: IEC 61850, Distributed Network Protocol (DNP3), and IEEE 1815
PEV as a DER: SAE J2847/3 and ISO 15118	Metering Model: ANSI C12.19, IEEE 1377
AC PEV smart charging standards: SAE J2847/1 and ISO 15118, DIN SPEC 70121:2014, IEEE 2030.5.	Multiple information models: CIM IEC 61970, IEC 61968, IEC 62325, and IEC 61850 data models
DC PEV charging standards: SAE J2847/2, ISO 15118, DIN SPEC 70122:2018, IEC 61851-24, and CHAdeMO (2030.1.1), GB/T 27930-2015	Communication protocols exist for charging station networks: OCPP, eMI3, OICP and proprietary protocols such as ChargePoint, Blink Network, SemaConnect, etc.
Wireless PEV charging standard: SAE J2847/6	
PEVs and customers: SAE J2847/5	

3.2.2 Features and Topologies

The chargers can be classified as unidirectional and bidirectional on the basis of the power flow, as single phase and three-phase chargers on the basis of the number of phases used, and as isolated and nonisolated chargers on the basis of the galvanic isolation provided. All the classification choices impact their capability to participate in V2G schemes and services. Each type is shown in Figure 3.2a–e. Typical single and two-stage unidirectional battery chargers with power factor correction (PFC) are shown in Figure 3.2a and b, respectively. The dashed line represents the third phase for three-phase chargers. Chargers can also be categorized as being single-stage or dual-stage based on their topology. The operations performed by the charger must meet the grid interface standards on the AC input side and control the DC current and/or voltage at the output to charge the battery.

3.2.3 Controls

The chargers shown in Figure 3.1 first have a PFC control loop that ensures the current drawn from the grid is at a controlled power factor, which, for unidirectional chargers, should be unity (or within acceptable limits as defined by the standards). There are typically two control objectives for any battery charger: (1) control the input current to be sinusoidal in shape preferably in phase with the input voltage and (2) control the charging current or output voltage as required by the charging algorithm. Single stage chargers perform both functions together, while dual stage chargers use the first stage to perform the first function while maintaining the DC link, and the second stage performs the second function independently. Typical control loops used for these functions differ for unidirectional and bidirectional chargers and are shown in Figure 3.3a–d. In the figures, the block labeled "*Mod*" represents the modulator with pulse width modulation being the most common modulator used.

EV Fleet Charging Management

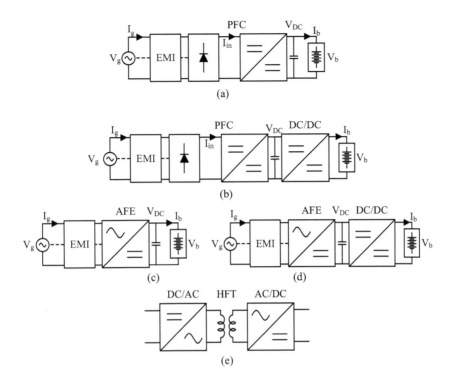

FIGURE 3.2 Battery charger topologies: (a) unidirectional single stage; (b) unidirectional dual stage; (c) bidirectional single stage; (d) bidirectional dual stage; and (e) isolated DC/DC converter.

3.2.4 Capabilities

The power electronic interfaces used in chargers can operate only in certain quadrants of the power plane. As per their capabilities, the chargers can be divided into the following three types (defined as Type I, II, and III for the sake of this chapter) as shown in Figure 3.4.

1. Type I Chargers

    Chargers that are fully bidirectional are either single stage with an active front end (AFE) or dual stage with a bidirectional DC/DC stage cascaded with an AFE. In theory, they can provide ±1 pu reactive power and ±1 pu active power independently. When providing both active and reactive power, the apparent power is limited to the charger kVA rating, S_{max}. The operating area of Type I chargers in the power plane is shown in Figure 3.4a.

2. Type II Chargers

    Unidirectional chargers with an AFE can provide ±1 pu reactive power, but since they cannot handle reverse power flow, the absorbed active power can only be varied from 0 to 1 pu. The operating point must remain within the kVA rating, the circle defined by S_{max}. The operating area of Type I chargers is shown in Figure 3.4b.

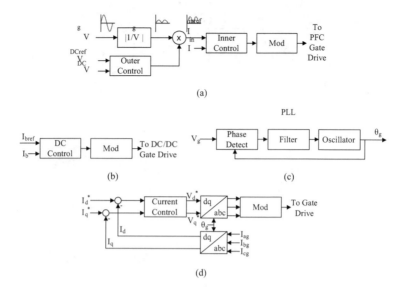

FIGURE 3.3 Battery charger control loops: (a) front end control; (b) DC/DC converter control; (c) angle detection phase locked loop (PLL); and (d) d-q control for front end three phase converters.

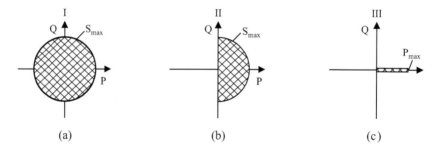

FIGURE 3.4 Charger active and reactive power capabilities for (a) bidirectional charger; (b) unidirectional charger with AFE; and (c) unidirectional charger with passive rectifier.

3. Type III Chargers

Unidirectional chargers that have a passive rectifier with a PFC preregulator on the front end will only be able to absorb 0–1 pu active power and are only capable of operating at unity power factor, i.e., $Q=0$ even if the second stage is a bi-directional DC/DC converter. The operating area of type III chargers is shown in Figure 3.4c.

3.3 EV AGGREGATION

Since EV chargers are directly interfaced with the batteries of EVs, they can be used to provide certain grid services, as long as they are capable of receiving external power setpoints in addition to the setpoint required to charge the EV battery. To be able to provide ancillary services to the grid, for example, to contribute to the grid primary frequency response, the available power should be at least 1 MW [37] (the number may be different depending upon the strength of the grid at the point of connection). A 1 MW interface is not realizable with a single charger; however, an entity responsible for aggregation of multiple chargers can negotiate a contract with the utility and provide a significant (e.g. 1 MW) controllable capacity while maintaining individual contracts with the vehicle owners [4]. This however requires predictable charging habits of the vehicle owners and large number of individual vehicle owners to minimize uncertainty caused by the variation in the participation of the individual EV owners [38]. Commercial fleet owners are more suitable for such applications, as the fleet availability can be predicted based on the fleet schedule. The charger interfaces for such fleets are also higher in power to ensure timely charging of high capacity batteries, when compared to passenger vehicles [39], thus requiring lesser number of EVs to aggregate large amounts of power.

An analysis was done to evaluate the number of vehicles required to aggregate 1 MW of power based on different types of EVs. Passenger EVs can be broadly classified into regular, extended range, and high-performance vehicles. Regular vehicles are either EVs or PHEVs, with a battery capacity ranging from 4 to 40 kWh. To completely charge the vehicle in 1 hour, the charger has to be rated at 4–40 kW and with chargers of these power ratings, the aggregator has to contract 25–250 vehicles depending on the power ratings of the individual chargers aggregated to accumulate 1 MW of available power. If the charging is allowed to be completed in 8 hours instead, the charger ratings change to 0.5–5 kW, and with these charger ratings, the aggregator will need to contract 200–2,000 vehicles depending on the power ratings of the individual chargers aggregated to accumulate 1 MW of available power. A similar analysis was carried out for extended range and high-performance passenger vehicles, light-duty, medium-duty, and heavy-duty vehicles. The results are shown in Table 3.6. It shows that the nature of aggregation changes with the type of vehicles used. For example, passenger vehicles usually have a defined availability pattern, i.e., they are expected to spend most of the time either in their garage (at night) or in a parking lot at work (during the day) for most urban and suburban settlements. When parked in their garages, the aggregator needs to ensure communication with various vehicles (assuming a contract exists between its owner and the aggregating entity) that may not be as physically close as in a parking lot or in a fleet garage.

Figure 3.5 shows the approximate power levels that can be achieved by aggregating vehicles with similar chargers and the grid services that they can be offered with such aggregation [40]. If only Level 2 chargers were aggregated, around 1 MW could be achieved with 89–160 vehicles which is a realistic number for fleet or parking lot operators. Depending on how many vehicles are aggregated, these vehicles can offer the available capacity for frequency regulation services or they can offer connected reserve service that can respond within minutes. To achieve higher power levels, the number of vehicles required will be unrealistic. DC fast chargers do provide an opportunity to accumulate a higher available power with fewer vehicles, but the purpose of installing fast chargers is to save time while charging, so managing the number of connected vehicles to obtain the required total power may be a challenge. However, DC fast stations with multiple DC fast chargers are integrated with energy

TABLE 3.6
Vehicle Types and the Typical Specifications

Vehicle Type	Battery Capacity (kWh)	Chargers Required for 1 hour charging	Total Vehicles for 1 MW	Chargers Required for 8 hour charging	Total Vehicles for 1 MW	Aggregating Entity
Passenger EV/ PHEV	4–40	AC Level 2 AC Level 3	25–250	AC Level 1 AC Level 2	200–2,000	Parking lot owners Commercial aggregators
Extended Range Passenger EV/ PHEV	15–50	15–50 kW AC Level 2 AC Level 3	20–67	AC Level 1 AC Level 2	160–530	-
High Performance Passenger EV	50–90	AC Level 2 AC Level 3	12–20	AC Level 2	89–160	-
Light Duty EV	11–90	AC Level 2 AC Level 3 DC Fast Charger	12–91	AC Level 1 AC Level 2	89–728	Parking lot owners Fleet owners
Medium Duty EV	100–150	AC Level 3 DC Level 2	7–10	AC Level 2	54–80	Fleet owners
Heavy Duty EV	200–400	AC Level 3 DC Level 2	3–5	AC Level 3 DC Level 1	20–40	

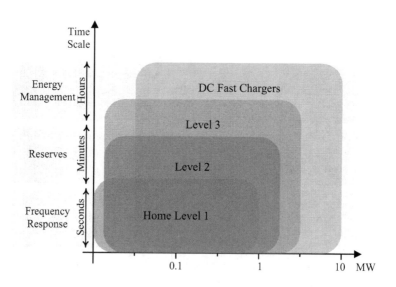

FIGURE 3.5 Charger levels and grid service capabilities.

FIGURE 3.6 EV aggregation, bidding, and disaggregation process.

storage and renewable energy resources form an aggregating entity, which can be treated as virtual power plant or microgrid, thus providing usable grid services [41].

There are generally two options for the relationship between the aggregator and EV owner/operator: a noncontractual and a contractual form [42]. The former involves a free participation and "pay-as-you-go" type of remuneration and the second a regular payment for the obligations of the service. Three types of contracts are defined based on the manner and degree that EV can be controlled: price-based, volume-based, and control-based [43]. Price-based contracts are associated with a price signal for the activation of V2G, by which the driver defines a minimum price to provide V2G service. Volume-based contracts involve the commitment of a predefined volume of energy within a certain time interval, while the driver cedes control to the aggregator as soon as the EV is plugged in with control-based contracts. The aggregated EVs can then bid in the service market based on the price and reward opportunities [44]. After market clearing, the aggregator is also in charge of the disaggregation process based on the service request considering the state of the EV and the expectation of EV owners, as illustrated in Figure 3.6.

3.4 AVAILABLE ANCILLARY GRID SERVICES WITH AGGREGATED EVs

Aggregated EVs can be used to offer various ancillary grid services. Ancillary services are defined by the US Federal Energy Regulatory Commission as the support services provided to the main electric power transmission function to maintain reliable operation of the electric power system. Various ancillary services have been defined and identified for third-party storage specifically [40] and for DER operators generally [45]. New ancillary services are also defined at the distribution grid level due to the deeper penetration of distributed renewable energy resources such as battery storage in EVs [46]. The services that can be provided, along with the examples found in the literature, by aggregated EVs are introduced and explained next.

3.4.1 Frequency Response and Regulation

Frequency response and regulation services are offered when a generation unit (or a group of units), aggregated EVs in this case, can change its output in response to the load fluctuations on a second to minute basis. EVs make a good candidate for this service as they are interfaced with power electronic converters, and the response rate of such converters can be very fast (only limited by the communication delays) as there is no inertia involved, unlike rotating generators. Typically, a 1%–2% of the peak load is required to perform regulation [45]. Since there is zero net energy exchanged for this service, it does not impact the overall battery charging process. Providing this service, however, limits the available charging power to the vehicle at a given time. This can be resolved by prioritizing the battery charging function at a compromised charging power and only participates with the remaining capacity resulting in a slow charging process or by participating in regulation only when the battery has reached 80%~90% of its charge. If regulation is performed using aggregated EVs, this can help offset more expensive fast response generators.

Various aggregator frameworks have been proposed in the literature that achieve frequency regulation. A framework that optimizes for aggregator revenue is proposed in Ref. [38] such that the available regulation capacity is modeled as a function of the battery state of charge of each vehicle. The work in Ref. [47] maximizes the available regulation capacity while scheduling the chargers for valley filling. The aggregator framework in Ref. [48] shows a method of aggregating unidirectional and bidirectional power chargers while considering their respective charging modes (i.e., constant current and constant voltage) and [49] presents an optimization approach for a fractional order controller used for load frequency control using aggregated EVs. The EVs in these papers are modeled as a large static battery, which has a collective SoC and power limits based on the aggregated individual EVs. This modeling approach neglects the dynamics of the effective available regulation capacity due to EV mobility.

A hierarchical model predictive control based aggregation framework with the EV mobility modeling to follow a regulation signal is proposed in Ref. [50]. The proposed framework works for all three levels of control, i.e., primary, secondary, and tertiary [51] and proposes an economic analysis strategy for the available regulation reserve (up and down) from EVs modeled with their dynamic mobility behavior, while [52] presenting a combined aggregation framework for plugin EVs, thermal loads, and CHPs for load frequency control with their individual uncertainties. The authors in Ref. [53] present a framework to follow a regulation signal that results in increased regulation capacity for the aggregator and therefore increased fleet utilization while meeting the consumer demand. The research in Ref. [54] presents a frequency regulation framework that either keeps the battery state of charge static or charges it while providing frequency support to the grid and for an interconnected area in Ref. [55]. Reference [56] proposes a regulation (up and down) capacity estimation algorithm for the aggregator by modeling it as an EV queuing network. The authors in Ref. [57] present contracting terms to maximize revenue for the aggregators with real regulation capacity estimation and [58] studies the effects of bidding in frequency regulation market that provides more compensation for a higher quality of regulation, i.e., a higher rate of ramping up and down.

EVs can be coordinated with other DERs for frequency regulation. An EV fleet which is coordinated with a high penetration of wind generators is presented in Refs. [59,60], while an economic analysis is presented in Ref. [61] for a system with high penetration of solar and wind generators. Integrated control of data centers and PEVs is demonstrated in Ref. [62].

A fair distribution of the regulation power among the participants is presented in Ref. [63] by introducing a fairness evaluation index and approaches for fair allocation. The results differ as the fairness basis is changed. Reference [64] proposes a

EV Fleet Charging Management

fair allocation charge scheduling for regulation for an aggregator that models the EV with its mobility uncertainties along with external energy sources (i.e. the vehicle getting charged from another entity).

The impact of communication delays and uncertainties on the provision of regulation service to the grid is assessed in Ref. [65]. As opposed to the central controller approach for frequency regulation, the controller can also be of a distributed nature such that communication requirements are minimized, and the control is performed with local droop control that is calibrated to meet the global objectives as shown in Ref. [39].

1. Charger Function for Frequency Regulation

    To provide regulation services, the charger should be capable of reacting to the power imbalance in the grid that is reflected as a change in the frequency of the network area. Therefore, the charger should be capable of varying its AC power output (or input) corresponding to a change in frequency at the node of connection. A general frequency vs. watt curve for this function is shown in Figure 3.7. Regulation can be provided by the charger either by curtailing the charging power when operating at rated power or by increasing the charging power when operating at lower than rated operating point [66]. The operating point of the charger limits the available regulation power for up and down regulation, respectively [55]. Falling or rising frequency can also have different slopes and different paths on the curve [67]. Type I chargers can provide a maximum of 2 pu (i.e. [−1 pu ↔ 1 pu]) regulation service due to their bidirectionality, while type II and type III chargers can only provide 1 pu (i.e. [0 ↔ 1 pu]) regulation, i.e., $-P_{max}=0$ in Figure 3.7.

3.4.2 Power Smoothing

Though not formally included in the existing ancillary service scheme, power smoothing services, especially high-frequency power smoothing, are proposed by many researchers to smooth out highly variable loads [60] or generators connected to the grid [68]. Due to the variable nature of the wind and solar energy resources, the output profile of these generators contains large peaks, and therefore, these sources

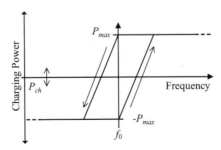

FIGURE 3.7 Charger control function for frequency–watt control.

need to be curtailed to ensure the stability of the network. Smoothing the output profile not only decreases the curtailed wind or solar power but also enhances the hosting capacity for such variable generation [69]. The service requires the provider to react in a seconds-to-minutes time range that is overlapping with or is slightly slower than the frequency regulation service.

Ramp rate-limiting services are introduced to compensate wind power fluctuations in the grid by power smoothing control at the point of common coupling (PCC) through demand response of EVs [70]. Power fluctuation smoothing of a wind farm located near a smart parking lot is achieved by an ADAptive Linear NEuron (ADALINE)-based method, which is more accurate than the conventional first-order low-pass filter-based approach [71]. A moving averaging-based wind power smoothing approach considering the constraints of wind ramping rate and the SoC of EV battery is proposed in Ref. [72]. A load power smoothing control using the EV storage capacity is also achieved by evaluating the load ramping rate and then allocating power output requirements to each EV based on an SoC-adaptive method [73]. When using EV batteries to smooth the voltage fluctuation due to renewable energy in the distribution system, a hull moving average (MMA) can eliminate the lag problem of the simple average method [74].

For real-time implementation and aggregation, a two-stage control strategy to smooth the wind power fluctuation is proposed in Ref. [75], with the first-order filter method adopted at the first stage to calculate the total power demand from aggregated EVs, while at the second stage, an aggregator management center is introduced to disseminate the power command to each load-management control center.

1. Charger Functions for Power Smoothing

    To offer a power smoothing service, a control loop or algorithm is required that reacts to the change in a generator power output [76] or a load power input [77]. This functionality is achieved by defining a curve that adds extra watts to the set output of the charger if the change in the renewable generator output is negative and vice versa. The change in P (ΔP) vs. watt curve for generation smoothing is shown in red and for load smoothing in black in Figure 3.8. A dead band is conventionally programmed to avoid reacting to small changes in the load.

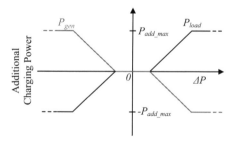

FIGURE 3.8 Charger control function for ΔP–watt control.

3.4.3 LOAD/GENERATION FOLLOWING

The following service, unlike frequency regulation, is an energy service that is required to make the profile of load or generation trackable by the dispatchable generators with limited ramping rate [78,79]. More general following services may also involve peak shaving and valley filling in the profiles, for which the following service can be treated as low-frequency power smoothing [80,81]. The following service is also termed as energy imbalance service in some literature [82]. Since this service requires a large amount of energy exchange, it depends on the state of charge of the individual EV batteries along with its total power capacity. Another issue with providing this service using aggregated EVs as a storage unit is that it requires the storage unit to be always connected to the grid. This is a difficult task to achieve for the aggregator as most EVs will have a pattern of availability. To realize this service, the aggregator will require significant coordination and optimization effort. The service requires reaction in a minute-to-hour time.

Various aggregator frameworks have been proposed in the literature that perform load and/or generation following. The probability of the location and charging state of EVs is calculated to determine the zonal available capacity to provide flexible ramping products for reliable power system operation [83]. A smart charge scheduling algorithm for the EVs was proposed in Ref. [84] that also coordinates with a wind generator such that the fleet discharges when the wind generator underproduces and charges when the wind generator overproduces. Network hosting capacity for wind was shown to be increased by optimal scheduling of vehicle charging in Ref. [85]. A coordination strategy between EV fleets and building energy management systems was presented showing that with optimal coordination, the peak demand on grid can be reduced in Ref. [86].

A scenario-based robust approach to tackle the uncertainty of EVs and wind power is described in Ref. [87] where the interdependency between the day-ahead market prices and the aggregator's bidding decisions is addressed using complementarity models. Vehicle uncertainty was modeled in Ref. [88], while a realistic battery model was included in the optimization formulation in Ref. [89]. A stochastic matching method for wind generation and EV charging load was proposed considering a sizeable fleet with a mobility pattern that complements the wind generation in Ref. [90] while [91] achieves the same using a three-level optimization problem for the energy management system without an aggregator. A decentralized scheduling algorithm for peak shaving and valley filling of the demand profile was proposed in Ref. [92]. Real-time charging control with distributed optimization to reduce communication overhead was presented in Ref. [93] and to reduce the computation burden on the central controller in Ref. [94].

A bidding strategy for the EVs to minimize the aggregator-owned renewable energy resource (RES) generation variation was proposed in Ref. [95], while [96] presents a bidding framework for aggregated EVs to bid for a downward reserve (i.e. the charging load is reserved for low price period). The authors in Ref. [97] present an aggregated control of wind power with EVs ensuring a match between the two by scheduling the EVs being charged and curtailing the wind. A distributed aggregation and coordination scheme was proposed [98] to increase renewable energy sources and EV hosting capacity that also improves the voltage profile of the network. Joint optimization for the aggregated EVs, carbon capture plant, and a fossil fuel plant was presented in Ref. [99] with an additional objective to limit emissions. Asset-aware aggregation frameworks that limit feeder capacity violations [100], transformer congestion [101], and feeder losses [102] and that limit the total load, voltage profile of the network, and the phase imbalances in the network due to single phase chargers were also proposed [103].

1. Charger Functions for Load/Generation Following

To offer load following or generation following services, a control loop or algorithm has to be added to the charger interface that reacts to the total power output by a generator [104] or the total power input by a load [105,106] that needs to be followed. The control needs a measurement signal from the generator (or load) being followed and the charger is required to proportionally change its output. The required watt vs. watt curve is shown in Figure 3.9. A dead band can be introduced to program a threshold after which the EVs start reacting to the watt signal being followed.

3.4.4 SPINNING RESERVE

Spinning reserves are required by the grid as online generation (or load) that can be ramped up (or ramped down) within 10 minutes of generation outage and sustain the operating point for typically up to 2 hours, the time required for the secondary reserves to takeover. It is demonstrated that EV providing spinning reserve service is more profitable than providing energy service only [107]. This service like the load/generation following function also requires net energy exchange during the time the service is provided. Therefore, it is also dependent on the state of charge of the individual EV batteries along with its total power capacity. To realize this service, the aggregator will also require significant coordination and optimization effort.

Various aggregator frameworks have been proposed in the literature that allow estimation of the available spinning reserve to bid in the reserve market. An autonomous distributed spinning reserve based on the expected EVs connected to the grid for charging is described in Ref. [67], while a distributed control strategy for the available reserve is proposed in Ref. [108]. A combined bidding framework for spinning

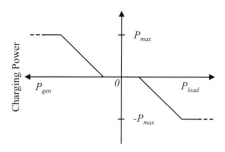

FIGURE 3.9 Charger control function for watt–watt control.

EV Fleet Charging Management

reserve and regulation for bidirectional chargers is proposed in Ref. [109] and for unidirectional chargers in Ref. [110], while a combined aggregation platform for regulation and spinning reserve is proposed in Ref. [111]. An estimation methodology for the true reserve available to the aggregator, factoring in the uncertainty in the mobility patterns of the vehicles, is proposed in Ref. [112]. Combined optimization of the cost of spinning reserve and the expected energy not supplied is performed [113]. A reserve provision for a network with a high wind penetration scenario using aggregated EVs is proposed in Ref. [114], while a synchronous machine emulating control for individual chargers and charging stations is proposed in Refs. [115,116] such that the chargers are able to implement the droop control with an emulated spinning rotor inertia reserve. The charger function already presented in Section 3.4.3 should also suffice for the implementation of spinning reserve applications as well.

3.4.5 Reactive Power Support

Reactive power support is required to increase the power quality and reliability of the grid [117,118]. It can be provided by the power electronic interfaces of the charger. The benefits of providing reactive power using aggregated EVs are that they provide it where it is needed, i.e., near the loads, reducing the power losses due to transmission of reactive power. It also replaces or reduces the requirements for capacitor banks on the distribution system. Since there is no energy exchange required, this service can be provided even when the batteries are fully charged; however, when the battery is being charged, the available reactive power is limited by the rating of the charger interface [119]. Reactive power is mostly required for voltage support and harmonic compensation applications. Since all these functions depend on the local operating variables, the communication burden is not increased significantly by offering this service.

Various aggregator frameworks have been proposed in the literature that allow reactive power support from the EV chargers. EV chargers capable of injecting reactive power to the grid as well as of bi-directional power flow are proposed in Ref. [120], while an integrated PV and battery charger system is proposed in Ref. [121] with a common bidirectional interface to the grid. The charging operation is controlled to compensate for the dips in the PV system power output while providing reactive power support to the grid which can be used to provide voltage support or other power quality services.

If the charger controls are designed to follow reactive power setpoints, provision of reactive power support can be supported by the charger.

3.4.6 Voltage Support

Typically, voltage support is provided by reactive power injection; however, in distribution grids (due to their resistive nature) voltage profiles rise due to the power injected by the distributed resources. Therefore, by curtailing the active power injected or by increasing the active power absorbed by the aggregated EVs, the voltage rise can be alleviated. Offering this service increases the hosting capacity of the grid for distributed generation.

Various aggregator frameworks have been proposed in the literature that provide voltage support service to the grid. A framework that improves the voltage profile of the network with the chargers to increase the EV hosting capacity of the network was presented in Ref. [122] with an assumption that EVs were always available for V2G service. A stochastic distributed optimization of reactive power with conditional ensemble predictions of V2G capability was introduced in Ref. [123]. A combined formulation for both voltage and frequency support was proposed in Ref. [124] for generalized reactive end-user devices including PEVs. A local and distributed voltage droop control was proposed in Ref. [125] that de-rates the charging power if the voltage goes down resulting in a better voltage profile, albeit at the cost of increased charging time for the vehicles. Another local voltage controller was proposed in Ref. [126] to reduce losses and communication requirements. An integrated EV and PV system was proposed in Ref. [127] that allows charging of the EV to support the PV output by charging when PV overproduces. Demonstration projects were presented in Ref. [128] for a fleet of EVs with unidirectional chargers that is able to provide V2G services like voltage support and frequency support.

1. Charger Functions for Voltage Support: Voltage–Watt Control

    To provide an active power voltage support service, a control loop that is capable of changing the charger maximum active power output in response to a change in the voltage of the node to which it is connected is discussed in Refs. [125,129]. The required voltage vs. watt control function is shown in Figure 3.10. It changes the maximum power limit in reaction to the voltage at the node it is connected to [130].

2. Charger Functions for Voltage Support: Voltage–VAR Control

    A similar control loop that changes the charger's reactive power output in response to a change in the voltage of the node it is connected to [131,132] is required for reactive power voltage support service. The required voltage vs. VAR control function is shown in Figure 3.11. This control is implemented as a supplementary service to the active power; therefore, the apparent power rating constraint must be respected. Falling or rising voltages can have different slopes and different paths on the curve. A dead band is usually programmed to avoid reaction to small changes in the voltage.

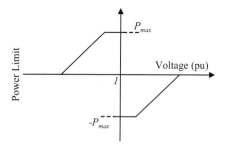

FIGURE 3.10 Charger control function for voltage–watt control.

EV Fleet Charging Management

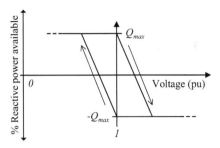

FIGURE 3.11 Charger control function for voltage–VAR control.

3.4.7 Discussion

Each service mentioned in Sections 3.4.1–3.4.6 requires a control loop that can be customized to provide a service based on local measurements of each EV [133]. The control loops can be implemented at the aggregator control center that applies the loop for each vehicle and then dispatches the set points [134] or they could be individually implemented in the EVs themselves. A purely centralized approach would require high processing power with the aggregator and a communication link with low latency, high bandwidth, and reliability. A purely decentralized approach, on the other hand, reduces the processing requirement and can operate with low bandwidth, high latency communication links [135]. A good compromise is when a hybrid hierarchical approach is used [136]. The aggregator may serve as a gateway to the rest of the vehicles, while only performing coordination and optimization functions, while each EV manages its own output.

The V2G services that can be offered using aggregated EVs, the functions required to be implemented on the chargers to offer the services, the reaction time required for the services, and their value propositions are summarized in Table 3.7. In the table, all except the last row are active power support functions and can be provided using type I chargers with maximum limits of 1 to −1 pu. Types II and III can provide the same function with limits of 0 and 1 pu. Similarly, the reactive power support function can be provided using a type I and II charger with limits equal to the rated charger power; however, type III charger does not have the capability of providing reactive power support.

3.5 CASE STUDIES

In this section, several case studies for the various types of ancillary services discussed in Section 3.4 are presented. The typical results selected are straightforward and illustrative to showcase different V2G services, while studies based on more advanced approaches can be found in the literature.

3.5.1 Frequency Regulation

In Ref. [137], the control strategy for EVs participating in voltage and frequency regulation is investigated. A frequency droop control is adopted, while the

TABLE 3.7
Types of V2G Services and the Typical Specifications

Grid Function	Control Function	Reaction Time	Value Proposition
Regulation/fast regulation	Frequency–watt control	Seconds–minutes	• Offset need for fast-response or very fast-response generation
Power smoothing	ΔP–Watt control	Seconds–minutes	• Increase the grid hosting capacity for highly variable generation • Increase the network reliability • Deferral of infrastructure upgrades to cater peaking loads
Load following	Watt–watt control	Minutes–hours	• Reduce generation variability • Extra generation sold elsewhere
Generation following	Watt–watt control	Minutes–hours	• Enhanced hosting capacity for highly variable renewable generation • Reduce generation curtailment for variable renewable generation
Energy arbitrage/load leveling/peak shaving	Direct P control	Hourly	• Allow peak power generation to be used else where
Spinning reserve	Frequency–watt control	Seconds–10 minutes	• Free up the generation reserve to be used elsewhere
Voltage active power support	V–watt control	Seconds	• Increase renewable sources hosting capacity at the distribution level
Voltage reactive power support	V–var control	Seconds	• Provide reactive power support near the loads • Increased power quality and system stability

available V2G regulation power is also constrained by the EV owners' transportation demand, initial and real-time battery SOC, and departure time. In case of voltage and frequency deviation beyond a threshold, the EVs will participate in regulation by fast adjustment of charging or discharging power without violating the constraints.

A case study of frequency response after 0.02 pu load increase was presented in Ref. [137]. Before EVs participate in frequency regulation, the maximum frequency deviation can reach up to 0.03 pu, which is reduced to 0.005 pu with the EVs participating in frequency regulation. The turbine output power of the conventional synchronous generators will also be affected. The overshoot of the mechanical power output can reach 0.025 pu without EVs' frequency regulation, which was almost completely eliminated due to EVs' control effect.

EV Fleet Charging Management

3.5.2 Power Smoothing

A real-time control strategy for aggregated EVs to smooth the fluctuation of wind-power is presented in Ref. [75]. The energy boundary model of EV battery is established to determine the charging/discharging margin of an EV at each moment. To smooth the power fluctuation mainly caused by wind variation, a first-order low-pass filter is used to determine the total power command of aggregated EVs, which is then allocated among individual EVs.

The effect of power smoothing control is demonstrated in the showcased results. In the disorderly charging mode, the maximum and minimum total load power is 4,440 kW and 2,050 kW, respectively, and the average total load power fluctuation rate is 0.0179. With the proposed control, the maximum and minimum are changed to 4,464 and 2,513 kW, respectively, and the average total load power fluctuation rate is reduced to 0.0078. The power fluctuation rate and the peak-valley difference are reduced by 56.4% and 18.4%, respectively. Therefore, not only the fluctuation rate of the total load is greatly decreased, but the peak-valley difference is reduced as well.

3.5.3 Load/Generation Following

A case study with EV used for load leveling (a general type of load following) is presented in Ref. [138]. Assuming that the target power curve with peak-shaving and valley-filling function is known, for example, determined based on grid efficiency, EV operation and its power, and the time-of-use power, a minimization function is designed to match the target power curve, and the V2G plan curve is proposed. The corresponding constraints include connected EV numbers, EV battery characteristics, and the EV user customized parameters such as charging period.

For a case study presented in Ref. [138], the valley-filling period approximately covers 23h00 to 7h00, and peak shaving occurs between 8h00 to 12h00 and 14h00 to 16h00. The maximum peak-shaving value is about 10% of the general power load at 10 am. The forecast load value is less than the target load value at 1 am when the grid is at the valley time, and its spare power can charge the EVs. The absolute value of the difference at this time is equal to the maximum power value of the grid supply and smaller than the total maximum charging demand of EVs. The charging power will then be limited to this value and distributed among individual EVs. On a daily scale, the V2G plan curve is close to the target curve except in the period between 19h00 and 21h00 because of the small number of connected EVs. As demonstrated, the load profile is greatly flattened, with the peak-shaving and valley filling control of EVs.

3.5.4 Spinning Reserve

The authors in Ref. [139] present a market-clearing mechanism with a probabilistic spinning reserve considering the uncertainty of wind and EVs. The system operator minimizes payments to generators and EV aggregators for providing energy and reserve, including energy cost, the start-up cost of generation units, reserve cost from conventional units, the cost of expected energy-not-supplied (EENS), and the cost of services from EV, with the presence of uncertain wind and EV participation. The proposed model is formulated and linearized so that it can be solved by means of available mixed-integer linear programming (MILP) solvers. It is shown that the EVs' participation in energy and reserve markets with competitive offer prices can counterbalance the increasing EENS and reserve costs of higher levels of uncertain wind profile.

Assuming that the price of V2G service is lower than the spinning reserve offered by generation units, the required amount of spinning reserve can be offset when EVs are used for this service. The results indicate the system operator schedules V2G only in the energy clearing process because of its low price. The spinning reserve and EENS are reduced by increasing V2G injected power. The operating cost is reduced by purchasing more V2G service and committing fewer conventional generation units.

3.5.5 VOLTAGE AND REACTIVE POWER SUPPORT

In Ref. [131], the sole effect of $P(V)$ voltage-watt droop charging (lower charging power at LV level), conventional voltage-var droop support $Q(V)$, and the combination of the two methods $[P\ Q](V)$ is examined for controlled EV charging with weak distribution grids. The control performance of the voltage profile due to a voltage dip at PCC is examined. Due to the high R/X ratio of the distribution feeder, the volt-var droop control from EV chargers alone is unable to meet the minimum voltage threshold, though the result is better than for the base case without any voltage support. In contrast, the $P(V)$ droop is more effective in improving the voltage, making voltage lie within the threshold. The combined voltage support method has the best performance regarding the effect of voltage regulation. The combined method can also partially relieve the adverse effect of the $P(V)$ droop charging, given that for the same voltage improvement, increased charging power and thus shorter charging time can be achieved by the additional voltage-var support.

3.6 CHALLENGES AND FUTURE RESEARCH DIRECTIONS

Despite the various frameworks proposed for the implementation of V2G technologies, the concept faces impediments in its wide acceptance leaving room for further research to be performed to develop practically implementable solutions. This section will elaborate on the challenges faced by the concept in the following subsections.

3.6.1 TECHNOLOGY INITIATIVES

1. Battery and Charger Technology

    One of the major drawbacks of providing V2G services is the additional degradation of the battery and its associated electronics due to additional cycling. Conventionally, EVs use high power lithium-ion batteries. Failure modes that reduce the battery lifetime include (1) overcharging; (2) overheating; (3) deep discharge; and (4) cycling. The battery lifetime is characterized by its residual usable capacity. By offering V2G services, the battery and its charger are cycled more often, which results in additional aging of the system reducing its lifetime, i.e., the usable capacity fades earlier than it would have if the battery was only used for vehicle propulsion. The provision of reactive power service does not cycle the battery. The charger, however, may be over-cycled in this case.

One way of addressing the issue of battery degradation is by online monitoring of the battery state of health [140] and proactively acting to avoid undue degradation of the battery. Another approach can be to include the battery degradation cost in the contract negotiated with the aggregator [141] or by adopting a framework that limits the services provided to the grid based on the cost of battery degradation [142,143]. A few solutions based on a second use of battery are also proposed [144]; however, the economic justification of such a use is unclear [145,146].

2. Communication Infrastructure

The communication requirements for fleet or parking space type vehicles are relatively simple because all the vehicles which have to be communicated with are in close physical proximity [39], and the required communication speeds are limited. For such cases, any available short-range wireless communication technology such as ZigBee, Bluetooth, and Wi-Fi or a wired technology such as GPIB and Ethernet is suitable. Sporadically contracted vehicles, however, require long-distance communication with distributed agents having large volumes of data exchange. This application requires the communication infrastructure to have a high bandwidth, low latency, high reliability, and greater security [147].

The communication infrastructure should ensure scalability, security, and privacy-preservation to allow integration of a large number of EVs by the aggregator. Currently, the concept of V2G has only been implemented in small scale demonstration projects using short-range communication networks. To be widely accepted, the concept entails a robust communication network that meets performance requirements [148,149]. Furthermore, the impact of aggregator–EV communication link uncertainties on the operation of the aggregator is yet to be ascertained.

3. Impact on Power Infrastructure

A high penetration of EVs may overload the distribution system when EVs are charged without any coordination measures implemented. Even with coordination, as the penetration of EV loads increases, an eventual overload on the network may result. This overload may not require immediate network upgrades as most of the distribution transformers are designed for cold pickup, i.e., for 200% load. On the other hand, if the growth of EV loads accelerate, a network upgrade will be required. Furthermore, not all EV users will participate in coordinated charging and providing V2G services. Integrating EVs with the grid will not only require a power network upgrade but will also require investment in additional energy generation. This additional energy can be provided by RESs; however, the stochastic nature of the generation must be matched with the stochastic nature of the EV load.

Additionally, since most V2G services require bidirectional power flow capability from the charger, a comprehensive recalibration and re-coordination of protection devices will be required to ensure adequate protection of the grid. With the trend toward fast charging and higher efficiency requirements, high-power chargers based on multilevel power electronic converters, wide-band-gap power switches, and solid-state transformers replacing the conventional bulky AC transformers can enhance the performance of EV charging stations and better enable V2G services to the grid [150,151].

3.6.2 Economical Aspects

1. Fair play

    There needs to be a strong economic incentive for the individual EV owner to allow the time-share use of their EV battery as a load leveling and load management device. For each ancillary service, the mechanism of market participation and payment should be designed to benefit both the grid and the EV owner without incurring economical losses to EV owner. This will mainly involve battery and charger life reduction and may result in disrupting travel plans and increased range anxiety. A complete business model of a V2G market design should take into account the critical cost and revenue elements of each player [152]. The cost elements include battery wear and replacement, charger over cycling, EVSE network and updates, communication hardware and software, and aggregator devices and software. The revenue elements, however, will involve a pool of aggregated vehicles, the EV plugged-in hours, and the available power and energy for V2G services. At the present stage, EV customers are motivated to participate in the most profitable service market such as frequency regulation. In the long term, they can also be encouraged to the services that are more urgent. For example, during an emergency with grid disconnection due to a major storm, the EVs within a community can provide emergency power for a limited period to selected devices, such as smartphones, lights, and refrigerators. In that case, the back-up power services will be given more importance and offered higher prices than services under normal circumstances. Furthermore, the issue of fair distribution of the service load to individual EVs is not easy to capture in contractual and operational terms. The concept of fairness varies depending on user, region, culture, and other socioeconomic factors. A step in this direction was regulation ensuring higher compensation for higher quality of reserve in the frequency regulation market [153].

2. Capacity estimation

    One of the issues that impedes the wide acceptance of the V2G concept is the ability of the aggregator to accurately estimate (or forecast if necessary) the total capacity available to provide specific V2G services. On the grid side, the estimation of the required capacity for given services becomes more challenging for the markets that have a lower time resolution, for example, the day ahead and hour ahead reserves and regulation market [154]. Furthermore, the structure of the regional energy and ancillary services market, along with the energy and power tariff scheme, may play an important role in the operation and control of the aggregated EVs. The market may allow for a viable business case (assuming the aggregation of adequate number of EVs) for the EV owners to participate in V2G services, as shown in Ref. [155] for California and in Ref. [61] for Florida. However, if the market price for such services is too low, due to various socio-political reasons, the economics for V2G services no longer hold [156]. The rates and prices of the grid services are not always set by economic factors. Socio-political issues significantly influence the electricity rates. For this reason, the solutions for removing such impediments are not straightforward.

3.6.3 Environmental Aspects

Climate change and air pollution are the two potential environmental concerns that V2G technology can help address. EVs can lower total emissions, compared to conventional gasoline-powered vehicles. By replacing high carbon electricity sources with renewable energy resources and replacing gasoline-powered vehicles with EVs, emission reduction and climate change benefits can be achieved. Moreover, the ancillary service markets can also be decarbonized (to an extent) by employing V2G technologies. However, research indicates that the energy mix in a utility power procurement, which might vary temporally and spatially, among other aspects, will also affect the decarbonization effect of V2G. It is pointed out that the penetration of EV participating in V2G services should exceed a certain ratio for the emission savings to exceed the additional electricity-based emissions as a result of charging [157]. By optimized charging schedules, V2G can achieve significant carbon emission reductions compared to conventional gasoline vehicles. However, for a grid with higher emission intensity, using V2G for load shifting may even increase the carbon emission [158]. In Ref. [159], environmental factors such as carbon reduction are incorporated in the impact of EV integration except for the conventional techno-economic benefits. Besides carbon reduction, the air pollution issues from gasoline oil such as NO_x, SO_x, etc. can also be relieved by the introduction of renewable energy and EVs.

However, EV battery manufacturing and disposal may also introduce some water and soil pollution from electrodes and electrolytes, especially heavy metals such as Pb, Cd, Ni, Co, and Mn and harmful acid and alkaline liquids. If EV battery life is significantly reduced in V2G services and more battery manufacturing is needed, such pollution issues will be more prominent.

3.6.4 Safety and Security

There are generally two types of safety issues in V2G services: (1) electrical safety when EV is used by the customer or is connected to EVSE. (2) Communication safety and cybersecurity, namely the privacy information shared by individual EV users with EV aggregator.

1. Electrical safety

    For AC charging, the AC arc can self-extinguish at the zero-crossing point, so the hazard of electrical shock and difficulty of protection will be lower. In contrast, DC-charging will not self-extinguish, and there is inadequate fault-sensing equipment and switchgear for protection. A DC arc does not have a zero-crossing point, rather a constant potential, if constant power is available. There are many causes for DC arcing, among which two types of arcing are most common for EV: a series arcing and a parallel arcing. Both types of DC arcs are difficult to detect but can be dangerous if not mitigated. Some of the standards on electrical safety and protection related to EV charging include IEC 61140, IEC 62040, IEC 60529, IEC 60529, ISO 6469-3, etc. [160].

2. Cybersecurity

    Grid power control signals and billing information will be transmitted over the communication channel. EV users will be concerned about the loss of data privacy such as the charging behavior or transaction details when using EV or providing V2G service [161]. Security for communication, or cybersecurity, must be guaranteed in the communication process. The need for security considerations can be seen in 802.11b and 802.1X security failures. It shows that poor implementations of an encryption algorithm can lead to security issues. The communication medium, either wired or wireless, will be the critical factor determining cybersecurity. Wireless networks, compared with wired networks, are more susceptible to security problems because the medium can be "tapped" from anywhere in the proximity of the network. Session hijacking, Denial of Service (DoS), and "rogue" access points are three common forms of attack on wireless networks. Wired networks can also be "tapped", but it requires a physical connection to the network. The standards on cyber-security include IEC 62351 and SAE J2931/7 [33].

3.6.5 Future Directions

1. Advanced EV charging networks with better placement and covering wider areas will be built [162]. These include slow EV chargers in home and urban environment and fast EV charging station integrating renewable energy resources and onsite energy storage for long-distance travel. Though V2G is mainly targeted at slow EV chargers, by aggregating fast chargers or using charging stations, V2G service may be also feasible, with more advanced coordination. Besides, new technologies such as wireless charging and unmanned vehicle driving will be matured to reduce the siting limitation of charging and make the V2G process smoother and less impactful to the EV users.
2. Internet and communication technology aided new business and operation modes such as blockchain and digital twin will boost V2G development. The blockchain technology adopting point-to-point (P2P) transaction mode

will digitalize and secure V2G business [163]. The digital twin or cloud-based platform can monitor the real-time status of EV and grid and ensure precise information procurement to control EV charging and V2G service provision [164]. The high-speed 5G communication technology will be a key driver in making these applications implementable. The merge of the Internet of Things (IoT), artificial intelligence (AI), sharing economy, smart grid, smart city, and cloud computing will make the integration of V2G service into the grid market more natural and the electrified transportation more intelligent [165].
3. More interdisciplinary studies will be carried out. As reviewed, the V2G integration will involve various factors, including technical, financial, socio-environmental, and individual/behavioral [166]. The previous studies mainly work on a single type of modeling or discipline, especially techno-economic assessment of V2G, but sophisticated behavioral and environmental models are not fully taken into account [167]. Consumers' notion, willingness, and expectation will be as important as technology in determining whether they will participate in V2G services. For this, more research based on multiple-method modeling approaches should be conducted.

3.7 CONCLUSION

The requirements, general charger topologies, and their control loops are reviewed in this chapter. It also presents the potential of EV batteries and their associated chargers to be used as distributed storage to provide ancillary grid services when aggregated. The services that can be provided depend on the power electronic grid interfaces of the chargers, the cumulative power of the chargers, and the cumulative energy of the batteries. The functions include fast frequency regulation, energy arbitrage and generation following in addition to the conventional ancillary support functions performed by synchronous generators such as reactive power compensation. The control functions required in the chargers to implement the services are also reviewed, and the impact of the charger topologies on their capability to provide the services is described. Challenges that impede wide adoption of the V2G services and the future research directions are also discussed.

REFERENCES

1. Statista. 2020. Worldwide number of electric vehicles in use from 2012 to 2019 (in 1,000s). Available: http://www.statista.com/statistics/270603/worldwide-number-of-hybrid-and-electric-vehicles-since-2009/.
2. E.D.T. Association. 2019. Electric drive sales dashboard. Available: http://electricdrive.org/index.php?ht=d/sp/i/20952/pid/20952.
3. A. Maitra, "Understanding the grid impacts of plug-in electric vehicles (PEV): Phase 1 study - distribution impact case studies" Electric Power Research Institute 000000000001024101, 2012.
4. M. A. Ortega-Vazquez, F. Bouffard, and V. Silva, "Electric vehicle aggregator/system operator coordination for charging scheduling and services procurement," *IEEE Transactions on Power Systems,* vol. 28, no. 2, pp. 1806–1815, 2013.

5. W. Kempton, J. Tomic, S. Letendre, A. Brooks, and T. Lipman, "Vehicle-to-grid power: Battery, hybrid, and fuel cell vehicles as resources for distributed electric power in California," 2001. University of California at Davis, Institute of Transportation Studies Report (UCD-ITS-RR-01-03), Davis, CA(2001), Available: https://itspubs.ucdavis.edu/publication_detail.php?id=360.
6. J. Y. Yong, V. K. Ramachandaramurthy, K. M. Tan, and N. Mithulananthan, "A review on the state-of-the-art technologies of electric vehicle, its impacts and prospects," *Renewable and Sustainable Energy Reviews,* vol. 49, pp. 365–385, 2015.
7. R. J. Bessa and M. A. Matos, "The role of an aggregator agent for EV in the electricity market," in 7th Mediterranean Conference and Exhibition on Power Generation, Transmission, Distribution and Energy Conversion (MedPower 2010), Agia Napa, Cyprus, 2010, pp. 1–9.
8. W. Kempton and J. Tomić, "Vehicle-to-grid power fundamentals: Calculating capacity and net revenue," *Journal of Power Sources,* vol. 144, no. 1, pp. 268–279, 2005.
9. W. Kempton and J. Tomić, "Vehicle-to-grid power implementation: From stabilizing the grid to supporting large-scale renewable energy," *Journal of Power Sources,* vol. 144, no. 1, pp. 280–294, 2005.
10. D. Wang, J. Coignard, T. Zeng, C. Zhang, and S. Saxena, "Quantifying electric vehicle battery degradation from driving vs. vehicle-to-grid services," *Journal of Power Sources,* vol. 332, pp. 193–203, 2016.
11. A. Khaligh and S. Dusmez, "Comprehensive topological analysis of conductive and inductive charging solutions for plug-in electric vehicles," *IEEE Transactions on Vehicular Technology,* vol. 61, no. 8, pp. 3475–3489, 2012.
12. S. Haghbin, S. Lundmark, M. Alakula, and O. Carlson, "Grid-connected integrated battery chargers in vehicle applications: Review and new solution," *IEEE Transactions on Industrial Electronics,* vol. 60, no. 2, pp. 459–473, 2013.
13. M. Yilmaz and P. T. Krein, "Review of battery charger topologies, charging power levels, and infrastructure for plug-in electric and hybrid vehicles," *IEEE Transactions on Power Electronics,* vol. 28, no. 5, pp. 2151–2169, 2013.
14. K. M. Tan, V. K. Ramachandaramurthy, and J. Y. Yong, "Integration of electric vehicles in smart grid: A review on vehicle to grid technologies and optimization techniques," *Renewable and Sustainable Energy Reviews,* vol. 53, pp. 720–732, 2016.
15. M. Yilmaz and P. T. Krein, "Review of the impact of vehicle-to-grid technologies on distribution systems and utility interfaces," *IEEE Transactions on Power Electronics,* vol. 28, no. 12, pp. 5673–5689, 2013.
16. L. Chunhua, K. T. Chau, W. Diyun, and G. Shuang, "Opportunities and challenges of vehicle-to-home, vehicle-to-vehicle, and vehicle-to-grid technologies," *Proceedings of the IEEE,* vol. 101, no. 11, pp. 2409–2427, 2013.
17. SAE International Surface Vehicle Recommended Practice. "J1773 - SAE electric vehicle inductively coupled charging," *SAE International J1773,* Rev. Jun 2014.
18. SAE International Surface Vehicle Recommended Practice. "J1772 - SAE electric vehicle and plug in hybrid electric vehicle conductive charge coupler," *SAE International J1772,* Rev. Oct. 2017.
19. V. Graf, "Requirements for Introduction of the 42V-PowerNet," in *42 V-PowerNets,* H. Wallentowitz and C. Amsel, Eds. Berlin, Heidelberg: Springer Berlin Heidelberg, 2003, pp. 193–204.
20. S. Q. Ali, D. Mascarella, G. Joos, and L. Tan, "Torque cancelation of integrated battery charger based on six-phase permanent magnet synchronous motor drives for electric vehicles," *IEEE Transactions on Transportation Electrification,* vol. 4, no. 2, pp. 344–354, 2018.
21. "Electromagnetic compatibility (EMC) - Part 3-4: Limits - Limitation of emission of harmonic currents in low-voltage power supply systems for equipment with rated current greater than 16 A," IEC/TS 61000-3-4, Oct.1998.

22. "Electromagnetic compatibility (EMC) - Part 3-2: Limits - Limits for harmonic current emissions (equipment input current ≤ 16 A per phase)," *IEC 61000-3-2,* 2005.
23. "IEEE recommended practice and requirements for harmonic control in electric power systems," *IEEE Std 519,* pp. 1–29, 2014.
24. "J2894-1 surface vehicle recommended practice - power quality requirements for plug-in electric vehicle chargers," *SAE International J2894-1,* 2011.
25. "IEEE standard for interconnection and interoperability of distributed energy resources with associated electric power systems interfaces," *IEEE Std 1547-2018 (Revision of IEEE Std 1547-2003),* pp. 1–138, 2018.
26. "IEEE approved draft standard for safety levels with respect to human exposure to electric, magnetic and electromagnetic fields, 0 Hz to 300 GHz," *IEEE PC95.1/D3.5,* pp. 1–312, 2018.
27. "J1113 - electromagnetic compatibility measurement procedures and limits for components of vehicles, boats (up to 15 m), and machines (Except Aircraft) (16.6 Hz to 18 GHz)," *SAE International J1113* 2013.
28. "Electric vehicle conductive charging system-Part 1: General requirements," *IEC Standard 61851-1,* 2017.
29. D. Hanauer, "Mode 2 charging—testing and certification for international market access," *World Electric Vehicle Journal,* vol. 9, no. 2, p. 26, 2018.
30. J. Bablo, "Electric vehicle infrastructure standardization," *World Electric Vehicle Journal,* vol. 8, no. 2, pp. 576–586, 2016.
31. L. Noel, G. Z. de Rubens, J. Kester, and B. K. Sovacool, "*Vehicle-to-Grid: A Sociotechnical Transition Beyond Electric Mobility*," Springer, Cham, Switzerland: Palgrave Macmillan, 1st ed. 2019.
32. C. Bo, K. S. Hardy, J. D. Harper, T. P. Bohn, and D. S. Dobrzynski, "Towards standardized vehicle grid integration: Current status, challenges, and next steps," in *2015 IEEE Transportation Electrification Conference and Expo (ITEC),* 2015, pp. 1–6.
33. T. Markel, A. Meintz, K. Hardy, B. Chen, and T. J. N. Bohn, Ed., "*Multi-lab EV Smart Grid Integration Requirements Study: Providing Guidance on Technology Development and Demonstration* (No. NREL/TP-5400-63963)" National Renewable Energy Laboratory (NREL), Golden, Colorado, 2015.
34. J. Hu, H. Morais, T. Sousa, and M. Lind, "Electric vehicle fleet management in smart grids: A review of services, optimization and control aspects," *Renewable and Sustainable Energy Reviews,* vol. 56, pp. 1207–1226, 2016.
35. J. Kester, L. Noel, X. Lin, G. Zarazua de Rubens, and B. K. Sovacool, "The coproduction of electric mobility: Selectivity, conformity and fragmentation in the sociotechnical acceptance of vehicle-to-grid (V2G) standards," *Journal of Cleaner Production,* vol. 207, pp. 400–410, 2019.
36. M. Kuzlu, M. Pipattanasompom, and S. Rahman, "A comprehensive review of smart grid related standards and protocols," in *2017 5th International Istanbul Smart Grid and Cities Congress and Fair (ICSG),* Istanbul, Turkey, 2017, pp. 12–16.
37. C. Yuen, A. Oudalov, and A. Timbus, "The provision of frequency control reserves from multiple microgrids," *IEEE Transactions on Industrial Electronics,* vol. 58, no. 1, pp. 173–183, 2011.
38. S. Han, S. Han, and K. Sezaki, "Development of an optimal vehicle-to-grid aggregator for frequency regulation," *IEEE Transactions on Smart Grid,* vol. 1, no. 1, pp. 65–72, 2010.
39. J. Tomić and W. Kempton, "Using fleets of electric-drive vehicles for grid support," *Journal of Power Sources,* vol. 168, no. 2, pp. 459–468, 2007.
40. "*Energy Storage: Possibilities for Expanding Electric Grid Flexibility*," ed: National Renewable Energy Laboratory (NREL), Golden, Colorado, 2015, Available: https://www.nrel.gov/docs/fy16osti/64764.pdf 2016.

41. I. Lymperopoulos et al., "Ancillary services provision utilizing a network of fast-charging stations for electrical buses," *IEEE Transactions on Smart Grid,* vol. 11, no. 1, pp. 665–672, 2020.
42. G. R. Parsons, M. K. Hidrue, W. Kempton, and M. P. Gardner, "Willingness to pay for vehicle-to-grid (V2G) electric vehicles and their contract terms," *Journal of Energy Economics,* vol. 42, pp. 313–324, 2014.
43. P. Lee, H. Esther, Z. Lukszo, and P. Herder, "Conceptualization of vehicle-to-grid contract types and their formalization in agent-based models," *Journal of Complexity,* vol. 2018, pp. 1–11, 2018.
44. S. Lefeng, L. Tong, and W. Yandi, "Vehicle-to-grid service development logic and management formulation," *Journal of Modern Power Systems Clean Energy,* vol. 7, no. 4, pp. 935–947, 2019.
45. B. Kirby, *"Ancillary Services: Technical and Commercial Insights,"* 2007. Available: http://www.consultkirby.com/files/Ancillary_Services_-_Technical_And_Commercial_Insights_EXT_.pdf
46. K. Oureilidis et al., "Ancillary services market design in distribution networks: Review and identification of barriers," *Energies,* vol. 13, no. 4, p. 917, 2020.
47. E. L. Karfopoulos, K. A. Panourgias, and N. D. Hatziargyriou, "Distributed coordination of electric vehicles providing V2G regulation services," *IEEE Transactions on Power Systems,* vol. 31, no. 4, pp. 2834–2846, 2016.
48. S. Izadkhast, P. Garcia-Gonzalez, and P. Frías, "An aggregate model of plug-in electric vehicles for primary frequency control," *IEEE Transactions on Power Systems,* vol. 30, no. 3, pp. 1475–1482, 2015.
49. S. Debbarma and A. Dutta, "Utilizing electric vehicles for LFC in restructured power systems using fractional order controller," *IEEE Transactions on Smart Grid,* vol. 8, no. 6, pp. 2554–2564, 2016.
50. F. Kennel, D. Görges, and S. Liu, "Energy management for smart grids with electric vehicles based on hierarchical MPC," *IEEE Transactions on Industrial Informatics,* vol. 9, no. 3, pp. 1528–1537, 2013.
51. D. Dallinger, D. Krampe, and M. Wietschel, "Vehicle-to-grid regulation reserves based on a dynamic simulation of mobility behavior," *IEEE Transactions on Smart Grid,* vol. 2, no. 2, pp. 302–313, 2011.
52. M. D. Galus, S. Koch, and G. Andersson, "Provision of load frequency control by PHEVs, controllable loads, and a cogeneration unit," *IEEE Transactions on Industrial Electronics,* vol. 58, no. 10, pp. 4568–4582, 2011.
53. K. Kaur, R. Rana, N. Kumar, M. Singh, and S. Mishra, "A colored petri net based frequency support scheme using fleet of electric vehicles in smart grid environment," *IEEE Transactions on Power Systems,* vol. 31, no. 6, pp. 4638–4649, 2016.
54. H. Liu, Z. Hu, Y. Song, and J. Lin, "Decentralized vehicle-to-grid control for primary frequency regulation considering charging demands," *IEEE Transactions on Power Systems,* vol. 28, no. 3, pp. 3480–3489, 2013.
55. H. Liu, Z. Hu, Y. Song, J. Wang, and X. Xie, "Vehicle-to-grid control for supplementary frequency regulation considering charging demands," *IEEE Transactions on Power Systems,* vol. 30, no. 6, pp. 3110–3119, 2015.
56. A. Y. S. Lam, K. C. Leung, and V. O. K. Li, "Capacity estimation for vehicle-to-grid frequency regulation services with smart charging mechanism," *IEEE Transactions on Smart Grid,* vol. 7, no. 1, pp. 156–166, 2016.
57. S. Han, S. Han, and K. Sezaki, "Estimation of achievable power capacity from plug-in electric vehicles for V2G frequency regulation: Case studies for market participation," *IEEE Transactions on Smart Grid,* vol. 2, no. 4, pp. 632–641, 2011.
58. E. Yao, V. W. S. Wong, and R. Schober, "Robust frequency regulation capacity scheduling algorithm for electric vehicles," *IEEE Transactions on Smart Grid,* vol. 8, no. 2, pp. 984–997, 2017.

59. W. Hu, C. Su, Z. Chen, and B. Bak-Jensen, "Optimal operation of plug-in electric vehicles in power systems with high wind power penetrations," *IEEE Transactions on Sustainable Energy,* vol. 4, no. 3, pp. 577–585, 2013.
60. C. Mu, W. Liu, and W. Xu, "Hierarchically adaptive frequency control for an EV-integrated smart grid with renewable energy," *IEEE Transactions on Industrial Informatics,* vol. 14, no. 9, pp. 4254–4263, 2018.
61. T. Ma and O. A. Mohammed, "Economic analysis of real-time large-scale PEVs network power flow control algorithm with the consideration of V2G services," *IEEE Transactions on Industry Applications,* vol. 50, no. 6, pp. 4272–4280, 2014.
62. S. Li, M. Brocanelli, W. Zhang, and X. Wang, "Integrated power management of data centers and electric vehicles for energy and regulation market participation," *IEEE Transactions on Smart Grid,* vol. 5, no. 5, pp. 2283–2294, 2014.
63. J. J. Escudero-Garzas, A. Garcia-Armada, and G. Seco-Granados, "Fair design of plug-in electric vehicles aggregator for V2G regulation," *IEEE Transactions on Vehicular Technology,* vol. 61, no. 8, pp. 3406–3419, 2012.
64. S. Sun, M. Dong, and B. Liang, "Real-time welfare-maximizing regulation allocation in dynamic aggregator-EVs system," *IEEE Transactions on Smart Grid,* vol. 5, no. 3, pp. 1397–1409, 2014.
65. H. Fan, L. Jiang, C. K. Zhang, and C. Mao, "Frequency regulation of multi-area power systems with plug-in electric vehicles considering communication delays," *IET Generation, Transmission & Distribution,* vol. 10, no. 14, pp. 3481–3491, 2016.
66. S. Izadkhast, P. Garcia-Gonzalez, P. Frías, and P. Bauer, "Design of plug-in electric vehicle's frequency-droop controller for primary frequency control and performance assessment," *IEEE Transactions on Power Systems,* vol. 32, no. 6, pp. 4241–4254, 2017.
67. Y. Ota, H. Taniguchi, T. Nakajima, K. M. Liyanage, J. Baba, and A. Yokoyama, "Autonomous distributed V2G (vehicle-to-grid) satisfying scheduled charging," *IEEE Transactions on Smart Grid,* vol. 3, no. 1, pp. 559–564, 2012.
68. A. Tuohy and M. O'Malley, "Wind Power and Storage," in *Wind Power in Power Systems* Hoboken, New Jersey: John Wiley & Sons, Ltd, 2012, pp. 465–487.
69. J. M. Maza-Ortega, J. M. Mauricio, M. Barragán-Villarejo, C. Demoulias, and A. Gómez-Expósito, "Ancillary services in hybrid AC/DC low voltage distribution networks," *Energies,* vol. 12, no. 19, p. 3591, 2019.
70. M. Raoofat, M. Saad, S. Lefebvre, D. Asber, H. Mehrjedri, and L. Lenoir, "Wind power smoothing using demand response of electric vehicles," *International Journal of Electrical Power Energy Systems,* vol. 99, pp. 164–174, 2018.
71. M. Jannati, S. Hosseinian, and B. J. R. E. Vahidi, "A significant reduction in the costs of battery energy storage systems by use of smart parking lots in the power fluctuation smoothing process of the wind farms," *Renewable Energy,* vol. 87, pp. 1–14, 2016.
72. S. Ziqi, P. Wei, D. Wei, and Q. Hui, "Capacity configuration of electric vehicle charging station for wind power smoothing," in *2014 IEEE Conference and Expo Transportation Electrification Asia-Pacific (ITEC Asia-Pacific)*, Beijing, China 2014, pp. 1–5, IEEE.
73. M. Wang et al., "Load power smoothing control of distribution network including photovoltaic generation with energy storage from electric vehicles," in *2017 IEEE Power & Energy Society General Meeting,* 2017, pp. 1–5.
74. A. Ali, D. Raisz, and K. Mahmoud, "Voltage fluctuation smoothing in distribution systems with RES considering degradation and charging plan of EV batteries," *Electric Power Systems Research,* vol. 176, p. 105933, 2019.
75. Z. Yu, P. Gong, Z. Wang, Y. Zhu, R. Xia, and Y. Tian, "Real-time control strategy for aggregated electric vehicles to smooth the fluctuation of wind-power output," *Energies,* vol. 13, no. 3, p. 757, 2020.
76. S. Gao, K. T. Chau, C. Liu, D. Wu, and C. C. Chan, "Integrated energy management of plug-in electric vehicles in power grid with renewables," *IEEE Transactions on Vehicular Technology,* vol. 63, no. 7, pp. 3019–3027, 2014.

77. L. Jian, H. Xue, G. Xu, X. Zhu, D. Zhao, and Z. Y. Shao, "Regulated charging of plug-in hybrid electric vehicles for minimizing load variance in household smart microgrid," *IEEE Transactions on Industrial Electronics,* vol. 60, no. 8, pp. 3218–3226, 2013.
78. N. Navid and G. Rosenwald, "Market solutions for managing ramp flexibility with high penetration of renewable resource," *IEEE Transactions on Sustainable Energy,* vol. 3, no. 4, pp. 784–790, 2012.
79. B. Zhang and M. Kezunovic, "Impact on power system flexibility by electric vehicle participation in ramp market," *IEEE Transactions on Smart Grid,* vol. 7, no. 3, pp. 1285–1294, 2016.
80. S. Khemakhem, M. Rekik, and L. Krichen, "A flexible control strategy of plug-in electric vehicles operating in seven modes for smoothing load power curves in smart grid," *Energy,* vol. 118, pp. 197–208, 2017.
81. W. Zhang, K. Spence, R. Shao, and L. Chang, "Grid power-smoothing performance improvement for PV and electric vehicle (EV) systems," in *2018 IEEE Energy Conversion Congress and Exposition (ECCE),* Portland, OR, USA 2018, pp. 1051–1057.
82. M. G. Vayá and G. Andersson, "Self Scheduling of plug-in electric vehicle aggregator to provide balancing services for wind power," *IEEE Transactions on Sustainable Energy,* vol. 7, no. 2, pp. 886–899, 2016.
83. D. Kim, H. Kwon, M.-K. Kim, J.-K. Park, and H. J. E. Park, "Determining the flexible ramping capacity of electric vehicles to enhance locational flexibility," *Energies,* vol. 10, no. 12, p. 2028, 2017.
84. M. Ghofrani, A. Arabali, M. Etezadi-Amoli, and M. S. Fadali, "Smart scheduling and cost-benefit analysis of grid-enabled electric vehicles for wind power integration," *IEEE Transactions on Smart Grid,* vol. 5, no. 5, pp. 2306–2313, 2014.
85. C. Shao, X. Wang, X. Wang, C. Du, C. Dang, and S. Liu, "Cooperative dispatch of wind generation and electric vehicles with battery storage capacity constraints in SCUC," *IEEE Transactions on Smart Grid,* vol. 5, no. 5, pp. 2219–2226, 2014.
86. J. E. Contreras-Ocana, M. R. Sarker, and M. A. Ortega-Vazquez, "Decentralized coordination of a building manager and an electric vehicle aggregator," *IEEE Transactions on Smart Grid,* vol. 9, no. 4, pp. 2625–2637, 2016.
87. C. L. Floch, E. Kara, and S. Moura, "PDE modeling and control of electric vehicle fleets for ancillary services: A discrete charging case," *IEEE Transactions on Smart Grid,* vol. 9, no. 2, pp. 573–581, 2016.
88. X. Bai and W. Qiao, "Robust optimization for bidirectional dispatch coordination of large-scale V2G," *IEEE Transactions on Smart Grid,* vol. 6, no. 4, pp. 1944–1954, 2015.
89. H. Liang, B. J. Choi, W. Zhuang, and X. Shen, "Optimizing the energy delivery via V2G systems based on stochastic inventory theory," *IEEE Transactions on Smart Grid,* vol. 4, no. 4, pp. 2230–2243, 2013.
90. Q. Huang, Q. S. Jia, Z. Qiu, X. Guan, and G. Deconinck, "Matching EV charging load with uncertain wind: A simulation-based policy improvement approach," *IEEE Transactions on Smart Grid,* vol. 6, no. 3, pp. 1425–1433, 2015.
91. C. T. Li, C. Ahn, H. Peng, and J. Sun, "Synergistic control of plug-in vehicle charging and wind power scheduling," *IEEE Transactions on Power Systems,* vol. 28, no. 2, pp. 1113–1121, 2013.
92. H. Xing, M. Fu, Z. Lin, and Y. Mou, "Decentralized optimal scheduling for charging and discharging of plug-in electric vehicles in smart grids," *IEEE Transactions on Power Systems,* vol. 31, no. 5, pp. 4118–4127, 2016.
93. V. d. Razo, C. Goebel, and H. A. Jacobsen, "Vehicle-originating-signals for real-time charging control of electric vehicle fleets," *IEEE Transactions on Transportation Electrification,* vol. 1, no. 2, pp. 150–167, 2015.

94. J. Rivera, C. Goebel, and H. Jacobsen, "Distributed convex optimization for electric vehicle aggregators," *IEEE Transactions on Smart Grid,* vol. 8, no. 4, pp. 1852–1863, 2017.
95. A. T. Al-Awami and E. Sortomme, "Coordinating vehicle-to-grid services with energy trading," *IEEE Transactions on Smart Grid,* vol. 3, no. 1, pp. 453–462, 2012.
96. R. J. Bessa, M. A. Matos, F. J. Soares, and J. A. P. Lopes, "Optimized bidding of a EV aggregation agent in the electricity market," *IEEE Transactions on Smart Grid,* vol. 3, no. 1, pp. 443–452, 2012.
97. P. Kou, D. Liang, L. Gao, and F. Gao, "Stochastic coordination of plug-in electric vehicles and wind turbines in microgrid: A model predictive control approach," *IEEE Transactions on Smart Grid,* vol. 7, no. 3, pp. 1537–1551, 2016.
98. E. L. Karfopoulos and N. D. Hatziargyriou, "Distributed coordination of electric vehicles providing V2G services," *IEEE Transactions on Power Systems,* vol. 31, no. 1, pp. 329–338, 2016.
99. Z. Li, Q. Guo, H. Sun, Y. Wang, and S. Xin, "Emission-concerned wind-EV coordination on the transmission grid side with network constraints: Concept and case study," *IEEE Transactions on Smart Grid,* vol. 4, no. 3, pp. 1692–1704, 2013.
100. J. M. Foster and M. C. Caramanis, "Optimal power market participation of plug-in electric vehicles pooled by distribution feeder," *IEEE Transactions on Power Systems,* vol. 28, no. 3, pp. 2065–2076, 2013.
101. B. Geng, J. K. Mills, and D. Sun, "Two-stage charging strategy for plug-in electric vehicles at the residential transformer level," *IEEE Transactions on Smart Grid,* vol. 4, no. 3, pp. 1442–1452, 2013.
102. O. Hafez and K. Bhattacharya, "Integrating EV charging stations as smart loads for demand response provisions in distribution systems," *IEEE Transactions on Smart Grid,* vol. 9, no. 2, pp. 1096–1106, 2016.
103. J. d. Hoog, T. Alpcan, M. Brazil, D. A. Thomas, and I. Mareels, "A market mechanism for electric vehicle charging under network constraints," *IEEE Transactions on Smart Grid,* vol. 7, no. 2, pp. 827–836, 2016.
104. F. R. Islam and H. R. Pota, "V2G technology to improve wind power quality and stability," in *Australian Control Conference (AUCC)*, Melbourne, VIC, Australia 2011, pp. 452–457.
105. P. Zhang, K. Qian, C. Zhou, B. G. Stewart, and D. M. Hepburn, "A methodology for optimization of power systems demand due to electric vehicle charging load," *IEEE Transactions on Power Systems,* vol. 27, no. 3, pp. 1628–1636, 2012.
106. M. Jun and A. J. Markel, "Simulation and analysis of vehicle-to-Grid operations in microgrid," in *2012 IEEE Power and Energy Society General Meeting*, San Diego, CA, USA 2012, pp. 1–5.
107. I. Pavić, T. Capuder, and I. Kuzle, "Value of flexible electric vehicles in providing spinning reserve services," *Applied Energy,* vol. 157, pp. 60–74, 2015.
108. M. R. V. Moghadam, R. Zhang, and R. T. B. Ma, "Distributed frequency control via randomized response of electric vehicles in power grid," *IEEE Transactions on Sustainable Energy,* vol. 7, no. 1, pp. 312–324, 2016.
109. E. Sortomme and M. A. El-Sharkawi, "Optimal combined bidding of vehicle-to-grid ancillary services," *IEEE Transactions on Smart Grid,* vol. 3, no. 1, pp. 70–79, 2012.
110. M. Ansari, A. T. Al-Awami, E. Sortomme, and M. A. Abido, "Coordinated bidding of ancillary services for vehicle-to-grid using fuzzy optimization," *IEEE Transactions on Smart Grid,* vol. 6, no. 1, pp. 261–270, 2015.
111. E. Sortomme and M. A. El-Sharkawi, "Optimal scheduling of vehicle-to-grid energy and ancillary services," *IEEE Transactions on Smart Grid,* vol. 3, no. 1, pp. 351–359, 2012.

112. H. Zhang, Z. Hu, Z. Xu, and Y. Song, "Evaluation of achievable vehicle-to-grid capacity using aggregate PEV model," *IEEE Transactions on Power Systems,* vol. 32, no. 1, pp. 784–794, 2016.
113. J. Zhao, C. Wan, Z. Xu, and K. P. Wong, "Spinning reserve requirement optimization considering integration of plug-in electric vehicles," *IEEE Transactions on Smart Grid,* vol. 8, no. 4, pp. 2009–2021, 2016.
114. J. R. Pillai and B. Bak-Jensen, "Integration of vehicle-to-grid in the western Danish power system," *IEEE Transactions on Sustainable Energy,* vol. 2, no. 1, pp. 12–19, 2011.
115. J. A. Suul, S. D. Arco, and G. Guidi, "Virtual synchronous machine-based control of a single-phase bi-directional battery charger for providing vehicle-to-grid services," *IEEE Transactions on Industry Applications,* vol. 52, no. 4, pp. 3234–3244, 2016.
116. C. Sun, S. Q. Ali, G. Joos, and F. Bouffard, "Virtual synchronous machine control for low-inertia power system considering energy storage limitation," in *2019 IEEE Energy Conversion Congress and Exposition (ECCE),* Baltimore, MD, USA 2019, pp. 6021–6028.
117. H. Tanaka, T. Tanaka, T. Wakimoto, E. Hiraki, and M. Okamoto, "Reduced-capacity smart charger for electric vehicles on single-phase three-wire distribution feeders with reactive power control," *IEEE Transactions on Industry Applications,* vol. 51, no. 1, pp. 315–324, 2015.
118. M. C. Kisacikoglu, M. Kesler, and L. M. Tolbert, "Single-phase on-board bidirectional PEV charger for V2G reactive power operation," *IEEE Transactions on Smart Grid,* vol. 6, no. 2, pp. 767–775, 2015.
119. J. Wang, G. R. Bharati, S. Paudyal, O. Ceylan, B. P. Bhattarai, and K. S. Myers, "Coordinated electric vehicle charging with reactive power support to distribution grids," *IEEE Transactions on Industrial Informatics,* vol. 15, no. 1, pp. 54–63, 2019.
120. H. Tanaka, F. Ikeda, T. Tanaka, H. Yamada, and M. Okamoto, "Novel reactive power control strategy based on constant DC-capacitor voltage control for reducing the capacity of smart charger for electric vehicles on single-phase three-wire distribution feeders," *IEEE Journal of Emerging and Selected Topics in Power Electronics,* vol. 4, no. 2, pp. 481–488, 2016.
121. M. Falahi, H. M. Chou, M. Ehsani, L. Xie, and K. L. Butler-Purry, "Potential power quality benefits of electric vehicles," *IEEE Transactions on Sustainable Energy,* vol. 4, no. 4, pp. 1016–1023, 2013.
122. F. Marra et al., "Improvement of local voltage in feeders with photovoltaic using electric vehicles," *IEEE Transactions on Power Systems,* vol. 28, no. 3, pp. 3515–3516, 2013.
123. H. V. Haghi and Z. Qu, "A kernel-based predictive model of EV capacity for distributed voltage control and demand response," *IEEE Transactions on Smart Grid,* vol. 9, no. 4, pp. 3180–3190, 2018.
124. M. Bayat, K. Sheshyekani, and A. Rezazadeh, "A unified framework for participation of responsive end-user devices in voltage and frequency control of the smart grid," *IEEE Transactions on Power Systems,* vol. 30, no. 3, pp. 1369–1379, 2015.
125. F. Geth, N. Leemput, J. V. Roy, B. J, R. Ponnette, and J. Driesen, "Voltage droop charging of electric vehicles in a residential distribution feeder," in *2012 3rd IEEE PES Innovative Smart Grid Technologies Europe (ISGT Europe),* 2012, pp. 1–8.
126. A. T. Al-Awami, E. Sortomme, G. M. A. Akhtar, and S. Faddel, "A voltage-based controller for an electric-vehicle charger," *IEEE Transactions on Vehicular Technology,* vol. 65, no. 6, pp. 4185–4196, 2016.
127. J. M. Foster, G. Trevino, M. Kuss, and M. C. Caramanis, "Plug-in electric vehicle and voltage support for distributed solar: Theory and application," *IEEE Systems Journal,* vol. 7, no. 4, pp. 881–888, 2013.

128. K. Knezovic, S. Martinenas, P. B. Andersen, A. Zecchino, and M. Marinelli, "Enhancing the role of electric vehicles in the power grid: Field validation of multiple ancillary services," *IEEE Transactions on Transportation Electrification,* vol. 3, no. 1, pp. 201–209, 2016.
129. A. Agrawal, M. Kumar, D. K. Prajapati, M. Singh, and P. Kumar, "Smart public transit system using an energy storage system and its coordination with a distribution grid," *IEEE Transactions on Intelligent Transportation Systems,* vol. 15, no. 4, pp. 1622–1632, 2014.
130. N. Leemput, F. Geth, J. V. Roy, A. Delnooz, J. Büscher, and J. Driesen, "Impact of electric vehicle on-board single-phase charging strategies on a flemish residential grid," *IEEE Transactions on Smart Grid,* vol. 5, no. 4, pp. 1815–1822, 2014.
131. S. Huang, J. R. Pillai, B. Bak-Jensen, and P. Thogersen, "Voltage support from electric vehicles in distribution grid," in *2013 15th European Conference on Power Electronics and Applications (EPE),* 2013, pp. 1–8.
132. M. C. Kisacikoglu, B. Ozpineci, and L. M. Tolbert, "EV/PHEV bidirectional charger assessment for V2G reactive power operation," *IEEE Transactions on Power Electronics,* vol. 28, no. 12, pp. 5717–5727, 2013.
133. B. Seal, *"Common Functions for Smart Inverters, Version 4,"* Electric Power Research Institute, Palo Alto, California 2016.
134. T. Masuta and A. Yokoyama, "Supplementary load frequency control by use of a number of both electric vehicles and heat pump water heaters," *IEEE Transactions on Smart Grid,* vol. 3, no. 3, pp. 1253–1262, 2012.
135. P. Richardson, D. Flynn, and A. Keane, "Local versus centralized charging strategies for electric vehicles in low voltage distribution systems," *IEEE Transactions on Smart Grid,* vol. 3, no. 2, pp. 1020–1028, 2012.
136. S. Vandael, B. Claessens, M. Hommelberg, T. Holvoet, and G. Deconinck, "A scalable three-step approach for demand side management of plug-in hybrid vehicles," *IEEE Transactions on Smart Grid,* vol. 4, no. 2, pp. 720–728, 2013.
137. X. Wang, Z. Y. He, and J. W. Yang, "Unified strategy for electric vehicles participate in voltage and frequency regulation with active power in city grid," *IET Generation, Transmission & Distribution,* vol. 13, no. 15, pp. 3281–3291, 2019.
138. Z. Wang and S. Wang, "Grid power peak shaving and valley filling using vehicle-to-grid systems," *IEEE Transactions on Power Delivery,* vol. 28, no. 3, pp. 1822–1829, 2013.
139. E. Mirmoradi and H. Ghasemi, "Market clearing with probabilistic spinning reserve considering wind uncertainty and electric vehicles," vol. 26, no. 3, pp. 525–538, 2016.
140. D. Liu, W. Xie, H. Liao, and Y. Peng, "An integrated probabilistic approach to lithium-ion battery remaining useful life estimation," *IEEE Transactions on Instrumentation and Measurement,* vol. 64, no. 3, pp. 660–670, 2015.
141. M. N. Mojdehi and P. Ghosh, "An on-demand compensation function for an EV as a reactive power service provider," *IEEE Transactions on Vehicular Technology,* vol. 65, no. 6, pp. 4572–4583, 2016.
142. A. Perez, R. Moreno, R. Moreira, M. Orchard, and G. Strbac, "Effect of battery degradation on multi-service portfolios of energy storage," *IEEE Transactions on Sustainable Energy,* vol. 7, no. 4, pp. 1718–1729, 2016.
143. K. Uddin, M. Dubarry, and M. B. Glick, "The viability of vehicle-to-grid operations from a battery technology and policy perspective," *Energy Policy,* vol. 113, pp. 342–347, 2018.
144. S. Tong, T. Fung, M. P. Klein, D. A. Weisbach, and J. W. Park, "Demonstration of reusing electric vehicle battery for solar energy storage and demand side management," *Journal of Energy Storage,* vol. 11, pp. 200–210, 2017.

145. C. Heymans, S. B. Walker, S. B. Young, and M. Fowler, "Economic analysis of second use electric vehicle batteries for residential energy storage and load-levelling," *Energy Policy,* vol. 71, pp. 22–30, 2014.
146. E. Martinez-Laserna et al., "Battery second life: Hype, hope or reality? A critical review of the state of the art," *Renewable and Sustainable Energy Reviews,* vol. 93, pp. 701–718, 2018.
147. V. C. Gungor et al., "A survey on smart grid potential applications and communication requirements," *IEEE Transactions on Industrial Informatics,* vol. 9, no. 1, pp. 28–42, 2013.
148. M. H. Eiza, Q. Shi, A. K. Marnerides, T. Owens, and Q. Ni, "Efficient, secure, and privacy-preserving PMIPv6 protocol for V2G networks," *IEEE Transactions on Vehicular Technology,* vol. 68, no. 1, pp. 19–33, 2019.
149. D. He, S. Chan, and M. Guizani, "Privacy-friendly and efficient secure communication framework for V2G networks," *IET Communications,* vol. 12, no. 3, pp. 304–309, 2018.
150. Q. Chen, N. Liu, C. Hu, L. Wang, and J. Zhang, "Autonomous energy management strategy for solid-state transformer to integrate PV-assisted EV charging station participating in ancillary service," *IEEE Transactions on Industrial Informatics,* vol. 13, no. 1, pp. 258–269, 2017.
151. J. Lu, L. Zhu, G. Liu, and H. Bai, "Device and system-level transient analysis in a modular designed sub-MW EV fast charging station using hybrid GaN HEMTs + Si MOSFETs," *IEEE Journal of Emerging and Selected Topics in Power Electronics,* vol. 7, no. 1, pp. 143–156, 2019.
152. D. M. Steward, *"Critical Elements of Vehicle-to-Grid (V2G) Economics,"* National Renewable Energy Lab (NREL), Golden, CO, 2017.
153. Federal Energy Regulatory Commission (FERC) "Frequency Regulation Compensation in the Organized Wholesale Power Markets," in *Order No. 755* vol. FERC 137, ed. 2011.
154. F. Varshosaz, M. Moazzami, B. Fani, and P. Siano, "Day-ahead capacity estimation and power management of a charging station based on queuing theory," *IEEE Transactions on Industrial Informatics,* vol. 15, no. 10, pp. 5561–5574, 2019.
155. N. Rotering and M. Ilic, "Optimal charge control of plug-in hybrid electric vehicles in deregulated electricity markets," *IEEE Transactions on Power Systems,* vol. 26, no. 3, pp. 1021–1029, 2011.
156. C. Quinn, D. Zimmerle, and T. H. Bradley, "An evaluation of state-of-charge limitations and actuation signal energy content on plug-in hybrid electric vehicle, vehicle-to-grid reliability, and economics," *IEEE Transactions on Smart Grid,* vol. 3, no. 1, pp. 483–491, 2012.
157. Y. Zhao, M. Noori, and O. Tatari, "Boosting the adoption and the reliability of renewable energy sources: Mitigating the large-scale wind power intermittency through vehicle to grid technology," *Energy,* vol. 120, pp. 608–618, 2017.
158. C. G. Hoehne and M. V. Chester, "Optimizing plug-in electric vehicle and vehicle-to-grid charge scheduling to minimize carbon emissions," *Energy,* vol. 115, pp. 646–657, 2016.
159. A. E. P. Abas, J. Yong, T. M. I. Mahlia, and M. A. Hannan, "Techno-economic analysis and environmental impact of electric vehicle," *IEEE Access,* vol. 7, pp. 98565–98578, 2019.
160. S. Vadi, R. Bayindir, A. M. Colak, and E. J. E. Hossain, "A review on communication standards and charging topologies of V2G and V2H operation strategies," *Energies,* vol. 12, no. 19, p. 3748, 2019.
161. C. Nitta, "System control and communication requirements of a vehicle-to-grid (V2G) network," in *Conference Proceedings: The 20th International Electric Vehicle Symposium and Exposition, EVS,* Nov 2003, Long Beach, CA 2003, vol. 20, p. 7.

162. D. Bowermaster, M. Alexander, and M. Duvall, "The Need for Charging: Evaluating utility infrastructures for electric vehicles while providing customer support," *IEEE Electrification Magazine,* vol. 5, no. 1, pp. 59–67, 2017.
163. Z. Zhou, B. Wang, M. Dong, and K. Ota, "Secure and efficient vehicle-to-grid energy trading in cyber physical systems: Integration of blockchain and edge computing," *IEEE Transactions on Systems, Man, and Cybernetics: Systems,* vol. 50, no. 1, pp. 43–57, 2020.
164. S. Park, S. Lee, S. Park, and S. J. S. Park, "AI-based physical and virtual platform with 5-layered architecture for sustainable smart energy city development," *Sustainability* vol. 11, no. 16, p. 4479, 2019.
165. L. Cai, J. Pan, L. Zhao, and X. Shen, "Networked electric vehicles for green intelligent transportation," *IEEE Communications Standards Magazine,* vol. 1, no. 2, pp. 77–83, 2017.
166. B. K. Sovacool, J. Axsen, and W. Kempton, "The future promise of vehicle-to-grid (V2G) integration: A sociotechnical review and research agenda," *Annual Review of Environment and Resources*, vol. 42, pp. 377–406, 2017.
167. R. Das et al., "Multi-objective techno-economic-environmental optimisation of electric vehicle for energy services," *Applied Energy,* vol. 257, p. 113965, 2020.

4 Challenges to Build a EV Friendly Ecosystem
Brazilian Benchmark

Ana Carolina Rodrigues Teixeira
University of São Paulo

CONTENTS

4.1 Introduction 73
4.2 Context and Brazilian Portrait 74
4.3 Challenges and Opportunities through the Brazilian Initiatives 76
    4.3.1 Economy and Production 76
    4.3.2 Public Policies 78
    4.3.3 Customer Acceptance 81
    4.3.4 Market, Logistics, Energy Matrix, and Environment 82
    4.3.5 Smart Grid 84
4.4 Case Study 85
    4.4.1 Public Perception 85
    4.4.2 Numeric Model 88
4.5 Summary and Conclusions 91
Acknowledgments 92
References 92

4.1 INTRODUCTION

Discussions about electric vehicle (EV) cannot be based just on automotive technologies. Many countries have been investing in fleet replacement using EV to reduce greenhouse gas (GHG) emissions and local air pollution. Different goals were proposed over the last years to replace the conventional fleet all around the globe, as the Norwegian case in which they had planned to change their entire fleet. This kind of target, however, cannot be equally used, and the whole ecosystem must be taken into consideration to stimulate and disseminate such technology. Hence, the introduction of EVs depends on many factors that should be considered, such as geography, public and private subsidies, energy, customer acceptance, investments, and infrastructure.

Many pilot projects have been coming to fruition to evaluate efficiency, costs, maintenance, and customer acceptance, although the extension and different characteristics of each region contribute to hinder the process. Therefore, this chapter covers these different characteristics of the Brazilian reality, since the country presents several economic problems in adittion to where the EVs started to be part of the country's reality over the last five years. Furthermore, this chapter also provides some examples to clarify how the country has been dealing with these issues.

Challenges of the EV's introduction using the Brazilian benchmark and the importance to develop the market before the actual introduction and possible fleet replacement will be discussed. Then, the chapter identifies the existing opportunities in Brazil that can be used as a differential to boost this kind of technology. This benchmark will be based on previous studies made by the author where the likelihood of replacement of the conventional taxi fleet to EV in different Brazilian cities was analyzed.

This chapter is divided into four parts. Section 4.2 brings the Brazilian context and its current situation, demonstrating how EVs can be influenced by different issues. Section 4.3 presents the challenges the country has been facing in this area, the opportunities to expand the market, and in what shape or form other characteristics such as energy matrix and environment can help us to reach a better ecosystem to introduce EV. Section 4.4 shows a case study applied to a municipality in Brazil about the replacement of conventional vehicles to EV in the taxi fleet. The last section concludes such chapter with lessons learned and final remarks.

4.2 CONTEXT AND BRAZILIAN PORTRAIT

Firstly, there is a chance that one would table the question: "why are you talking about ecosystem while EVs are being explained?", and the answer is: because the word "ecosystem" is commonly related to biological issues. Nevertheless, according to the Cambridge dictionary [1], an ecosystem is "all the living things in an area and the way they affect each other and the environment". This word, however, can have another meaning related to the business scope such as "a group of business activities that affect each other and work well together" [1] or "all the different activities, companies, systems, etc. that are involved in a particular area of business, especially new technology" [2]. These definitions explain why it is important to consider different aspects when we talk about the introduction of EVs in a specific place, wherein geography, subsidies, energy, customer acceptance, investment, and infrastructure will always influence the success of this new product. Thus, to build a friendly ecosystem for the EV's introduction, it is necessary to consider all these multiplicities.

Different countries require distinct strategies to insert a new product in the market; it is important, though, and it adds knowledge to observe what other subjects are doing over the world. It is helpful not only to build actions to deal with future problems but also to learn with flawed solutions. Geography and country extension, for example, will impact, directly the use of EVs, especially when the battery electric vehicles (BEVs) are considered. In Brazil's case, for example, the country depends on different modal types (i.e. road, rail, waterway, airway, and pipeline), but more than 60% stands for the road modal. Brazil has around 1.7 million km of road networks and a total fleet of 100 million vehicles, which makes it common to travel more than 300 km during holidays and vacation periods. Consequently, there is a need for higher-autonomy batteries and a strong recharging-complex infrastructure to supply the electricity demand.

Challenges to Build a EV Friendly Ecosystem

Another issue to be taken into consideration is the electric matrix, which is different in each country, although extremely important to the following discussion: (1) it is the base of the supply chain related to the demand required for EVs, and its management is a crucial key to reach intermittent energy quality; (2) the lack of a strong electric matrix contributes to failures and energy outages.

Furthermore, EV has been considered a way to reduce GHG emissions and local air pollutants. Hence, a clean electric matrix is important to reach this goal, since, in 2018, 83% of the electricity produced in Brazil came from renewable sources, which include hydropower, wind energy, solar energy, and biomass [3]. Nonetheless, it is necessary to manage supply and demand during drought periods to avoid cases of energy outage in Brazil. Lastly, even with an electric matrix based on fossil fuel sources, EVs can contribute to reducing local air pollution, directly improving the air quality in cities and contributing to better public health by reducing the attributable deaths to air pollution. Figure 4.1 shows the Brazilian electric matrix composition.

As mentioned above, the infrastructure is a key for the introduction of EV in the market since without recharging infrastructure, there is no possibility to maintain BEVs in any place. However, this issue might increase in developing countries, such as Brazil, due to high EV purchase costs, which may lead to a fewer number of such vehicles on streets and consequently minor need for public infrastructure.

In this way, some companies have been building recharging points in specific areas and highways in Brazil, but there is still a minimal number of fast recharge stations. Taking the opportunity for this issue, government subsidies are necessary to push and incentivize the market, both from the customers' and companies' point of view. Public policies are another topic discussed in this chapter, and it is presented as a challenge faced in Brazil since EVs are still a new product, and there are a few subsidies to push the market and stimulate costumers to the conventional vehicles.

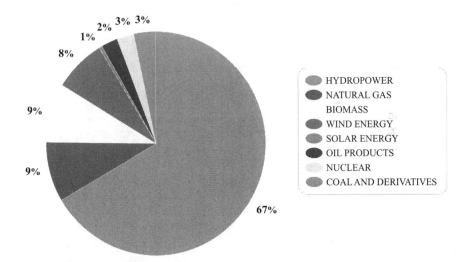

FIGURE 4.1 Brazilian electric matrix considering the data for 2018. Data from EPE 2019 [3].

Brazil accounts with 210 million people divided into five regions with different needs in terms of energy demand, consumption and availability, economy, education, and environmental problems; besides that, another issue that must be considered is customers' acceptance since they surely would be the key for success. Each place has a specific culture that affects the way people buy and deal with innovation. Moreover, a new product in the market always comes with many doubts, in which the case of EV is no different, such as purchasing price, maintenance costs plus its availability, and energy price. Owing to economic struggles, especially, the EV acquisition represents several uncertainties for Brazilians.

Indeed, in several countries, EV's history started in the 19th century with ups and downs over the centuries and just by the end of the 20th century that it was stimulated. Some initiatives started to appear in Brazil in this period as the case of Itaipu Electric and Itaipu e400, which were developed by Gurgel S.A. Nonetheless, due to high costs and low autonomy, EV came back to the Brazilian market only over the last years. Aside from many advantages of EVs, there are uncertainties related to costs, maintenance, and lack of knowledge about the new product, which is responsible for restricting its introduction in the Brazilian market. Throughout this chapter, this affair will be discussed and applied for a case study with taxi drivers in Brazil.

4.3 CHALLENGES AND OPPORTUNITIES THROUGH THE BRAZILIAN INITIATIVES

Considering the Brazilian portrait, which was briefly presented, it is possible to introduce some challenges and opportunities faced in Brazil about the EV insertion. It is worth mentioning that to consider all these issues together as an ecosystem, it is necessary that each aspect, in this case, illustrated as a gear (Figure 4.2), converses with each other. Any changes made in the production area, for example, could affect all the chain. The same occurs with the others (Figure 4.2).

4.3.1 Economy and Production

Considering the EV's insertion in the Brazilian market, the main challenge faced by the country is the economic issue. According to Rubens [4], the price is the major key to reach the customers, followed by maintenance costs and energy prices (fuel/electricity) [5]. Based on research conducted by Teixeira & Sodré [6], the purchase cost would influence the purchase decision the most. From this perspective, the Brazilian context is not favorable to the EV introduction. In 2020, the national average income is USD 430, and the basic salary is USD 191 (with the currency of 1 USD/5.433 BRL – 31 April 2020). On the other hand, as we can see in Figure 4.3, the best-selling vehicle in the country can be found by USD 9,000 (with the currency of 1 USD/5.433 BRL – 31 April 2020), and the cheaper EV on the market is being sold by the amount of USD 22,000 (with the currency of 1 USD/5.433 BRL – 31 April 2020). Therefore, the only way people can afford any kind of vehicle is through credit loans or using the vehicle as a way to generate income, the same way taxi drivers do, aiming to reduce the payback period [7]. However, even with this possibility, taxi drivers struggle to

Challenges to Build a EV Friendly Ecosystem

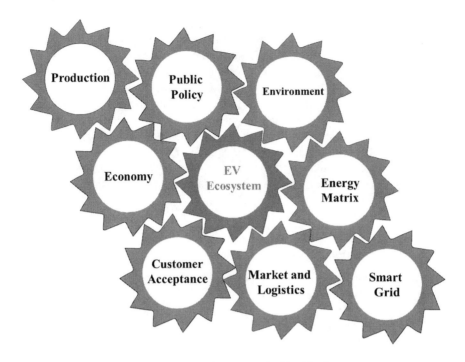

FIGURE 4.2 Challenges and opportunities approached in this chapter.

maintain an EV in Brazil (Figure 4.3). Additionally, the gasoline price is high in the country when compared to the basic and average incomes – along with other countries with a similar economy, which is estimated at about US$ 0.72/liter [8].

This purchase cost – despite being high – is only possible (and it is not higher) because the country invests in massive production in different states. For example, companies like FIAT Chrysler, Volkswagen, Chevrolet, Ford, and Honda have

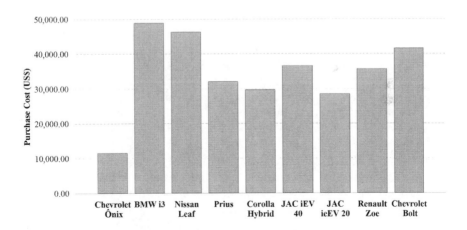

FIGURE 4.3 EVs' purchase price in Brazil. Data from Angelo 2019 [67].

industrial plants in the country, and they produced, altogether, almost 3 million vehicles in 2019 [9]. It is worth noting that Brazil does not have any kind of pure EV production nowadays, which can impact the purchase costs.

Nevertheless, this is only possible because of the incentives they receive to stay in the country, the labor costs, and the supply chain that is developed around those manufacturers. It is worth mentioning that despite the production of vehicles, the government also invests in fuel production through its state-owned company Petrobras, which is also responsible for the distribution and part of the gas stations in the country. Understanding such a structure is a paramount issue because talking about the production is not a strict conversation on how to manufacture vehicles but also about the whole structure that needs to be created to guarantee the vehicle's full operation through the year, i.e., it is important to discuss the production of components, fuel production, and even the production of asphalt that will interfere with the useful life of the vehicles.

4.3.2 Public Policies

Since we are talking about production and public incentives, it is crucial to mention public policies. Over the last decade, the use of EVs was considered a way to hold the low-carbon technologies and reach sustainability [10]. Many countries have been promoting the EV market through investments. Due to the quest to reach emission targets and standards, distinct countries determined goals about the EV introduction in the market and have encouraged people to use this technology. In the USA, incentives vary according to the state that can include a subsidy for purchase, home charge aid, preferential route, free or reduced-price parking, and discounts on licensing [11,12]. The Netherlands [13] and Norway [14] have national policies such as fee waivers, reduction on the energy prices, and road priority. China has several targeted policies that affect less than 50% of inhabitants, which cover fee waivers, reduction on the energy prices, road priority, and access to restricted traffic zones [15,16]. In another direction, Brazilian public policies and incentives are not enough to boost the EV market. Figure 4.4 shows a timeline with public policies carried out in Brazil regarding different types of fuels considering the automotive area.

In Brazil, the government usually offers tax reduction and financial aids to those brands willing to open factories – which is also usual all over the world. Additionally, it is usual to offer a tax reduction in the final prices to ensure an increase in the demand based on price reduction. As can be seen in Figure 4.4, there were different types of policies in the automotive area to boost low-carbon fuel technologies. However, comparing EV with other fuel technologies, a small number of public policies cover legislation and programs to incentivize EVs in Brazil. In this context, public policies can be presented as a challenge to be faced in the country.

The history with fuel incentives in Brazil started with ethanol, and the addition of this fuel to gasoline became mandatory, one of the government measures to guarantee its competitiveness in the market. In 1975, the National Alcohol Fuel Program PROALCOOL was launched, which had different phases, and it was responsible for increasing the number of ethanol plants and establishing the obligation of the ethanol-gasoline mixture. Another consequence of this program was the launch of vehicles powered by 100% ethanol to maintain this biofuel in the Brazilian market [17].

Challenges to Build a EV Friendly Ecosystem

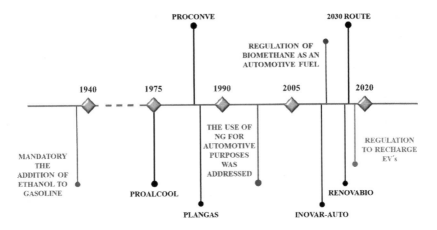

FIGURE 4.4 Timeline with Brazilian policies considering the automotive approach.

However, the natural gas was introduced in the country from 1970 to 1980 with National Anticipation Planning for the Gas Production (PLANGAS), which was the project to increase investment in the fuel supply and develop the market [18]. These initiatives reduced the investments on the PROALCOOL program and sales of vehicles powered by ethanol fuel. In the same period, the government launched the Air Pollution Control Program by Motor Vehicles (PROCONVE) to reduce and control air pollutant emissions [11], and percentage of 22% of ethanol added to gasoline became mandatory throughout the national territory. Besides that, the PROCONVE program suggested new technologies to reduce especially nitrogen oxides, sulfur oxides, and particulate matter emissions [19–21].

Other public policies over the years through the legislation implemented help to increase the natural gas for automotive purposes, the addition to biodiesel instead of diesel, and regulation of biomethane as an automotive fuel. The INOVAR-AUTO program was established in 2012 and lasted until 2017 by the Brazilian government to stimulate competition and increase efficiency and productivity in the automotive chain, considering the whole manufacturing, technological, and commercial service network [22]. The results can be summarized as investments increasing, larger production capacity, better automotive energy efficiency through new technologies, and the development of sustainability through new indicators [23].

The sequence of this program was followed by the 2030 Route Program, which was created to spread the global insertion of the Brazilian automotive sector, through the export of vehicles and auto parts. Besides, the program aims to increase competitiveness through technology improvement, reach new standards related to sustainability, and stimulate the research and development (R&D) to help the industry to reach its goals [24]. This program was the first to look for EVs in Brazil and has some measures such as the prediction of tax reduction over industrialized products as the case of EVs, which may reduce the acquisition cost and, thus, encourage their purchase. Also, the government has regulated the commercialization of energy in recharging infrastructures [25], which facilitates the provision of this service in distributors, gas stations, and shopping centers.

Instead of the 2030 Route established over the last years in Brazil, there are still few policies related to EVs, which represents a challenge for its insertion in the country. There is a need for other public policies strictly for this type of vehicle in the same way that PROALCOOL was for ethanol. According to Consoni et al [26], national efforts to increase the market and improve the possibilities of the EV implementation are represented by recent actions, mainly in the academic area. At the same time, some actions toward EV implementation represent an opportunity for the market. Considering the research scope, the 13,243/2016 National Law [27] establishes measures to encourage innovation, scientific and technological research focusing on training, the achievement of technological autonomy, and the development of national/regional productive systems in the country. An example of an R&D project, related to EV, from 2010 is the Smart Grid Brazilian Program funded by Brazilian Electric Energy Agency (ANEEL). The project proposed, for the Smart Grid Brazilian Plan, the transition of the current national electricity sector to full adoption of the smart grid concept throughout the country. Several areas as governance, energy measurement, distribution automation, and the insertion of EVs were taken into consideration [6,28,29].

Regarding the production scope, the first large-scale initiatives started in the 1980s, when Gurgel S.A., in partnership with Furnas Centrais Elétricas S.A., developed two models of purely electric cars – Itaipu Elétrico and Itaipu e400. However, these models did not obtain market space due to their high costs and low autonomy, as well as nationalization and oil substitution measures, such as PROALCOOL [30]. Another initiative took place in 2004, when the company Itaipu Binacional and Kraftwerke Oberhasli AG (KWO), the Swiss hydro controller, signed a technology cooperation agreement for the design and construction of electric cars. The initiative was also supported by FIAT, as well as other institutions in Brazil, Paraguay, and Switzerland [31]. Renault is also a partner of the Itaipu Binacional plant, which facilitates and creates favorable conditions for the acquisition of the models. Additionally, Renault has also delivered several units of the Zoe model 100% electric compact hatch, which are used in impact studies of the new technology in connection to the power grid.

Considering the infrastructure scope, the partnership between ANEEL, National Bank for Economic Development (BNDES), and Technology's Research and Projects Financing institution (FINEP) called *"Inova Energia"* was launched from 2013 to 2016 to hold the smart grid implementation and technological development of solar and wind energy devices, promote the production of EV components, and improve its energy efficiency [32,33]. In 2018, ANEEL approved the legislation NR 819/2018 [25], which settled the conditions for public recharging infrastructure to EV, representing an opportunity to the area, and other companies can be attracted to invest in Brazil. Through this measure, companies responsible for supply stations can charge vehicle owners for the electricity, the same way gas stations can charge for the fuel provided and negotiate energy prices. It is noteworthy that the electricity from EVs cannot be transferred to the distribution grid according to the Normative Resolution 482/2012 [34], which means that the bi-directional flow of energy as expected in a smart grid is not allowed yet. Some companies such as EDP and BMW are investing in EV public infrastructure along highways between the cities of São Paulo and Rio de Janeiro, creating the possibility to use EVs for long distances and solving problems with autonomy.

In the consumption scope, some municipal regulations contribute to the use of low carbon vehicles. The 6,545/2017 Act [19] for the city of São Bernardo do Campo (SP) provides municipal policy to encourage the use of electric, hybrid, and hydrogen-powered vehicles and other future alternative technologies that do not pollute the environment according to the municipal 6,163/2011 Act [20]. The incentive corresponds to a return of 25% (twenty-five percent) of the value paid for motor vehicle property tax (IPVA) [21] considering the limits: first five years of taxation; vehicles licensed in the city; and vehicles with a value equal or less than an equivalent of USD 31,120 (Exchange Rate – USD/5.433 BRL – 31 April 2020). The city of São Paulo also has the 15,997/2014 Act, which establishes the municipal policy to encourage the use of electric/hybrid vehicles and hydrogen-powered vehicles. The incentive corresponds to a partial refund of the IPVA considering the limits: first five years of taxation; vehicles licensed in the city; and vehicles with a value equal or less than an equivalent of USD 27,700 (Exchange Rate – USD/5.433 BRL – 31 April 2020). Regarding the motor vehicle property tax, seven Brazilian states (Rio Grande do Sul, Maranhão, Piauí, Ceará, Rio Grande do Norte, Pernambuco, and Sergipe) have its exemption.

To improve the benefits for users/buyers, it is important to highlight some partnerships in Brazilian cities. From 2012 through 2016, Rio de Janeiro and São Paulo received the Nissan Leaf model as a pilot project to be used in the taxi fleet. In general, taxis circulated at airports and could be recharged on quick recharging points at some specific points. The results of the project showed a reduction of about 13 tons of CO_2 and saving around USD 1,800 (USD/5.433 BRL – 31 April 2020) per year for each taxi. In Curitiba, the project called "Eco-Elétrico" started in 2014, as a result of partnerships between Itaipu Binacional, Renault, Nissan, and Portugal's Center for Excellence and Innovation in the Automotive Industry, aimed to integrate EVs into the city public service fleet. In Belo Horizonte – MG, for example, the Toyota Prius hybrid vehicle has been inserted in the taxi fleet in 2017. This is the result of a partnership between Belo Horizonte City Council, BHTrans (Belo Horizonte Transportations and Traffic Company), and the Development Bank of Minas Gerais (BDMG), which intends to incorporate 600 vehicles into the city fleet. It is worth mentioning that, in 2017, the government presented the BYD e6 to also be used as a clean technology alternative for taxi drivers. These initiatives agree with the mobility plan and are part of the project designed to implement a low carbon mobility system in Minas Gerais.

Despite the initiatives that arose in Brazil during the last years, there is still much to be done to ensure that EVs can be used on a large scale. Besides that, the government ought to have goals and priorities to increase the possibilities, investments, and development of public policies regarding EVs to push the sector.

4.3.3 Customer Acceptance

Another issue to be discussed when we consider a new product is the customer acceptance, which shows if the new product could be absorbed by the market and could contribute to improving the technologies [35], especially in a country with different types of regions and cultures, which represent a challenge for the EV adoption. As previously mentioned, Brazil always faces periods with economic difficulties, directly impacting the way products are consumed, especially EVs – due to many issues already discussed, such as the lack of national EV production, aside from the different types of uncertainties that come with EVs. First and foremost, the lack of knowledge about EVs – considering the costs, maintenance, life cycle, efficiency, and autonomy – is an important aspect that justifies low adherence. Additionally, this issue is connected to another one aforementioned: if the government does not have any goals set to develop and promote EV technology, the dissemination of the information becomes compromised. Second, EV prices are extremely high when compared to conventional vehicle prices as shown in Figure 4.3. Therefore, the population has doubts about high investments in an innovative product. Then, autonomy is always a concern in Brazil since the country is extended and highly dependable of highways. Last but not least, the lack of public recharging infrastructure is a concern once it is common to travel for more than 300 km many times per year.

Considering the high EV purchase costs, an option in Brazil is investing in this type of vehicle for the taxi's fleet, as presented by some pilot projects. Owing to the revenue related to this job, the short distances generally traveled per trip, the high mileage over the years, and intermediate waiting times, this makes taxis the ideal candidates to be replaced by the EVs. This adoption has been observed in countries such as China, Korea, and the USA [36–38]. Besides that, the increase in EVs in the taxi's fleet contributes to improving the diffusion of electric mobility and encourage other customers to purchase EVs [39]. Some studies such as the one introduced by Ozaki and Sevastyanova [40] tell us that the main motivational factors that influence the purchase of an EV are the government subsidies for these vehicles. According to Rudolph [41], the likelihood for a driver to exchange his internal combustion engine vehicle to an EV is 73.4% when we have an increase in the availability of charging infrastructure and 61.7% when a discount is offered to reduce the EV's purchase price. In the study presented by Krause et al. [42], subsidies for the purchase of the EV would help 82% of the interviewees to consider the purchase of the EV. A study with taxi drivers developed by Teixeira [43] in Belo Horizonte, Brazil, showed that 70% of the respondents would change the conventional vehicle by an EV, and regarding around 33% of them, it would depend on the purchase cost. Some of these results will be presented in the case study section.

4.3.4 MARKET, LOGISTICS, ENERGY MATRIX, AND ENVIRONMENT

The first three topics presented in this section cover economic aspects, production, public policies, and customer acceptance. Although these topics represent many of the challenges faced in Brazil, the country had a different type of development in those areas, also representing opportunities to the industry. Considering the topics listed as opportunities in Figure 4.2, another important issue to be discussed is the Brazilian market and logistics. As mentioned, the country has almost 2 million km of road networks mainly used by the logistics sector, which is extremely dependable of this structure to supply to different regions. It means that if there is a technology with good efficiency and autonomy, low cost, and minimal emissions, this could be thoroughly applied in Brazil, especially considering ground vehicles.

Brazilian road freight is responsible for about 60% of the food and fuel goods that circulate throughout the country and is especially important due to the small rail network in Brazil. According to data from the CNT Annual Report [44], the percentage of cargo vehicles comprising trucks, vans, trucks, trailers, and semi-trailers corresponds to about 17% of the Brazilian fleet. In 2018, the Brazilian population faced a strike of the cargo service, having a great impact on the distribution of essential goods and especially on the fuel's distribution. This fact was also a watershed which led to a change in the thought about road transport, promoting research and developing technologies capable of diversifying the Brazilian logistics and energy matrix. In a context where freight transport is mostly performed by diesel-powered cargo vehicles, any service shutdown will always have major impacts. Consequently, the use of other mobility technologies needs to be analyzed and considered.

Studies regarding buses and electric trucks have been developed in many countries [5,45–47] as the case of the United States and in the European Union. However, other studies [29,48] show that impacts on GHG emissions need to be evaluated when the introduction of EV since the energy to supply this technology is dependent on the country's electrical matrix and the life-cycle boundary analyzed. The energy matrix diversification represents a step to reduce the dependence of any type of source and increase the possibility to develop new strategies for energy supply, using different technologies. As it was mentioned, Brazil has a cleaner energy matrix when compared to other countries, since the country has more than 80% of energy from renewable sources. Thus, the low CO_2 emission factor from electricity production is a favorable component for the introduction and use of EVs in the country.

Another point of view considering the use of EV is not only the benefits from GHG reduction but also the reduction of local air pollution in the cities, which contributes to improving air quality and public health. It is interesting to note that in the São Paulo State [49] and the city of Rio de Janeiro [50] (Brazil), air pollution had a significant reduction during the partial lockdown period due to COVID-19 pandemic – according to the World Health Organization (WHO) [51], premature deaths around the world can be associated with outdoor air pollution. In city centers, intensive road traffic contributes to increasing the concentration of pollutants and the number of deaths due to cardiovascular, respiratory, and lung cancer diseases. Several studies show the relationship between particulate matter emissions and those diseases [52–54]. Therefore, the possibility to reduce pollutant concentration can contribute to improving human health. Hence, the synergy between all the factors already quoted and the impacts that each one has on the globe should be highlighted.

Another type of business that has been gaining ground in the Brazilian market and has a high potential of dissemination is car sharing with EVs as the core initiative for this type of service. Some factors as the case of the reduction of the number of vehicles in the streets and the parking stress [55], the possibility to have a vehicle for a short period and when it is really necessary [56], and the lack of all fixed costs (insurance and maintenance, i.e.) associated to the vehicle [57] lead to the success of this business model. Some government actions and private companies have been investing in this kind of service in different cities in Brazil such as the city of São

Paulo with the Beep-beep startup, the city of Belo Horizonte with car sharing for taxi fleet, and in the city of Brasília with car sharing for government approach.

4.3.5 Smart Grid

Considering the EV scenario in Brazil, it is also important to highlight the concept of a smart grid that is being developed in the country with some pilot projects, seeking to incorporate technologies for sensing and monitoring network performance. The current Brazilian electric grid has a unidirectional flow, which means that the energy is produced from different sources and then is distributed to the customers, with no possibility to insert energy back in the system from a distributed generation, for example. Differently, in a smart grid, the flow of information is bidirectional, which means that there is communication with all areas considering the smart grid system, helping to increase the economy with loss reduction, improving the system efficiency, and increasing the credibility and safety for all customers [58]. In this case, the energy efficiency can be increased due to this whole communication, and customers can manage the consumption in real-time, being able to trade energy through microgeneration. In general, the first step to reach the smart grid system is to change the measurement equipment to monitor consumption, allowing customers to understand energy spending according to each device.

Many advantages can be observed with smart grid implementation. Considering the customer's point of view, there is a cost reduction with different types of rates, energy efficiency improvement, access to new energy sources, and the possibility of participating in the market in a different role – as a producer. Considering the dealer's position, there is an increase in operational efficiency, narrowing the relationship between company and customer, in addition to the optimization of such investments. These improvements can promote business recovery and increase partnership. From the government's perspective, the smart grid system can contribute to cost reduction, promote new taxes according to the demand, and improve the energy quality and its respective quality index. For other areas, the new system may increase the research, develop the productive chain, and create new jobs.

A smart grid system promotes an easier EV introduction in the market considering its evolution, interaction, and monitoring since it provides capable management of charging processes. It increases the project's feasibility by providing a better distribution of electricity and preventing energy outages. Furthermore, EVs can be used as energy sources to provide electricity for the grid in periods of the day with high demand, and this procedure can be managed by the consumer. Particularly in Brazil, this energy market provided by the smart grid can contribute to enticing the population to invest in EVs, reducing the total costs associated with this product. This is the intersection between both themes.

Despite the issues faced by many countries with smart grid implementation and this new system, they can be settled as a challenge and an opportunity in Brazil. The first one is related to its geographic extension, which makes the smart grid implementation in the whole country an issue and hard to come to fruition equally for all regions. On the other hand, since there are some pilot projects in Brazil, it could help the EV insertion in the market. In 2010, as previously mentioned, ANEEL launched an R&D project to carry out the technological migration of the Brazilian current electric for the full adoption of the concept of smart grids nationally [59]. Some pilot projects can be highlighted such as the case of the "Cities of the Future" developed by CEMIG (Minas Gerais State Energy Company) in Sete Lagoas (MG/Brazil), "Smart Grid Light" developed by Light in Rio de Janeiro (RJ/Brazil), and "InovCity" developed by EDP Bandeirante, which has been implemented through the installation of smart meters, cybersecurity, distributed generation from solar energy, EV's consideration, and customer awareness [60]. Even though these pilot projects have been developed over the years, there is still a lack of public policies to hold and regulate all these changes in the system [61,62].

4.4 CASE STUDY

In order to illustrate the topics previously discussed and display the synergy in an EV ecosystem, a case study related to the Brazilian scenario will be presented – the results [43] are gathered in this section, which has two parts: public perception from the taxi drivers and numeric model. Figure 4.5 shows a general flowchart about the case study steps.

4.4.1 Public Perception

The case study was developed in the municipality of Belo Horizonte – the capital of the second largest Brazilian state (Minas Gerais) [63] – aiming to analyze the possibility to introduce EVs in the taxi fleet in 2017. To begin with, the taxi driver's profile was evaluated through an interview process since understanding taxi drivers' needs and issues was the starting point of the study.

The interview was performed with 238 respondents, considering all regions in the municipality. One of the concerns in this research was to evaluate basic knowledge

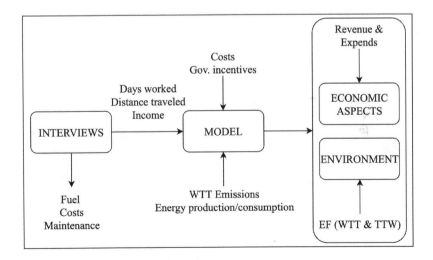

FIGURE 4.5 Case study flowchart.

about EVs, such as advantages and disadvantages regarding this technology since it is not totally disseminated in the country. Figure 4.6 shows the number of EVs licensed in Brazil from 2011 to 2020. As can be seen, the number of EVs is still lower than conventional, but it has been increasing in the last years, especially considering 2019, in which the number of EVs being licensed has tripled. This fact shows the potential of EV's introduction in the country even with a slow process when compared to the rest of the world.

To evaluate the customer acceptance, taxi drivers were invited to answer some questions regarding the advantages of EVs instead of conventional vehicles. The level of instruction of the interviewees was requested, enabling to infer the relation between the level of education and basic knowledge about EVs.

Figure 4.7 shows the relationship between the respondents' education level and knowledge about the advantages of the VE. As can be seen, the lack of information decreases with the increase in schooling. For those with elementary education, about 40% of the respondents do not know any EV advantage. However, these percentages are reduced for those with high school education (about 27%) and even lower for those with higher education (about 4%). Besides, for those with only primary education, the savings that EV could bring would be the main advantage. For the respondents with high school education, the first advantage would be the reduction of pollutants and air quality improvement, and for those who have higher education, both pollutant reduction and savings would be the major advantages.

Public policies to promote basic education and increase insertion in higher education could expand the access to information, contributing to improving knowledge about alternative transportation. Therefore, enhancing the information diffusion

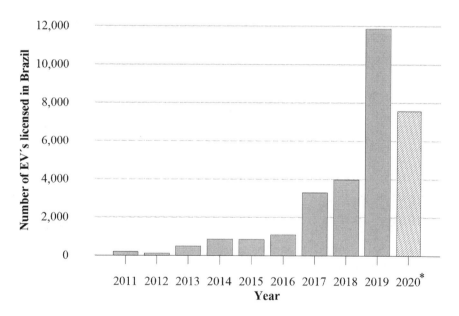

FIGURE 4.6 Brazilian EV licensed from 2011 to 2020 (*Data until July 2020). Data from ANFAVEA 2020 [9].

Challenges to Build a EV Friendly Ecosystem

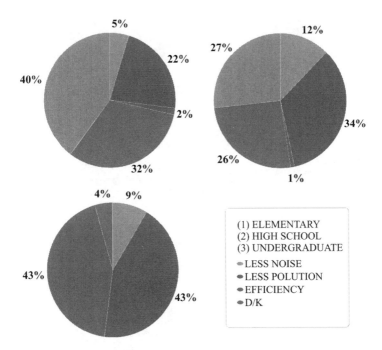

FIGURE 4.7 EVs' advantages according to the taxi drivers.

about collective and alternative transport models could help the population to have the opportunity to understand the advantages, whether social, economic, and environmental, of using these types of vehicles. Another concern evaluated with the interviews was the efforts to replace the conventional vehicle by an EV, the results of which are represented in Figure 4.8. The main concern reported by the taxi drivers was the purchase cost, which is one factor responsible for preventing vehicle replacement. As previously mentioned, these values are way higher when compared to the conventional vehicle prices, and the income provided by the taxi service is not sufficient to afford the EV initial costs. Besides, the lack of government incentives and knowledge was also the most cited factor, by the taxi drivers, as a barrier to change their vehicles since there are many uncertainties related to infrastructure, electricity prices, politics, and others.

Figure 4.9 represents the data reported by the taxi drivers as their monthly income and the costs due to operating and maintaining their vehicles. Considering these data, 11% of the respondents have a lower income than the costs to maintain the conventional car, which makes it difficult to change to an electric option. In this case, it means that taxi drivers are facing losses with the service. Furthermore, 41% of the respondents have the operation and maintenance costs representing 50% of their monthly income, which means that the profit from the taxi service is minimal, restricting the possibility of changing the vehicle. In this context, the importance of government incentives is mentioned again to support and facilitate EV access, especially in public transportation (taxis and buses).

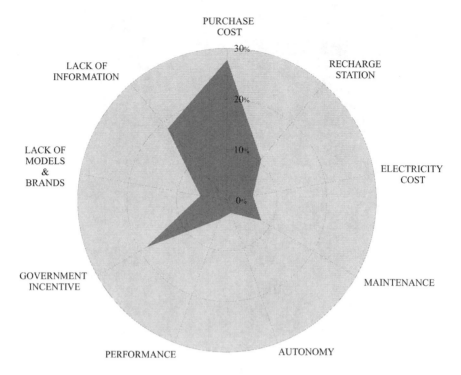

FIGURE 4.8 Challenges about replacement of conventional vehicles by EVs.

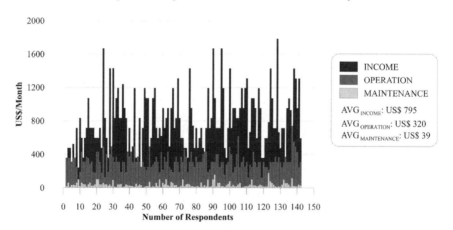

FIGURE 4.9 Income and costs associated with operation and maintenance reported by the taxi drivers.

4.4.2 Numeric Model

Based on the answers from the interviews, a numeric model was developed using Matlab software to analyze several aspects that might influence the EV insertion in the Brazilian market. Different alternative vehicles were compared to the conventional one, the Toyota Etios – which was considered as the basis for the economic analysis due to taxi drivers' preferences and its efficiency. For the electric/hybrid vehicles, three different models were compared: the Nissan Leaf, Toyota Prius, and BYD e6. These alternative vehicles were chosen since Prius started to be tested in the Belo Horizonte's taxi fleet; Leaf is the bestselling EV in the world; and e6 is very common within taxi drivers in China. The maintenance items were considered according to the Preventive Maintenance Plan of each type of vehicle, and costs associated with each one were considered through a market research. Energy costs, whether with gasoline, ethanol, or electricity, were also considered depending on the type of vehicle evaluated.

Regarding the environment analysis, it was developed considering the CO_2e (CO_2 equivalent) emissions from the fuel/electricity production – Well-to-Tank (WTT) – and from the vehicle's operation phase – Tank-to-Wheel (TTW). Both together bring the Well-to-Wheels analysis (WTW). Emission factors for the WTT phase were used based on data from SEEG [64], and CO_2e emissions from the ethanol, gasoline, and electricity production in the country were considered. Figure 4.10 shows the values from the period from 1970 to 2018. It can be seen that the emission factor from the electricity production has increased in last years due to the use of energy from thermoelectric power system to support the dry period, which brings the need for a diversification of the energy/electric matrix. Brazil has 67% of the electric matrix from hydropower, as mentioned before, which is directly impacted by the weather. Thus, there is a need for diffusion of other renewable sources such as wind and solar, since both of them have a huge potential for expansion in the country.

The inventory data of licensed taxis were provided by Belo Horizonte's Transportation and Traffic Company (BHTRANS) [65] and are grouped according to brand, model, type of fuel, and year of manufacture, which can interfere in the emission calculation. In Brazil, there is the possibility to choose the fuel between gasoline

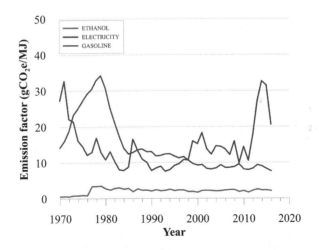

FIGURE 4.10 CO_2e emission factor from the fuel and electricity production. Data from SEEG 2018 [64].

and hydrous ethanol, and to consider it in the model, there is a function developed by Goldemberg [66], which relates the fuel choice and the ratio between hydrated ethanol and gasoline prices. For the State of Minas Gerais, this ratio is around 70%, which was considered in the model. Fuel consumptions from the vehicles were used to calculate the emissions for the vehicles analyzed.

Figure 4.11 shows the costs associated with each vehicle under consideration for fuel prices, maintenance, purchase price, and taxes. Due to higher gasoline and ethanol prices compared to the electricity in Brazil, the Etios model showed the highest percentage (18.79%) of operating costs compared to other vehicles analyzed. The purchase cost was the most representative (55.19%), and taxes represent about 18% of total costs, and maintenance accounts for around 7%. For electric and hybrid vehicles, the cost of purchasing is still the most representative (around 70% of total costs). Interestingly, for Leaf, e6, and Prius, tax-related costs (industrialized products tax and licensing fees) are higher than the operating costs of these vehicles, and they can vary from 4% to 6% of the total costs. In general, EVs and the hybrid ones have maintenance costs of approximately 1%–3% against 7% of conventional vehicles.

Figure 4.12 shows the emissions for each vehicle per kilometer traveled. Considering WTT phase, pure EVs (Leaf and e6) have higher CO_2e emissions compared to Etios and Prius, due to the higher emission factor from electricity production. However, the absolute values are quite similar. On the other hand, in the TTW phase, Toyota Etios has the higher CO_2e emission factor compared to the other vehicles due to the automotive technology and fuel, representing 90% of the total emissions. For the EV side, during the operation phase, there are no emissions. In this context, it is important to show the importance of the possibility to use EVs to reduce local emissions, helping to improve the air quality.

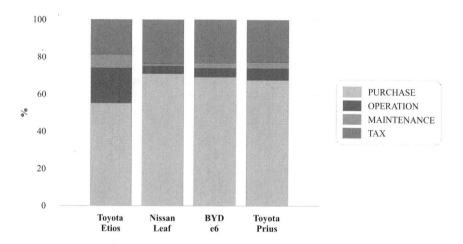

FIGURE 4.11 Percentage costs regarding different vehicles.

Challenges to Build a EV Friendly Ecosystem

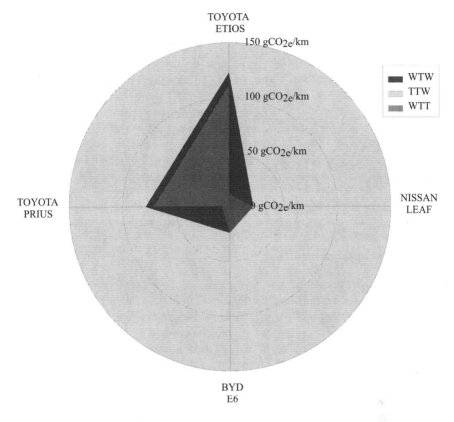

FIGURE 4.12 CO_2e emission factor.

4.5 SUMMARY AND CONCLUSIONS

This chapter brought the main topics considering challenges and opportunities to create a friendly EV ecosystem to facilitate the introduction of EVs in the market and presented a case study to show this synergy. Some Brazilian actions were shown as challenges that the country has been facing over the years, which are related to barriers to market diffusion, technologies, economic, environmental, and energy issues related to this area. Among these, it is possible to quote high purchase costs, lack of recharging infrastructure, the need for specific public policies to support the EV introduction, and customer acceptance. From the opportunities side, which are associated with the EV introduction, it was mentioned that the market and logistics are key components and the potential areas, since there is a country dependency on road networks; energy matrix and the environment, once it is based on renewable sources and can contribute to reducing local air pollution; and also smart grid system, which can be considered as both: challenge and opportunity, while Brazil has some pilot projects, which can help the EV diffusion, although its continental size can make it difficult for such project to be implemented.

Some measures that help a new product boost in the market are public policies through specific programs such as the experience mentioned about the ethanol topic in Brazil. From the production perspective, the national industry can promote EV and its components with low costs instead of importing them, and it can also increase its feasibility and the competition with conventional vehicles. From the customer acceptance side, another challenge to be faced is the necessity to disseminate correct and helpful information about EVs since the lack of data represents a barrier to change the conventional vehicle fleet. Therefore, the idea of starting this dissemination with heavy-duty drivers is a way to implement EVs from a different perspective, since the Brazilian road network is extended and have a great potential for investments. Finally, as long as all aspects of the ecosystem are not taken into consideration simultaneously, introducing it in the country might be a harsh reality to be faced on a large scale, and the same goes for the rest of the world, in which successful initiatives in different countries can show the synergy between these aspects.

ACKNOWLEDGMENTS

The author gratefully acknowledges the Pontifical Catholic University of Minas Gerais, Higher Education Improvement Coordination (Coordenação de Aperfeiçoamento de Pessoal de Nível Superior - CAPES), the University of São Paulo, and support from SHELL Brazil and São Paulo Research Foundation (Fundação de Amparo à Pesquisa do Estado de São Paulo – FAPESP), through the Research Centre for Gas Innovation (RCGI) hosted by the University of São Paulo (FAPESP Grant Proc. 2014/50279-4 and 2019/09242-3).

REFERENCES

1. Cambridge, "Dictionary," 2020. [Online]. Available: https://dictionary.cambridge.org/pt/dicionario/ingles-portugues/ecosystem.
2. Macmillan, "Dictionary," 2020. [Online]. Available: https://www.macmillandictionary.com/dictionary/british/ecosystem.
3. EPE, "BEN - Balanço Energético Nacional," 2019.
4. G. Zarazua de Rubens, "Who will buy electric vehicles after early adopters? Using machine learning to identify the electric vehicle mainstream market," *Energy*, vol. 172, pp. 243–254, 2019.
5. R. E. da Silva, P. M. Sobrinho, and T. M. de Souza, "How can energy prices and subsidies accelerate the integration of electric vehicles in Brazil? An economic analysis," *Electr. J.*, vol. 31, no. 3, pp. 16–22, 2018.
6. A. C. R. Teixeira and J. R. Sodré, "Simulation of the impacts on carbon dioxide emissions from replacement of a conventional Brazilian taxi fleet by electric vehicles," *Energy*, vol. 115, pp. 1617–1622, 2016.
7. A. C. R. Teixeira and J. R. Sodré, "Impacts of replacement of engine powered vehicles by electric vehicles on energy consumption and CO_2 emissions," *Transp. Res. Part D Transp. Environ.*, vol. 59, pp. 375–384, Mar. 2018.
8. Global Petrol Prices, "Gasoline Price," 2020. [Online]. Available: https://pt.globalpetrolprices.com/gasoline_prices/.
9. ANFAVEA, "Vehicles - Production, licensing and exports data," 2020. [Online]. Available: http://anfavea.com.br/estatisticas.
10. Barton; Schütte, "Electric vehicle law and policy: a comparative analysis.," *J. Energy Nat. Resour. Law,* vol. 35, no. 2, pp. 147–170, 2017.

11. Y. Zhou, M. Wang, H. Hao, L. Johnson, H. Wang, and H. Hao, "Plug-in electric vehicle market penetration and incentives: a global review," *Mitig. Adapt. Strateg. Glob. Chang.*, vol. 20, no. 5, pp. 777–795, 2015.
12. S. Wee, M. Coffman, and S. La Croix, "Data on U.S. state-level electric vehicle policies, 2010–2015," *Data Br.*, vol. 23, p. 103658, 2019.
13. W. Sierzchula, S. Bakker, K. Maat, and B. Van Wee, "The influence of financial incentives and other socio-economic factors on electric vehicle adoption," *Energy Policy*, vol. 68, pp. 183–194, 2014.
14. A. C. Mersky, F. Sprei, C. Samaras, and Z. S. Qian, "Effectiveness of incentives on electric vehicle adoption in Norway," *Transp. Res. Part D Transp. Environ.*, vol. 46, pp. 56–68, 2016.
15. Z. Wan, D. Sperling, and Y. Wang, "China's electric car frustrations," *Transp. Res. Part D Transp. Environ.*, vol. 34, pp. 116–121, 2015.
16. A. R. Gopal, W. Y. Park, M. Witt, and A. Phadke, "Hybrid- and battery-electric vehicles offer low-cost climate benefits in China," *Transp. Res. Part D Transp. Environ.*, vol. 62, no. March, pp. 362–371, 2018.
17. J. A. Puppim de Oliveira, "The policymaking process for creating competitive assets for the use of biomass energy: The Brazilian alcohol programme," *Renew. Sustain. Energy Rev.*, vol. 6, no. 1–2, pp. 129–140, 2002.
18. E. Moutinho dos Santos, M. T. W. Fagá, C. B. Barufi, and P. L. Poulallion, "Gás natural: A construção de uma nova civilização," *Estud. Avancados*, vol. 21, no. 59, pp. 67–90, Jan. 2007.
19. D. T. Hountalas, G. C. Mavropoulos, and K. B. Binder, "Effect of exhaust gas recirculation (EGR) temperature for various EGR rates on heavy duty DI diesel engine performance and emissions," *Energy*, vol. 33, no. 2, pp. 272–283, 2008.
20. M. H. M. Yasin et al., "Study of a diesel engine performance with exhaust gas recirculation (EGR) system fuelled with palm biodiesel," *Energy Procedia*, vol. 110, no. December 2016, pp. 26–31, 2017.
21. M. Koebel, G. Madia, and M. Elsener, "Selective catalytic reduction of NO and NO_2 at low temperatures," *Catal. Today*, vol. 73, no. 3–4, pp. 239–247, 2002.
22. V. Lazzarotti, R. Manzini, L. Pellegrini, and E. Pizzurno, "Open innovation in the automotive industry: Why and how? Evidence from a multiple case study," *Int. J. Technol. Intell. Plan.*, vol. 9, no. 1, pp. 37–56, 2013.
23. BRASIL, "INOVAR-AUTO Avaliação de Impacto do Programa Inovar-Auto Dezembro de 2019," 2019.
24. BRASIL, "Provisional Measure no 843/2018. Establishes mandatory requirements for the commercialization of vehicles in Brazil, establishes the Rota 2030- Mobility and Logistics Program and provides for the tax regime of non-produced auto parts." 2018.
25. ANEEL, "RES. NORM. 819 de 19 de junho de 2018," 2018. [Online]. Available: http://www2.aneel.gov.br/cedoc/ren2018819.pdf.
26. F. Consoni, "Estudo de Governança e Políticas Públicas para Veículos Elétricos Estudo de Governança e Políticas Públicas para Veículos Elétricos," 2018.
27. BRASIL, "Lei 13,243 de 11 de janeiro de 2016." 2016.
28. A. C. R. Teixeira, D. L. Da Silva, L. D. V. B. MacHado Neto, A. S. A. C. Diniz, and J. R. Sodré, "A review on electric vehicles and their interaction with smart grids: The case of Brazil," *Clean Technol. Environ. Policy*, vol. 17, no. 4, 2015, pp. 841–857.
29. E. A. M. Falcão, A. C. R. Teixeira, and J. R. Sodré, "Analysis of CO_2 emissions and techno-economic feasibility of an electric commercial vehicle," *Appl. Energy*, vol. 193, pp. 297–307, May 2017.
30. B. S. M. C. Borba, "Modelagem integrada da introdução de veículos leves conectáveis à rede elétrica no sistema energético brasileiro." 2012.

31. C. R. B. Novais, "Modalidade Elétrica: Desafios e Oportunidades," *FGV Energ.*, p. 10, 2016.
32. R. Marx and A. M. De Mello, "New initiatives, trends and dilemmas for the Brazilian automotive industry: The case of Inovar Auto and its impacts on electromobility in Brazil," *Int. J. Automot. Technol. Manag.*, vol. 14, no. 2, pp. 138–157, 2014.
33. G. Masiero, M. H. Ogasavara, A. C. Jussani, and M. L. Risso, "The global value chain of electric vehicles: A review of the Japanese, South Korean and Brazilian cases," *Renew. Sustain. Energy Rev.*, vol. 80, no. May, pp. 290–296, 2017.
34. ANEEL, "RES. NORM. 482 de 17 de abril de 2012," 2012. [Online]. Available: http://www2.aneel.gov.br/arquivos/PDF/Resolução Normativa 482, de 2012-bip-junho-2012.pdf.
35. F. Liao, E. Molin, H. Timmermans, and B. van Wee, "Consumer preferences for business models in electric vehicle adoption," *Transp. Policy*, vol. 73, no. November 2017, pp. 12–24, 2019.
36. J. Kim, S. Lee, and K. S. Kim, "A study on the activation plan of electric taxi in Seoul," *J. Clean. Prod.*, vol. 146, pp. 83–93, 2017.
37. N. Wang, Y. Liu, G. Fu, and Y. Li, "Cost–benefit assessment and implications for service pricing of electric taxies in China," *Energy Sustain. Dev.*, vol. 27, pp. 137–146, 2015.
38. E. Park and S. J. Kwon, "Renewable electricity generation systems for electric-powered taxis: The case of Daejeon metropolitan city," *Renew. Sustain. Energy Rev.*, vol. 58, pp. 1466–1474, 2016.
39. J. Asamer, M. Reinthaler, M. Ruthmair, M. Straub, and J. Puchinger, "Optimizing charging station locations for urban taxi providers," *Transp. Res. Part A Policy Pract.*, vol. 85, pp. 233–246, 2016.
40. R. Ozaki and K. Sevastyanova, "Going hybrid: An analysis of consumer purchase motivations," *Energy Policy*, vol. 39, no. 5, pp. 2217–2227, May 2011.
41. C. Rudolph, "How may incentives for electric cars affect purchase decisions?," *Transp. Policy*, vol. 52, pp. 113–120, 2016.
42. R. M. Krause, S. R. Carley, B. W. Lane, and J. D. Graham, "Perception and reality: Public knowledge of plug-in electric vehicles in 21 U.S. cities," *Energy Policy*, vol. 63, no. 2013, pp. 433–440, 2013.
43. A. C. R. Teixeira, "Análise numérica da inserção de veículos elétricos na frota de táxis de belo horizonte," Ponitfícia Universidade Católica de Minas Gerais, 2018.
44. CNT, "Anuário CNT do transporte – Estatísticas Consolidadas," 2018.
45. D. C. Quiros et al., "Real-world emissions from modern heavy-duty diesel, natural gas, and hybrid diesel trucks operating along major California freight corridors," *Emiss. Control Sci. Technol.*, vol. 2, no. 3, pp. 156–172, 2016.
46. Y. Zhang, Y. Jiang, W. Rui, and R. G. Thompson, "Analyzing truck fleets' acceptance of alternative fuel freight vehicles in China," *Renew. Energy*, vol. 134, pp. 1148–1155, 2019.
47. P. Kluschke, T. Gnann, P. Plötz, and M. Wietschel, "Market diffusion of alternative fuels and powertrains in heavy-duty vehicles: A literature review," *Energy Reports*, vol. 5, pp. 1010–1024, 2019.
48. E. F. Choma and C. M. L. Ugaya, "Environmental impact assessment of increasing electric vehicles in the Brazilian fleet," *J. Clean. Prod.*, vol. 152, pp. 497–507, 2017.
49. L. Y. K. Nakada and R. C. Urban, "COVID-19 pandemic: Impacts on the air quality during the partial lockdown in São Paulo state, Brazil," *Sci. Total Environ.*, vol. 730, no. PG-139087-139087, p. 139087, 2020.
50. G. Dantas, B. Siciliano, B. B. França, C. M. da Silva, and G. Arbilla, "The impact of COVID-19 partial lockdown on the air quality of the city of Rio de Janeiro, Brazil," *Sci. Total Environ.*, vol. 729, p. 139085, 2020.

51. WHO, "Ambient (outdoor) air pollution," 2018.
52. C. E. Reid, E. M. Considine, G. L. Watson, D. Telesca, G. G. Pfister, and M. Jerrett, "Associations between respiratory health and ozone and fine particulate matter during a wildfire event," *Environ. Int.*, vol. 129, no. March, pp. 291–298, 2019.
53. J. Rovira, J. L. Domingo, and M. Schuhmacher, "Air quality, health impacts and burden of disease due to air pollution (PM10, PM2.5, NO2 and O3): Application of AirQ+ model to the Camp de Tarragona County (Catalonia, Spain)," *Sci. Total Environ.*, vol. 703, p. 135538, 2020.
54. E. Engström and B. Forsberg, "Health impacts of active commuters' exposure to traffic-related air pollution in Stockholm, Sweden," *J. Transp. Heal.*, vol. 14, no. August, p. 100601, 2019.
55. T. H. Stasko, A. B. Buck, and H. Oliver Gao, "Carsharing in a University setting: Impacts on vehicle ownership, parking demand, and mobility in Ithaca, NY," *Transp. Policy*, vol. 30, pp. 262–268, 2013.
56. K. Kortum, "Driving smart: Carsharing mode splits and trip frequencies," *93rd Annu. Meet. Transp. Res. Board*, no. January 2014, pp. 1–11, 2014.
57. F. Jin, E. Yao, and K. An, "Analysis of the potential demand for battery electric vehicle sharing: Mode share and spatiotemporal distribution," *J. Transp. Geogr.*, vol. 82, no. June 2019, p. 102630, 2020.
58. G. Dileep, "A survey on smart grid technologies and applications," *Renew. Energy*, vol. 146, pp. 2589–2625, 2020.
59. ANEEL, "Redes Inteligentes do Brasil," 2020. [Online]. Available: http://redesinteligentesbrasil.org.br/.
60. K. G. Di Santo, E. Kanashiro, S. G. Di Santo, and M. A. Saidel, "A review on smart grids and experiences in Brazil," *Renew. Sustain. Energy Rev.*, vol. 52, pp. 1072–1082, 2015.
61. G. G. Dranka and P. Ferreira, "Towards a smart grid power system in Brazil: Challenges and opportunities," *Energy Policy*, vol. 136, no. September 2019, 2020.
62. G. de A. Dantas et al., "Public policies for smart grids in Brazil," *Renew. Sustain. Energy Rev.*, vol. 92, no. May, pp. 501–512, 2018.
63. IBGE, "IBGE," 2019. [Online]. Available: https://ibge.gov.br/. [Accessed: 17-Oct–2019].
64. SEEG, "Sistema de Estimativa de Emissões de Gases de Efeito Estufa," 2018.
65. BHTRANS, "Taxi inventory," 2018. [Online]. Available: https://prefeitura.pbh.gov.br/bhtrans.
66. J. F. Goldemberg, F. E. B. Nigro, and S. T. Coelho, "Situação atual, Perspectivas, Barreiras e Propostas," *Bioenergia no Estado São Paulo*, 2008.
67. B. Angelo, "Carros elétricos no Brasil: veja todos os modelos e preços," 2019. [Online]. Available: https://autopapo.uol.com.br/noticia/carros-eletricos-no-brasil-modelos-precos/.

5 Coordinated Operation of Electric Vehicle Charging and Renewable Power Generation Integrated in a Microgrid

Alberto Borghetti, Fabio Napolitano, Camilo Orozco Corredor, and Fabio Tossani
University of Bologna

CONTENTS

5.1 Introduction	97
5.2 The Stochastic Optimization Model	100
5.2.1 Model of the Microgrid	100
5.3 Scenarios and Tree Generation Procedure	102
5.3.1 Scenario Generation for the V2G Parking Lot	102
5.3.2 Scenario Generation for the PV Unit and Local Load	103
5.3.3 Tree Generation by Using k-Means	103
5.4 Microgrid Simulation Results	105
5.4.1 Description of the Case Study	105
5.4.2 Scenario-Based Tree Generation	105
5.4.3 Solution of the Multistage Stochastic Model	108
5.5 Conclusions	114
Acknowledgments	116
Nomenclature	116
References	117

5.1 INTRODUCTION

The literature on charging load modelling is becoming quite large, as shown in the recent survey presented in Ref. [1]. An analysis of the advantages and drawbacks of different approaches to the integration of electric vehicles (EVs) is presented in Ref. [2]. Additionally, a study of the state of art of fast-charging stations, including experimental test, has been introduced in Ref. [3].

The current transition to electric mobility and the integration of distributed energy generation into microgrids both support the employment of bidirectional public charging stations of plug-in electric vehicles (PEVs). Moreover, the appearance of clusters of rapid charging stations in parking lots has propitiated the development of mainly two types of interactions between PEVs and the power distribution network, according to their dispatchability: grid-to-vehicle (G2V) charging stations, in which PEVs arrive in the stations to be charged as fast as possible, and vehicle-to-grid (V2G) scenarios, which represent typical parking lots where PEVs stay for a significantly long time. For the first scenario (G2V), one of the key aspects is the queuing model that can represent the arrival and departure times of EVs, as illustrated, e.g., in Ref. [4] and reference therein. For the second scenario, the V2G parking lot can be operated so as to achieve either or both of the two typical objectives of storage systems: load flattening and balancing of renewable generation [5].

Typically, in the V2G scenario, the owner of the charging facility aims to obtain economic benefits for the services, while offering to the PEV owner the option to charge the vehicle at the lowest possible cost. To achieve these objectives, smart charging approaches are implemented to align the charging and dispatching processes of storage systems with the optimization objectives. For example, Ref. [6] propose an optimization model for the assessment of the contribution of V2G systems. The method described in Ref. [7] addresses the integrated operational planning of a distribution system, by using an aggregator conceived as an intermediate agent between end-users and distribution system operators (DSOs).

The integrated operation of parking facilities with renewable energy resources (RESs) has been extensively studied, as shown in Ref., e.g., [8]. An evaluation of the integration of PEVs with photovoltaic (PV) systems, in order to cope with the fluctuation of solar irradiance, has been performed in Ref. [9]. An additional approach that takes into account the uncertainties of PEV's arrival and grid power price has been presented in Ref. [10]. A generation scheduling method for the coordinated operation of an industrial microgrid, which considers electricity and heat generation, electrical loads, PV units, and PEVs, is presented in Ref. [11].

Among the various schemes typically adopted for regulating the participation of distributed energy resources (DERs), fixed or time-of-use (TOU) tariffs, designed by retailers or DSOs, are often adopted. A study of the feasibility of premium tariff rates for V2G services similar to feed-in-tariff (FIT) programs for RES has been presented in Ref. [12].

This chapter focuses on the operation of the parking lot equipped with bidirectional charging stations inside the energy management system of a microgrid. A typical scenario corresponding to the integration of G2V and V2G services, battery energy storage (BES) systems, and DERs in a microgrid site is illustrated in Figure 5.1. In the represented case, a central dispatching system solves the optimization problem that minimizes the energy procurement costs of the considered site.

As an example, the model described in this chapter represents the day-ahead optimization of the global charging and discharging of the batteries of the EVs connected to the charging stations in order to reduce the energy procurement cost. The optimization problem is solved by the dispatching centre of the microgrid. The optimization model is based on the application of stochastic linear programming, where the

EV Charging and Renewable Power Generation

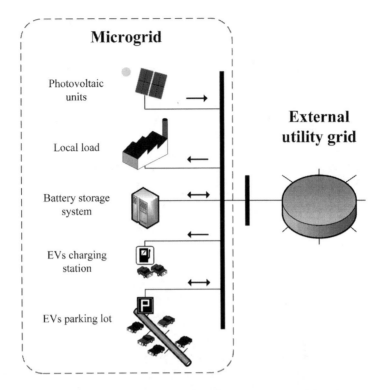

FIGURE 5.1 Scheme of the microgrid with both local generation and EV charging stations.

uncertainty associated with the operation of the system during the day is modelled by stochastic processes.

The contributions of this chapter are:

- The presentation of an approach for the day-ahead scheduling of a microgrid, including a parking lot, that facilitates the integration of the production from renewable resources;
- The presentation of a scenario-based approach to describe the uncertainties associated with the departures and arrivals of the EVs, as well as with the power production from renewable energy and load profile;
- The development of a clustering procedure that allows the solution of the scenario-based multistage optimization model with a reasonable computational effort.

The following sections describe the tree generation procedure for the case of a microgrid that includes both distributed generation (e.g., PV units) and a parking lot equipped with charging stations for PEVs with V2G technology. For this purpose, we make reference to the linear programming model presented in Ref. [13], although the procedure can be suitably adapted to be used with different stochastic models for the optimal operation of EV's parking lots in microgrids (e.g., those presented in Refs. [6] and [14]) and in power distribution systems (e.g., Refs. [15] and [16]).

A peculiar characteristic of the linear programming model presented in Ref. [13] is the representation of specific operating rules relevant to the initial energy available in the cars entering the parking lot. While in Ref. [13] a two-stage stochastic programming model has been adopted, this chapter describes the extension of the model to the multistage case, following the approach of Ref. [17]. For such a purpose, a scenario tree is needed to define each scenario as a root-leaf path through all the stages, representing the conditional probabilities with links between nodes. Each path corresponds to a set of values of stochastic parameters with a specific history up to the relevant node.

The structure of this chapter is as follows: Section 5.2 reviews the main characteristics of the stochastic optimization model of the parking lot within the microgrid. Section 5.3 describes the generation of the scenarios and the relevant tree implemented in the AIMMS modelling environment. Section 5.4 presents some optimization results. Section 5.5 concludes this chapter.

5.2 THE STOCHASTIC OPTIMIZATION MODEL

This section describes the specific linear programming model applied to the parking lot with EV's charging stations. In order to perform the analysis of the model, we consider the day-ahead optimization problem of the dispatching centre of a microgrid that is aimed at reducing the total cost for electric energy procurement. The microgrid is connected to the external utility grid and includes the EV's parking lot with V2G technology, as well as a PV unit and local loads.

5.2.1 Model of the Microgrid

In the multistage stochastic optimization model, each scenario ω is represented by the path that connects the consecutive nodes from the root of the tree to the relative leaf node. All the variables associated with the scenarios that pass through the same intermediate node are identical. In general, a stage consists of multiple time periods t; therefore, all the variables associated with the periods of the same stage coincide.

The optimization problem considered by the dispatching centre is represented by (5.1), where we consider a day-ahead optimization horizon of 24 hours:

$$\min \sum_{\omega} \pi_\omega C_\omega \qquad (5.1)$$

with

$$C_\omega = \sum_{t} \left[\rho_t^T \left(E_{t\omega}^- - r E_{t\omega}^+ \right) + C_{t\omega}^S \right]. \qquad (5.2)$$

For the V2G parking lot, the decision variables $P_{t\omega}$ and $\mu_{t\omega}$ are defined as follows: $P_{t\omega}$ is the power exchange of the V2G parking lot calculated for each node of the scenario tree (positive if provided to the grid), and $\mu_{t\omega}$ is the nonnegative utilization coefficient of the total energy $E_{t\omega}^{S+}$ initially stored in the EVs that arrive at time t in the parking lot.

The constraints that represent the behaviour of the V2G parking lot are:

$$P_{t\omega}^{\min} \leq P_{t\omega} \leq P_{t\omega}^{\max} \tag{5.3}$$

$$E_{t\omega}^{S} = (1-\delta)E_{(t-1)\omega}^{S} - P_{t\omega}\Delta t + E_{t\omega}^{S+} - E_{t\omega}^{S-} \tag{5.4}$$

$$E_{t\omega}^{S} \leq E_{t\omega}^{S\max} \tag{5.5}$$

$$0 \leq \mu_{t\omega} \leq \frac{\left(E_{t\omega}^{S+} - e_{\min}N_{t\omega}^{in}E_{nom}^{EV}\right)}{E_{t\omega}^{S+}} \cdot u\left(E_{t\omega}^{S+} - e_{\min}N_{t\omega}^{in}E_{nom}^{EV}\right) \tag{5.6}$$

$$E_{t\omega}^{V2G} = (1-\delta)E_{(t-1)\omega}^{V2G} - P_{t\omega}\Delta t + \mu_{t\omega}E_{t\omega}^{S+} + \sum_{j=0}^{t}(1-\mu_{t\omega})E_{(j,t)\omega}^{\mu} - E_{t\omega}^{S-} \tag{5.7}$$

$$l_{t\omega} \geq \begin{cases} (1-\eta^r)P_{t\omega} \\ \left(1-\dfrac{1}{1-\eta^s}\right)P_{t\omega} \end{cases} \tag{5.8}$$

$$C_{t\omega}^{S} \geq \begin{cases} c^r P_{t\omega}\Delta t + c^{\mu}\mu_{t\omega}E_{t\omega}^{S+} \\ -c^s P_{t\omega}\Delta t + c^{\mu}\mu_{t\omega}E_{t\omega}^{S+} \end{cases}. \tag{5.9}$$

Constraint (5.3) limits the maximum and minimum power exchange of the V2G parking lot.

The energy stored by each battery at the end of period t is determined by (5.4), and constraint (5.5) limits the maximum value of the storage capability.

When EVs reach the parking lot, their energy adds to the total energy of the V2G parking lot $E_{t\omega}^{S}$. Constraint (5.6) limits the maximum value of $\mu_{t\omega}$ so the initial charge of the battery may be used only for the amount exceeding a predefined minimum fraction ($e_{\min}$) of the rated energy size E_{nom}^{EV}. This constrain avoids deep discharge conditions. In (5.6), function $u(\cdot)$ represents the step function with value 0 for negative arguments and 1 for positive arguments.

Constraint (5.7) is aimed to calculate the available energy for dispatching services $E_{t\omega}^{V2G}$. For such purpose, the dispatching centre determines the exploitation factor $\mu_{t\omega}$ of the total energy $E_{t\omega}^{S}$. This scheme permits the implementation of the battery-to-battery charging strategy in a V2G system. The total cost $C_{t\omega}^{S}$ considers the cost of this retrieval according to specific price (i.e., c^{μ}). Such a cost is considered at the arriving time of the EVs, while the associated energy can be retrieved in all of the periods during the parking time.

Disregarding network power losses, the energy balance in each period of the microgrid is

$$\sum_u P_{ut\omega} \Delta t = E_{t\omega}^+ - E_{t\omega}^- \qquad (5.10)$$

The net power of the V2G parking lot, calculated considering the losses, is given by

$$P_{ut\omega} = P_{t\omega} - l_{t\omega} \qquad (5.11)$$

Both $l_{t\omega}$ and $C_{t\omega}^S$ (nonnegative variables according to equations 5.8 and 5.9, respectively) are minimized as a result of the minimization of the objective function in equation 5.1.

According to this model, the power values associated with the PV unit and the local loads cannot be dispatched.

5.3 SCENARIOS AND TREE GENERATION PROCEDURE

To build the scenario tree described in the Introduction, a set of equiprobable scenarios is at first generated. This section deals with the procedure for the scenario generation of the V2G parking lot. For completeness, this section of this chapter will also describe the scenario generation of the PV unit output power and the local loads.

Once the scenarios have been generated, the scenario tree is obtained by a routine based on a recursive k-means clustering procedure applied to the total set of initial scenarios, with the scenario reduction technique described in Ref. [18]. The relevant parameters considered by the procedure of aggregation are the number of parked EVs, PV generation, and the non-dispatchable load in the microgrid.

5.3.1 SCENARIO GENERATION FOR THE V2G PARKING LOT

The scenarios for the V2G parking lot are generated as follows. For each scenario ω, a population of N_{tot} EVs willing to enter in the parking lot is generated according to chosen statistical distributions (e.g., a normal distribution with the mean value equal to the forecasted number of potential users of the parking lot). In general, not all the N_{tot} EVs can be connected to a charging station if all the charging stations available in the parking lot are engaged.

Each i-th EV that is allowed to be recharged is characterized by the parameters t_i^+, E_i^0, and s_i. In the simulation example reported in this chapter for illustrative purposes, also these parameters are generated following normal distributions with the mean value equal to the forecasted value. Although the rated capacity of the batteries is considered constant and equal for all the EVs, it could also be assumed as a variable.

The energy stored E_i^- is equal to 1 pu if the time of stay s_i is long enough to get a full recharge; otherwise, it is set to ratio between s_i and the time needed for a full recharge.

EV Charging and Renewable Power Generation

For each time t and scenario ω, the procedure builds two sets: $S_{t\omega}^+$ that consider the EVs incoming at time t and $S_{t\omega}^-$ that consider the EVs leaving at time t.

On the basis of these sets, the procedure calculates the parameters needed in equations 5.3–5.9: the increase in the energy stored in the parking lot due to arrivals at time t $\left(E_{t\omega}^{S+} = \sum_{i \in S_{t\omega}^+} E_i^0 \right)$, the decrease in the energy stored in the parking lot due to departures at time t $\left(E_{t\omega}^{S-} = \sum_{i \in S_{t\omega}^-} E_i^- \right)$, and the decrease in the energy stored in the parking lot at time t due to the initial charge of the EVs entered at time j and leaving at t $\left(E_{(j,t)\omega}^{\mu} = \sum_{i \in \{S_{j\omega}^+ \cap S_{t\omega}^-\}} E_i^0 \right)$.

5.3.2 Scenario Generation for the PV Unit and Local Load

The scenarios of the PV output and the local loads are generated on the basis of the day-ahead forecasts and the probability distributions that characterize the expected deviations with respect to the forecasts.

For the PV units, the scenarios are generated by using the procedure described in Ref. [19] based on a first-order autoregressive Markov process that provides a time series $x_{t\omega}$ that represents the one-lag autocorrelation

$$x_{t\omega} = \phi x_{(t-1)\omega} + \varepsilon, \quad (5.12)$$

where ε is the white noise with zero mean and standard deviation $\sigma = \sqrt{1-\phi^2}$. In order to obtain the generation of the PV outputs $P_{t\omega}^{PV}$, $x_{t\omega}$ is added to the normalized profile of the forecast, and then the probability transformation, which includes the cumulative density functions of both the normalized time series and the forecasted profile, is applied.

In order to avoid unrealistic scenarios, all the $P_{t\omega}^{PV}$ profiles do not differ by more than 10% from the forecasted profile for at least 90% of the time.

In the numerical test presented in this chapter, the profiles of the local loads, besides the EVs at the unidirectional charging stations, are obtained by multiplying the forecasted profile by $1+k_t$, where k_t is generated by using a normal distribution with a mean value equal to 0 and standard deviation $\sigma = \sqrt{1-\psi(t)^2}$, where $\psi(t)$ is a decreasing function of t in order to represent the increase of the incertitude with time.

5.3.3 Tree Generation by Using k-Means

As described in Ref. [18], after the generation of the set of initial scenarios, the application of the k-means clustering method allows to build the scenario tree. For such a purpose, the 24-hour optimization horizon is divided into four stages of six hours each. Figure 5.2 illustrates the main steps of the implemented method.

Each scenario $\xi_{t\omega}$ corresponds to a realization of the stochastic parameters of number of EVs in the parking lot, PV generation, and total load (i.e., considering local load and charge of EVs at the charge station) in the form:

$$\xi_{t\omega} = \left[N_{t\omega}^{EV}, P_{t\omega}^{PV}, P_{t\omega}^{Load} \right] \quad (5.13)$$

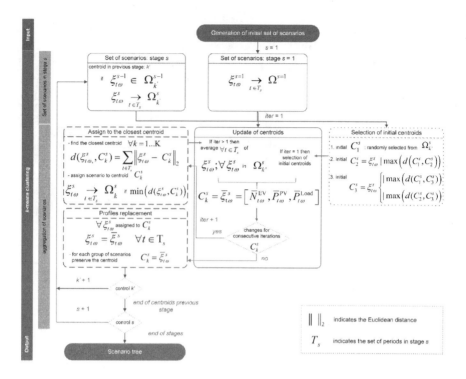

FIGURE 5.2 Implemented *k*-means algorithm.

At stage $s = 1$ (i.e., the very first six hours of the day), all scenarios have the same value of parameter, i.e., $\xi_{t\omega} = \bar{\xi}_{t\omega}$.

At stage $s = 2$ (i.e., from 7 am to 12 pm), the set of scenarios is aggregated in a number K of desired groups represented by centroid C_k^s; for the numerical tests included in this chapter, a number of three clusters have been predefined ($K = 3$). Firstly, the initial K cluster centres are selected (i.e., centroids); the first cluster centre C_1^s is randomly chosen from the relevant set of scenarios in the stage. Then, the following two initial cluster centres, i.e., C_2^s and C_3^s, correspond to the farthest two scenarios $\xi_{t\omega}$ obtained, on the basis of the Euclidean distance.

Then, in order to identify the closest centroid C_k^s to the relevant scenario $\xi_{t\omega}$, distance d is calculated as

$$d\left(\xi_{t\omega}, C_k^s\right) = \sum_{t \in T_s} \left\| \xi_{t\omega} - C_k^s \right\| \quad \forall k = 1,...,K, \tag{5.14}$$

where $\| \ \|_2$ indicates the Euclidean distance, and the individual scenario is assigned to the cluster with the corresponding minimal distance.

Next, the centroid of each cluster C_k^s is updated as the mean value of all the scenarios aggregated in each cluster. This process is iteratively carried on, until there are not changes in the centroid values.

EV Charging and Renewable Power Generation

Starting from an original set of equiprobable scenarios at the considered stage s, the probability of each cluster π_k^s corresponds to the total sum of the probabilities of the scenarios aggregated in the same cluster.

The profiles of all the scenarios belonging to the same cluster C_k^s are replaced by the profile of the relevant centroid, i.e., $\xi_{t\omega} = \bar{\xi}_{t\omega} \; \forall t \in T_s$ if $\xi_{t\omega}^s \in \Omega_k^s$.

At the subsequent stages, the k-means clustering routine is carried on for each centroid k obtained in the previous stage.

Finally, the implemented k-means clustering algorithm delivers the scenario tree that represents the realization of the considered stochastic model, while considering the relevant non-anticipativity constraints.

At each stage defined, each scenario is grouped to one of the centroids of the algorithm on the basis of the average value of the number of parked EVs, PV output, and total non-dispatchable load in the six hours.

All the values of the parameters of the scenarios grouped together are averaged in order to assign a unique value to each of them for each hour of the stage. At each period j, the values of $E^\mu_{(j,t)\omega}$ of all the scenarios in the same group are averaged for all $t > j$. Moreover, a matrix that contains the number of EVs that arrive in period j and leave in period t is defined and averaged as done for $E^\mu_{(j,t)\omega}$. In order to guarantee the existence of a feasible solution of the optimization problem of the V2G parking lot, this matrix is used to define the average values of the number of parked EVs, and therefore of $E^{S\max}_{t\omega}$, while $E^\mu_{(j,t)\omega}$ is used to define the average values of $E^{S+}_{t\omega}$ and $E^{S-}_{t\omega}$.

5.4 MICROGRID SIMULATION RESULTS

5.4.1 Description of the Case Study

The case study refers to a microgrid composed by a parking lot with 100 bidirectional charging points - each with 7kW rated power -, by a 3.5MWp-PV system, local loads up to 3 MW and a connection point to the external utility grid.

In the presented simulation results, the parking lot is empty at time $t = 0$, the energy capacity of the EV's batteries is 24 kWh, efficiencies η^r and η^s are equal to 0.96, c^μ is 1.8 €/MWh, the minimum energy value $e_{\min}$ is equal to 0.2 pu, price ρ_t^T is equal to 72.39 €/MWh from 7 am to 11 pm and is equal to 51.62 €/MWh in the other hours, ratio r between selling price and buying price of electric energy is equal to 0.5, and δ is neglected.

The stochastic linear programming model has been implemented in the AIMMS development environment and tested by using the Cplex solver version 12.9 on a PC equipped with a 2-GHz Intel-i7vPro with 8 GB of RAM, running 64-bit Windows 10.

5.4.2 Scenario-Based Tree Generation

The number of generated scenarios, before the grouping procedure is applied, is 60 for all the stochastic variables of the microgrid.

As mentioned, the stochastic events for the parking lot are obtained starting from the random generation of N_{Tot}, t_i^+, E_i^0, and s_i according to normal distribution functions with the mean values and the standard deviations as reported in Table 5.1. Figure 5.3 shows the 60 profiles of $N_{t\omega}^{EV}$.

The profiles of the PV output are generated assuming $\phi = 0.999$ and are shown in Figure 5.4. The load profiles are generated by assuming ψ linearly decreasing from 1 at $t = 1$ to 0.99 at $t = 24$. Figure 5.5 shows the profiles of the total load of the microgrid.

Figure 5.6 shows the scenario tree with 26 leaf nodes obtained by the implemented k-means, with the indication of the probability associated with each leaf node.

TABLE 5.1

Parameters of the Normal Distribution Functions for the V2G Scenario Generation

Population Parameters	Mean Value	Standard Deviation
Time of entrance t	9	2
Initial charge (pu) E_i^0	0.3	0.3
Time of stay s_i	8	2
Number of entrances N_{Tot}	100	10

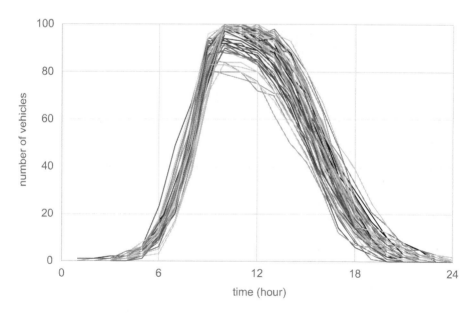

FIGURE 5.3 Number of parked EVs in the parking lot.

EV Charging and Renewable Power Generation

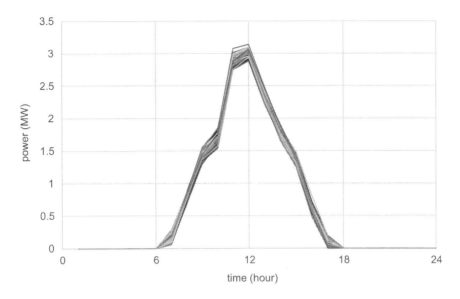

FIGURE 5.4 PV output profiles

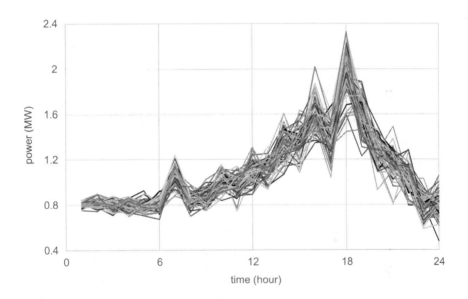

FIGURE 5.5 Total load profiles

Figures 5.7 to 5.11 illustrate the 26 aggregated scenarios of Figure 5.6: Figure 5.7 shows the number of parked EV's $N_{t\omega}^{EV}$, Figure 5.8 shows $E_{t\omega}^{S+}$, Figure 5.9 shows $E_{t\omega}^{S-}$, Figure 5.10 shows the PV power outputs, and Figure 5.11 shows the local load profiles.

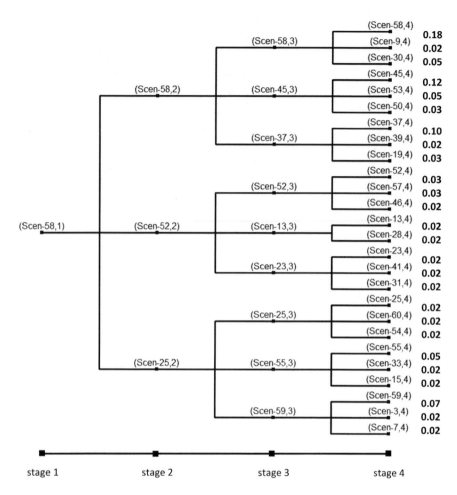

FIGURE 5.6 Multistage scenario tree with the indication of the probabilities associated with each leaf node.

5.4.3 Solution of the Multistage Stochastic Model

The optimization procedures have been implemented in AIMMS Developer modelling environment and solved in around 15 seconds.

Figure 5.12 shows the dispatched power of the V2G parking lot. The results of Figure 5.12 can be compared with the power profiles at the EV's unidirectional charging station as shown in Figure 5.13, which correspond to an uncontrolled charge of the EVs for each of the 60 scenarios before grouping. In particular, Figure 5.12 illustrates the effect of the energy stored during the hours of maximum production of the PV unit and the power contribution of the parking lot during the evening load-peak hours.

Figure 5.14 shows the profile of $\mu_{t\omega}$ for the scenario in the tree of Figure 5.6. The initial energy of the incoming EVs is used in the early morning in order to increase

EV Charging and Renewable Power Generation

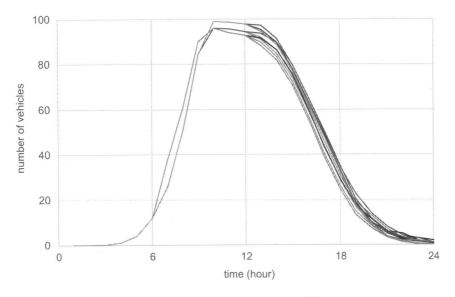

FIGURE 5.7 Selected scenarios of number of parked EV's $N_{t\omega}^{EV}$.

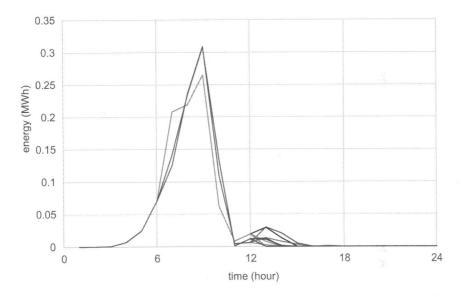

FIGURE 5.8 Selected scenarios of total energy $E_{t\omega}^{S+}$ entering the parking lot due to EV's arrivals.

the storage of the PV output in the parking lot during the following hours (taking into account the low value of the ratio between sale and purchase utility tariffs) and in late afternoon although the effect is limited due to the few EVs entering in those hours.

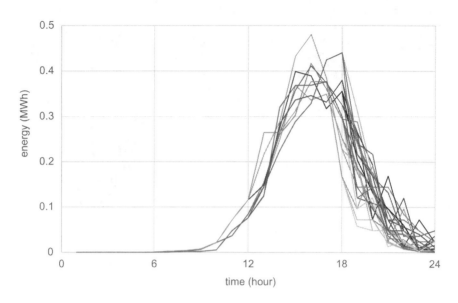

FIGURE 5.9 Selected scenarios of total energy $E_{t\omega}^{S-}$ leaving the parking lot due to EV's departures.

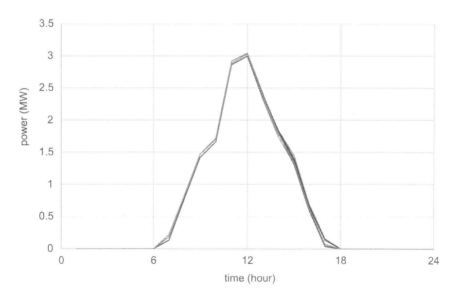

FIGURE 5.10 Selected scenarios of power delivered by the photovoltaic system.

Figure 5.15 shows the total energy stored in the parking lot, and Figure 5.16 shows the power exchange between the microgrid and the external utility grid.

The value of stochastic solution (*VSS*) and the expected value of perfect information (*EVPI*) are typical performance metrics. According to, e.g., Ref. [20], the definitions of *VSS* and *EVPI* are as follows:

EV Charging and Renewable Power Generation

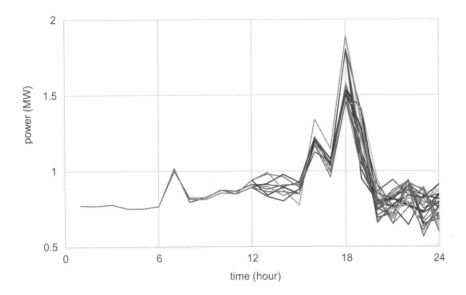

FIGURE 5.11 Selected scenarios of power consumptions of the local loads of the microgrid.

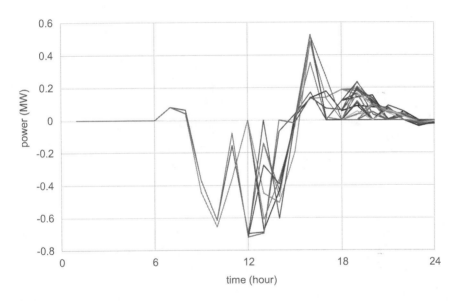

FIGURE 5.12 Power delivered by the parking lot calculated by the stochastic optimization model in the selected scenarios.

- *VSS* is the difference between *EEV* and *RP*, where *EEV* is the expected value solution and *RP* is the solution of the recourse problem, i.e., the cost value of the multistage stochastic problem (5.1). In order to calculate *EEV*, at first, the values of the decision variables for each time period t are obtained by the solution of the deterministic model in which all random variables are

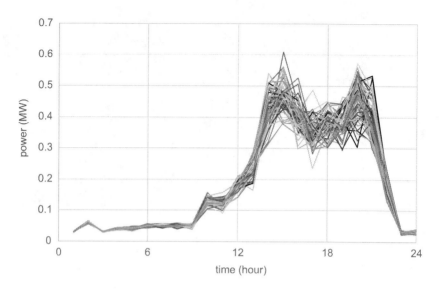

FIGURE 5.13 Uncontrolled charge at the EV's charging station for the 60 initial scenarios.

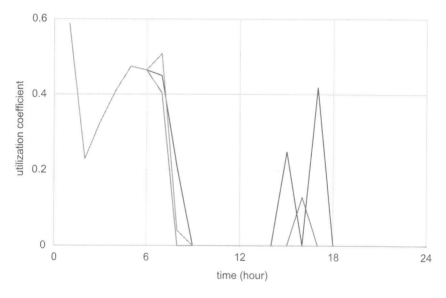

FIGURE 5.14 Utilization coefficient calculated by the stochastic optimization model in the selected scenarios ($e_{min} = 0.2$ pu).

replaced by their expected values; then, *EEV* is the solution of the stochastic problem in which the decision variables are fixed parameters.
- *EVPI* is the difference between *RP* and the wait-and-see (*WS*) solution given by the calculation of the expected value of the set of deterministic solutions, each relevant to one of the tree scenarios.

EV Charging and Renewable Power Generation

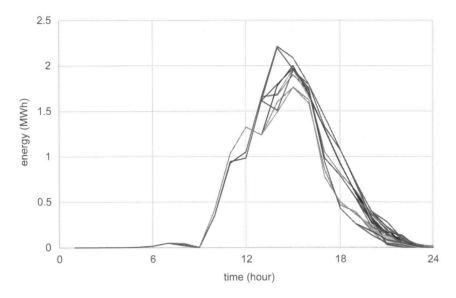

FIGURE 5.15 Energy stored in the parking lot.

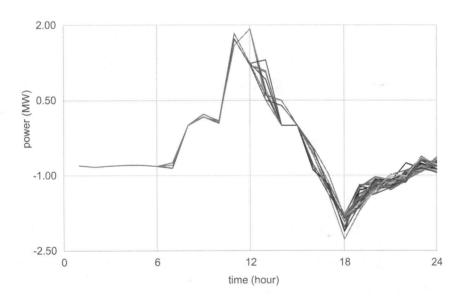

FIGURE 5.16 Power exchanged between the microgrid and the external grid (positive if exported from the microgrid).

The *VSS* and *EVPI* calculated for the considered case studies are presented in Table 5.2.

For a further comparison, the value of the objective function obtained for the deterministic problem that uses as input the average values of the stochastic parameters $N_{t\omega}^{EV}$, $P_{t\omega}^{PV}$, and $P_{t\omega}^{Load}$ is € 907.55. The average value of the deterministic solution

TABLE 5.2
Performance Evaluation for the Multistage Stochastic Solution

Solution	Cost Value (€)	VSS (€)	EVPI (€)
EEV	870.79	6.04	0.05
RP	864.74		
WS	864.69		

TABLE 5.3
Objective Function of the Stochastic Solution for Different Values of e_{min}

e_{min}	0.1 pu	0.15 pu	0.2 pu	0.3 pu	0.4 pu	0.5 pu
RP (€)	864.38	864.52	864.74	865.46	867.29	867.30

for an initial set of 60 scenarios (which does not provide a unique profile of charge, discharge decision, and utilization coefficient $\mu_{t\omega}$) is € 851.62.

In view of these results, the adoption of the multistage stochastic model is expected to allow a significant improved flexibility and cost reduction for many scenarios with respect to the adoption of the solution provided by the deterministic model.

We compare now the *RP* cost values obtained for different values of the minimum energy value e_{min}, which is used in equation 5.6 in order to limit utilization coefficient $\mu_{t\omega}$ of the initial charge of the vehicle arriving in the parking lot. As expected, Table 5.3 shows that larger values of e_{min} cause the increase in the objective function values. Furthermore, Figure 5.17 shows the effect of the different e_{min} values on the profile of utilization coefficient $\mu_{t\omega}$. For e_{min} values equal and exceeding 0.5 pu, the resulting $\mu_{t\omega}$ is always 0.

5.5 CONCLUSIONS

This chapter deals with the operation of a microgrid with a parking lot that allows bidirectional charging services. The optimization problem minimizes the daily cost of the microgrid by means of a multistage stochastic problem. For such a purpose, the uncertainties associated with the number of parked EVs, PV generation, and non-dispatchable loads in the microgrid are represented by the relevant scenario tree.

The employed *k*-means clustering procedure for the scenario tree generation provides appropriate results even with a limited number of cluster centres. In order to evaluate the performance of the multistage stochastic optimization, *VSS* and *EVPI* have been calculated, confirming the advantage of the multistage scenario-based approach over the solution given by a deterministic model based only on the forecast of the stochastic parameters. In addition, the computational effort is reasonable for the considered four-stage stochastic problem.

The use of the optimization model has been illustrated for the day-ahead dispatch of an industrial microgrid with a significant production from renewable resources, but is expected to be useful also to study the services of V2G parking lots in other

EV Charging and Renewable Power Generation

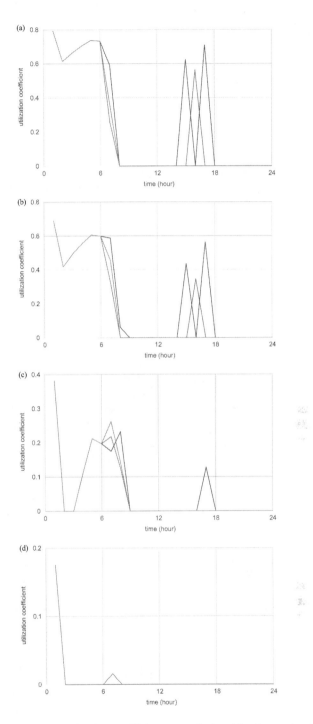

FIGURE 5.17 Profiles of utilization coefficient $\mu_{t\omega}$ calculated by the stochastic optimization model for different values of e_{min}: (a) 0.1 pu; (b) 0.15 pu; (c) 0.3 pu; (d) 0.4 pu.

context (e.g., the minimization of the procurement energy costs of a single or multiple EV parking lots and the congestion reduction analysis in urban distribution networks).

ACKNOWLEDGMENTS

This work was funded by the Italian Ministry for Education, University and Research under the grant PRIN-2017K4JZEE "Planning and flexible operation of micro-grids with generation, storage and demand control as a support to sustainable and efficient electrical power systems: regulatory aspects, modelling and experimental validation" and by Regione Emilia-Romagna through the European Regional Development Fund 2014–2020 - "Energynius - Energy networks integration for urban systems" project (CUP E31F18001040007). These activities have been carried out in collaboration with Saeed Rahmani Dabbagh and Mohammad Kazem Sheikh-El-Eslami (co-authors of the papers that present some of the obtained results). The authors are indebted to Sara Chiriatti and Paolo Prevedello for their help in performing the simulations. The comments and suggestions provided by Prof. Carlo Alberto Nucci are gratefully acknowledged.

NOMENCLATURE

Symbols relevant to the optimization model of the microgrid:

π_ω:	Probability associated with scenario ω
C_ω:	Cost associated with scenario ω
ρ_t^T:	Time-of-use tariff for purchasing from the grid in period t
r:	Ratio between sale and purchase tariffs
$E_{t\omega}^+$ and $E_{t\omega}^-$:	Nonnegative variables that correspond to the hourly energy sold and purchased from the utility grid
$P_{t\omega}$:	power exchange of the parking with the external provider
$P_{t\omega}^{\max}$ and $P_{t\omega}^{\min}$:	available total maximum power for discharging and charging processes of the parking lot, respectively, which depend on the total set of parked EVs.
Δt:	duration of period t
$E_{t\omega}^S$:	energy stored in the parking lot
$E_{t\omega}^{S\max}$:	maximum storage capability that depends on the number of parked EVs
E_{nom}^{EV}:	rated energy size of the EV batteries
$N_{t\omega}^{\text{in}}$:	number of EV's arrivals
$E_{t\omega}^{V2G}$:	energy available for vehicle-to-grid (V2G) services
$E_{t\omega}^{S-}$:	decrease in the energy stored in the parking lot due to EV's departures
$E_{t\omega}^{S+}$:	increase in the energy stored in the parking lot due to EV's arrivals
$\mu_{t\omega}$:	is the nonnegative utilization coefficient of the total energy $E_{t\omega}^{S+}$ initially stored in the EVs that arrive at time t in the parking lot
$E_{(j,t)\omega}^\mu$:	the initial energy of the EVs arriving in period j and leaving in time period t ($j = 0$ indicates the EVs already parked when the optimization horizon begins)
$l_{t\omega}$:	charging/discharging power losses
η^r and η^s:	average efficiencies of the EV batteries and charging stations during discharging and charging processes, respectively

δ:	self-discharging rate
$C_{t\omega}^{S}$:	total cost associated with the use of the V2G parking lot according to prices c^r and c^s of the discharging and charging processes, respectively, and to the price c^μ of the utilization of $E_{t\omega}^{S+}$
$P_{ut\omega}$:	net active power provided or absorbed by unit u (positive if provided to the microgrid), i.e., in our case, the parking lot, the PV unit, and the local loads

Symbols relevant to the scenario generation for parking lot:

N_{tot}:	population of EVs willing to enter in the parking lot
t_i^+:	time of entrance of the ith EV
E_i^0:	initial charge of the ith EV
s_i:	time of stay of the ith EV
E_i^-:	charge at departure of the ith EV
$S_{t\omega}^+$:	set of EVs incoming at time t in scenario ω
$S_{t\omega}^-$:	set of EVs leaving at time t in scenario ω

Symbols relevant to the scenario generation for the PV unit and local load:

$x_{t\omega}$:	time series that represents the one-lag autocorrelation
ϕ:	autocorrelation parameter
ε:	additive white noise employed for the scenario generation technique

Symbols relevant to the scenario tree generation:

Ω:	set of scenarios $\xi_{t\omega}$
$N_{t\omega}^{EV}$:	number of EVs in the parking lot in period t and scenario ω
$P_{t\omega}^{PV}$:	photovoltaic generation in period t and scenario ω
$P_{t\omega}^{Load}$:	total load (i.e., considering local load and charge of EVs at the charge station) in period t and scenario ω
C_k^s:	set of K centroid clusters at stage s
T_s:	set of periods in stage s
$\bar{\xi}_{t\omega}$:	mean value of all the scenarios $\xi_{t\omega}$ assigned to the same cluster for time t.

REFERENCES

1. Y. Xiang, S. Hu, Y. Liu, X. Zhang, and J. Liu, "Electric vehicles in smart grid: A survey on charging load modelling," *IET Smart Grid*, vol. 2, no. 1, pp. 25–33, 2019, doi: 10.1049/iet-stg.2018.0053.
2. J. García-Villalobos, I. Zamora, J. I. San Martín, F. J. Asensio, and V. Aperribay, "Plug-in electric vehicles in electric distribution networks: A review of smart charging approaches," *Renew. Sustain. Energy Rev.*, vol. 38, pp. 717–731, Oct. 2014, doi: 10.1016/j.rser.2014.07.040.
3. D. Sbordone, I. Bertini, B. Di Pietra, M. C. Falvo, A. Genovese, and L. Martirano, "EV fast charging stations and energy storage technologies: A real implementation in the smart micro grid paradigm," *Electr. Power Syst. Res.*, vol. 120, pp. 96–108, Mar. 2015, doi: 10.1016/j.epsr.2014.07.033.

4. F. Varshosaz, M. Moazzami, B. Fani, and P. Siano, "Day-ahead capacity estimation and power management of a charging station based on queuing theory," *IEEE Trans. Ind. Informatics*, vol. 15, no. 10, pp. 5561–5574, 2019, doi: 10.1109/TII.2019.2906650.
5. C. Develder, M. Strobbe, K. De Craemer, and G. Deconinck, "Charging electric vehicles in the smart grid," *Smart Grids from a Global Perspective*, pp. 235–248, 2016.
6. C. Battistelli, L. Baringo, and A. J. Conejo, "Optimal energy management of small electric energy systems including V2G facilities and renewable energy sources," *Electr. Power Syst. Res.*, vol. 92, pp. 50–59, 2012, doi: 10.1016/j.epsr.2012.06.002.
7. A. Zakariazadeh, S. Jadid, and P. Siano, "Integrated operation of electric vehicles and renewable generation in a smart distribution system," *Energy Convers. Manag.*, vol. 89, pp. 99–110, Jan. 2015, doi: 10.1016/j.enconman.2014.09.062.
8. F. Mwasilu, J. J. Justo, E.-K. Kim, T. D. Do, and J.-W. Jung, "Electric vehicles and smart grid interaction: A review on vehicle to grid and renewable energy sources integration," *Renew. Sustain. Energy Rev.*, vol. 34, pp. 501–516, Jun. 2014, doi: 10.1016/j.rser.2014.03.031.
9. J. Traube et al., "Mitigation of solar irradiance intermittency in photovoltaic power systems with integrated electric-vehicle charging functionality," *IEEE Trans. Power Electron.*, vol. 28, no. 6, pp. 3058–3067, 2013, doi: 10.1109/TPEL.2012.2217354.
10. T. Zhang, W. Chen, Z. Han, and Z. Cao, "Charging scheduling of electric vehicles with local renewable energy under uncertain electric vehicle arrival and grid power price," *IEEE Trans. Veh. Technol.*, vol. 63, no. 6, pp. 2600–2612, 2014, doi: 10.1109/TVT.2013.2295591.
11. S. Y. Derakhshandeh, A. S. Masoum, S. Deilami, M. A. S. Masoum, and M. E. Hamedani Golshan, "Coordination of generation scheduling with PEVs charging in industrial microgrids," *IEEE Trans. Power Syst.*, vol. 28, no. 3, pp. 3451–3461, Aug. 2013, doi: 10.1109/TPWRS.2013.2257184.
12. D. B. Richardson, "Encouraging vehicle-to-grid (V2G) participation through premium tariff rates," *J. Power Sources*, vol. 243, pp. 219–224, 2013, doi: 10.1016/j.jpowsour.2013.06.024.
13. S. R. Dabbagh, M. K. Sheikh-El-Eslami and A. Borghetti, "Optimal operation of vehicle-to-grid and grid-to-vehicle systems integrated with renewables," *2016 Power Systems Computation Conference (PSCC)*, Genoa, Italy, 2016, pp. 1–7, doi: 10.1109/PSCC.2016.7540933.
14. M. Honarmand, A. Zakariazadeh, and S. Jadid, "Integrated scheduling of renewable generation and electric vehicles parking lot in a smart microgrid," *Energy Convers. Manag.*, vol. 86, pp. 745–755, Oct. 2014, doi: 10.1016/j.enconman.2014.06.044.
15. M. E. Khodayar, L. Wu, and M. Shahidehpour, "Hourly coordination of electric vehicle operation and volatile wind power generation in SCUC," *IEEE Trans. Smart Grid*, vol. 3, no. 3, pp. 1271–1279, 2012, doi: 10.1109/TSG.2012.2186642.
16. M. Shafie-khah et al., "Optimal behavior of electric vehicle parking lots as demand response aggregation agents," *IEEE Trans. Smart Grid*, vol. 7, no. 6, pp. 2654–2665, Nov. 2016, doi: 10.1109/TSG.2015.2496796.
17. A. Borghetti, F. Napolitano, S. Rahmani-Dabbagh, and F. Tossani, "Scenario tree generation for the optimization model of a parking lot for electric vehicles," in *2017 AEIT International Annual Conference: Infrastructures for Energy and ICT: Opportunities for Fostering Innovation, AEIT 2017*, 2017, vol. 2017-Jan, doi: 10.23919/AEIT.2017.8240519.
18. C. Orozco, A. Borghetti, S. Lilla, G. Pulazza and F. Tossani, "Comparison Between Multistage Stochastic Optimization Programming and Monte Carlo Simulations for the Operation of Local Energy Systems," *2018 IEEE International Conference on Environment and Electrical Engineering (EEEIC Europe)*, Palermo, Italy, 2018, pp. 1–6, doi: 10.1109/EEEIC.2018.8494563.
19. G. J. Osório, J. M. Lujano-Rojas, J. C. O. Matias, and J. P. S. Catalão, "A new scenario generation-based method to solve the unit commitment problem with high penetration of renewable energies," *Int. J. Electr. Power Energy Syst.*, vol. 64, pp. 1063–1072, 2015, doi: 10.1016/j.ijepes.2014.09.010.
20. L. F. Escudero, A. Garín, M. Merino, and G. Pérez, "The value of the stochastic solution in multistage problems," *Top*, vol. 15, no. 1, pp. 48–64, 2007, doi: 10.1007/s11750-007-0005-4.

6 Energy Storage Sizing for Plug-in Electric Vehicle Charging Stations

*I. Safak Bayram, Ryan Sims, Edward Corr,
Stuart Galloway, and Graeme Burt*
University of Strathclyde

CONTENTS

6.1 Introduction 119
6.2 Literature Review 121
    6.2.1 Literature on Smart Charging and Impacts of PEV Charging 121
    6.2.2 Literature on Charging Station Design... 122
    6.2.3 Literature on Probabilistic Modelling of PEV
          Charging Infrastructures 122
    6.2.4 Contributions 123
6.3 Demonstration and Testing Platform of a PEV Charging Infrastructure 124
    6.3.1 Overview of PEV Research and Testing Projects at PNDC............. 124
    6.3.2 Summary of Results 126
6.4 System Model 129
    6.4.1 Markov-Modulated Poisson Process 130
    6.4.2 Matrix Geometric Approach 132
    6.4.3 Algorithmic Solution Technique... 133
6.5 Numerical Evaluations..134
    6.5.1 Computation of Station Parameters 134
    6.5.2 Charging Station Economic Analysis... 135
6.6 Conclusions 139
Acknowledgement 139
Bibliography 139

6.1 INTRODUCTION

The future of electric power grids is currently shaped by two major advancements, namely higher use of renewables on the supply side and increasing adoption of plug-in electric vehicles (PEVs) on the demand side. These advancements aim to decarbonize electricity and transportation networks since more than half of the global energy-related carbon emissions are attributed to these two sectors. The push towards PEVs is supported by legislations and regulations to encourage PEV uptake. For instance, a number of countries, including the UK, France, and Norway, plan to phase out fossil fuel cars by introducing a ban on the sale of such vehicles and increase the coverage of charging network within the next two decades (Bayram and Tajer, 2017). On the other hand, reaching net zero goals would require an exponential adoption of PEVs; for instance, in the UK, there are currently two hundred thousand PEVs on the road, and this number needs to be around four million by 2030 to meet government policies (Haslett, 2019). Similarly, the State of California has a mandate to acquire one-and-half million PEVs by 2025 and generate half of its electricity with renewables by 2030. Net zero policy impacts are visible in France as there is a spring-back effect on the year-on-year (January 2019–2020) PEV market share jump from 2.7% to 11%. To that end, after being considered as a fringe technology, PEV market is getting closed to a tipping point (Sperling, 2018). This can be viewed in PEV sales and forecasts, as shown in Figure 6.1.

To support electrification of transportation, there is a need to deploy charging nodes to meet various customer needs shaped by time, location, and duration of service. In this chapter, we present a large-scale PEV charging station architecture equipped with an on-site storage unit. The primary goal is to develop a probabilistic method to optimally size storage unit and show how on-site storage can be effective in reducing peak demand and operational costs. Furthermore, we present an overview of demonstration studies conducted at Power Networks Demonstration Centre (PNDC is a research and testing hub founded by government, industry, and academia and is part of the University of Strathclyde) charging infrastructures.

At the moment, there are three typical charging options for PEVs (Falvo, 2014). First option is level 1 charging, which takes place in customer's premises. Level 1 charging uses the existing, typically single-phase, electrical circuit at residential units (2–3 kW) and fills the PEV battery during the night. Second charging option uses AC level 2 chargers, which are typically located at public parking lots (e.g.,

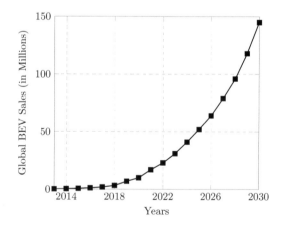

FIGURE 6.1 Battery electric vehicle (BEV) sales and forecasts (International Energy Agency).

workplaces, shopping malls). This type of chargers typically supplies 6–7 kW power to stationary vehicle. Third option is called fast or ultrafast charging, which can transfer DC power at a rate of 50–350 kW (Srdic and Lukic 2019). Note that a typical fast model (50 kW) can deliver enough charge for a 100-mile trip with 30 minutes of charging, while ultrafast models are more preferred by high-end PEVs with large batteries. In this chapter, the proposed model employs level 2 chargers to serve cars parked at a large-scale lot. The related literature can be classified into three groups. First group of studies is related to smart charging (see Section 6.2.1), which is aimed at mitigating disruptive impacts of PEV demand on the power grid by exploiting the demand flexibility of PEVs. Second group of studies is related to design of charging stations and is categorized according to technology and economic operation domains (see Section 6.2.2). Third group is related to probabilistic modelling of charging stations (see Section 6.2.3).

6.2 LITERATURE REVIEW

6.2.1 Literature on Smart Charging and Impacts of PEV Charging

Smart charging of PEVs is critical in transition towards electric transportation. Existing electric power grids are not designed to serve large PEV loads, and concurrent charging of PEVs will lead to major technical challenges on the distribution, transmission, and generation components. At the distribution level, clusters of PEV load during peak hours can lead to premature aging of transformers, increase distribution system losses, and deteriorate power quality (García-Villalobos et al., 2014). Increased stress and voltage fluctuations will further risk the consistency and safety of the network. According to a field study conducted in the UK (Cross, 2016), one-third of the low-voltage (LV) feeders will require intervention when 40%–70% of residents have PEVs. At the transmission level, PEV load increases transmission congestion level, which is a major challenge as the investments towards new transmission lines have been declining. To support increasing electrification demand, there is an urgent need to expand transmission network capabilities (Jurgen Weiss, 2019). Finally, at the generation level, uncontrolled PEV load could lead to an increase in peak system load, which requires a system additional deployment of new system upgrades (Wang and Tehrani, 2015). To overcome the aforementioned disruptive impacts, design and operation of charging facilities play a key role in transition to electric transportation.

PEVs are considered as a new kind of electric loads that have both temporal and charging power flexibility. When a PEV is connected to a charger, charging session starts at a constant power. In the case of multiple PEVs connected simultaneously, they collectively increase peak loading and potentially trigger the aforementioned disruptions. Smart charging is the optimization of charging power by exploiting PEV flexibility to maximize one or more benefits such as reducing peak load, increasing renewable energy utilization, lowering the cost of PEV charging, or deferring infrastructural upgrades (García-Villalobos et al., 2014). Smart charging can be implemented through standards such as IEC 61851 and ISO 15110 that enable control and communication between the charger and the vehicle. In addition, a group of PEV owners, coordinated by an aggregator, can participate in ancillary energy markets to stabilize electricity grids and, in return, receive payments for services rendered (Han et al., 2010). Vehicle-to-grid (V2G) applications are particularly important to smoothen what is known as solar "duck curves", which are used to define the net electricity generation curve when there is a significant solar generation. In this case, PEVs' charging rates are adjusted in a way to minimize the ramping-up requirements of traditional power generators and lower financial losses (Lee et al., 2019).

6.2.2 Literature on Charging Station Design

The approach described in this chapter focuses on economic operation of charging stations and energy storage sizing (Negarestani et al., 2016; Sarker, 2018). In this type of works, a critical component is local storage unit, which is typically employed to shave peak load, reduce demand charges, and provide an additional income via energy market participation. In the study by Negarestani et al., (2016), an optimal sizing approach is proposed for energy storage systems (ESSs) in fast charging stations. In this work, PEV demand is calculated based on driving patterns and optimal storage size is determined based on cost minimization. In the study by Sarker et al., (2018), an optimization framework is presented for an optimal bidding strategy in day-ahead electricity markets for a PEV charging station with an on-site storage. In the current charging station applications, one of the main issues is related to expensive demand charges that constitute a sizable portion of monthly electricity bills and reduce station profits (The 50 States of Electric Vehicles, 2018). Demand charges are pricing tools to limit peak consumption of large customers by inducing a fee commensurate to the peak consumption during any fifteen minutes during each month. In June 2016, using a charging facility with two fast chargers, the following bill was issued (Bayram and Ismail, 2019). The energy charge was 284 USD, and the demand charge for the peak power was more than 2900 USD, representing 91% of the total operational cost. High demand charges both compromise business models and negatively impact PEV sales if prices are reflected to customers.

6.2.3 Literature on Probabilistic Modelling of PEV Charging Infrastructures

Since the experimentation of capital-intensive PEV charging stations is not possible, analytical modelling of PEV charging demand and infrastructure is used to provide insights to system planners in how different system components interact with each other. In line with the previous discussion, economic operation of charging infrastructures has been the topic of several mathematical modelling and optimization research works. Stochastic modelling and queuing systems have been widely used as such methods capture the probabilistic nature of problem related to different battery packs, technologies, weather parameters, and customer arrival and departure processes (Hu et al., 2016). Moreover, station may have uncertainties related to renewable energy output, and storage unit can be modelled as a linear for simplification or nonlinear "buffer" if battery's chemical dynamics are taken into account. Some of the related studies can be enumerated as follows. In the study by Aveklouris et al., (2017), fluid approximation of queueing models is adopted to calculate charging

Energy Storage Sizing for PEVs

station overloading probabilities. In the study by Ucer et al., (2019), a queueing model is employed to calculate waiting times and service quality for a number of charging stations located in Ohio by using actual traffic traces. In the study by Fan et al., (2015), a charging station is modelled using a queuing model and captured the effect of constant-current and constant-voltage charging on customer waiting times in the station. Customer arrival and charging demand statistics are the important system parameters in charging stations. In the study by Fotouhi et al., (2019), using actual PEV charging data (level 2 chargers) from a major North American University Campus between 2010 and 2015, a Markovian model for representing the charging behaviour of PEV owners is presented. The results show that PEV owners not necessarily fully charge their batteries; hence, service duration is shorter than expected. In the study by Bayram et al., (2014), shared-based ESS located at residential units is modelled using fluid dynamic approach and storage sizing problem is solved by computing outage probability of the system, which is defined as the event when the load is higher than the supply.

6.2.4 Contributions

The contributions of this chapter can be enumerated as below:

- Firstly, an actual demonstration and testing platform of a PEV charging infrastructure is introduced to show a detailed overview of a PEV charging infrastructure and sample measurement results.
- Secondly, a probabilistic system model for the large-scale PEV charging station equipped with an on-site energy storage is presented (shown in Figure 6.2). By considering the probabilistic nature of the customer demand, the proposed architecture is modelled by a Markov-modulated Poisson process (MMPP).
- Thirdly, a matrix geometric-based algorithm is presented in detail and used to solve the associated capacity planning problem to find optimal energy storage size and station capacity with respect to customer demand statistics.
- Fourthly, practical case studies are developed to show that (1) by accounting for the statistical variations in customer demand, the power required for the station is significantly less than the sum of chargers' rated power and (2) on-site storage units can help station operators to significantly reduce their electricity bills.

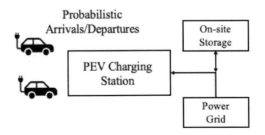

FIGURE 6.2 System overview.

6.3 DEMONSTRATION AND TESTING PLATFORM OF A PEV CHARGING INFRASTRUCTURE

Before presenting an analytical model for a large-scale charging station, an overview of an actual small-scale PEV charging station with associated hardware and software components is presented to provide a better understanding of actual system components. The PNDC has completed several research projects relating to the electrification of transport and charging infrastructure. This includes electrical impact assessments of wireless inductive charging, on-street pop-up charger performance testing, and active power quality compensation of single-phase harmonic and load imbalance impacts of electric vehicle (EV) charge points. The PNDC was founded with the goal of accelerating the penetration of disruptive technologies from early-stage research into business as usual adoption by the electricity industry. The facility comprises a fully representative distribution network, including the capabilities summarized in Table 6.1. This enables the research, test, and demonstration of hardware, software, and integrated systems solutions in a safe, controlled environment.

6.3.1 Overview of PEV Research and Testing Projects at PNDC

In 2019/2020, the PNDC supported Power Line Technologies Ltd. (PTL), Chronos Technology Ltd, and the University of Strathclyde with the development of the ENERSYN platform, which enables the hosting of partner-developed applications. The platform monitors the LV network via high-fidelity voltage and current measurements, making this data available to hosted "apps". Two apps developed were a micro-phasor measurement unit (PMU) and a non-intrusive load monitoring (NILM) algorithm, to detect the connection of EVs to their chargers. The NILM algorithm uses machine-learning techniques applied to LV network data to detect unique features related to EV charger operation. The test setup was varied in two ways. The first variation involved the Enersyn platform monitoring a rapid EV charging load isolated from background noise. The second variation involved monitoring a rapid

TABLE 6.1
Selected Hardware and Software Assets of PNDC

Asset	Rating/Comments
11-kV overhead/underground distribution	Up to 60 kM of representative 11-kV network
400-V low-voltage distribution	Up to 6 kM of representative LV network
Controllable motor-generator (MG) set	1-MW Motor/5-MVA generator
Controllable load banks	600-kVA controllable resistive/inductive
Real-time digital simulator (RTDS)	6 racks of RTDS execution hardware
Power hardware in the loop (PHIL)	540-kVA bidirectional power converter
Distributed energy resources (DERs)	E.g., EV charge points, PV inverters, loads
Distribution management system	Operational GE PowerOn SCADA
Data acquisition system	Fluke & Beckhoff monitoring and logging

EV charging load with other background loads supplied from the same distribution circuit. A high-level representation of the setup is illustrated in Figure 6.3. Rogowski coil current transducers are installed on the incoming cables to the distribution board, and voltage transducer measurements taken off terminals inside the building's distribution board. Conducting testing with and without background load permitted the assessment of the NILM algorithm and its ability to disaggregate system noise from loads of interest.

The ENERSYN platform and an off-the-shelf data acquisition system were deployed in parallel to ensure the accuracy of data capture. Data was recorded using a National Instruments CompactRIO (NI CRIO) data acquisition system and a LabVIEW Virtual Instrument (VI) hosted on a PC. Three analogue input modules were deployed, with 4 channels per module. The other module employed in this setup was a global positioning system (GPS) time synchronization module. Data was sampled at a rate of 100 kHz to provide a high-resolution data set for the development of machine-learning features that underpin the NILM algorithm.

Concurrently, PTL, Chronos, and University of Strathclyde deployed the developed ENERSYN monitoring platform. The developed system incorporates current and voltage measurements, and records high-speed waveform events (100 kHz sampling rate) which are time-stamped using GPS. The ENERSYN platform uses a long-range (LoRa) GPS timing module supplied by Chronos. Captured waveform events are then analysed by the on-board PMU and passed to the NILM algorithm (depicted in Figure 6.4). Noteworthy waveform events are flagged and forwarded onto the ENERSYN server for further analysis.

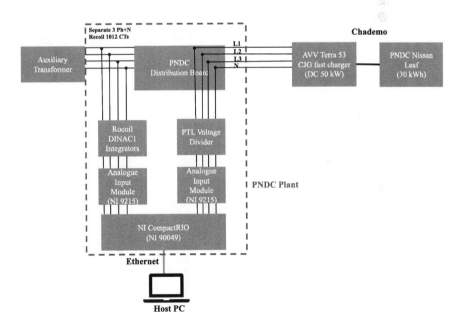

FIGURE 6.3 Iteration of test equipment and facility setup at the PNDC for NILM testing.

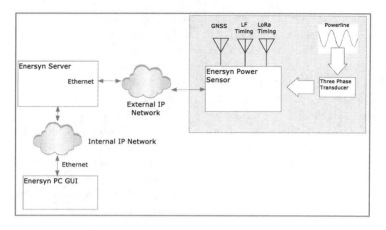

FIGURE 6.4 ENERSYN platform block diagram.

6.3.2 SUMMARY OF RESULTS

High-resolution load signatures for a range of PEV charging profiles were analysed using data gathered by the National Instruments CRIO monitoring system. Previous studies (Zhang et al., 2011) have established a set of general usage patterns for charging PEVs, enabling four prescribed charging schedules to be derived, as illustrated in Figures 6.5–6.8. These prescribed profiles were recreated, using PNDC-owned EVs, and used as inputs to the development of the NILM algorithm.

PEV charger behaviour for non-prescribed charging profiles was also investigated by logging the electrical parameters on a public rapid 50-kW charger. The logged

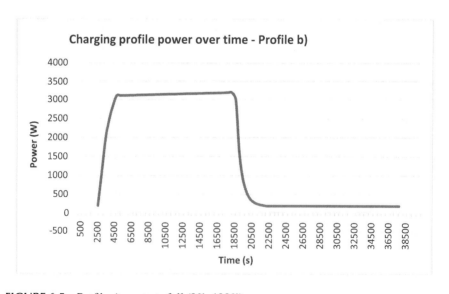

FIGURE 6.5 Profile a) empty to full (0%–100%).

Energy Storage Sizing for PEVs

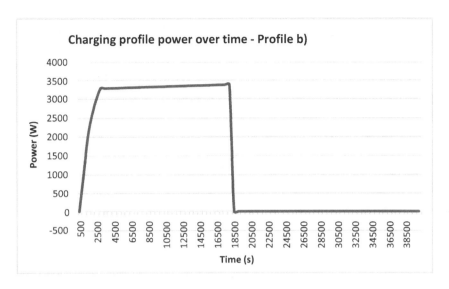

FIGURE 6.6 Profile b) empty to part full (0%–75%)

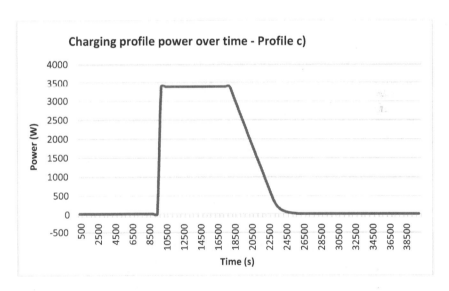

FIGURE 6.7 Profile c) part full to full (25%–75%);

data was correlated with voluntary questionnaire responses by charge point users about the start and end state of charge of their charging session. Figure 6.9 outlines the responses received, each point corresponding to a charge start and stop percentage.

Based on public responses, the majority of charging sessions started at 10%–30% state of charge (SOC) and stopped in the range of 70%–100% SOC. This additional step provided inputs to ensure that the data for training the NILM algorithm was

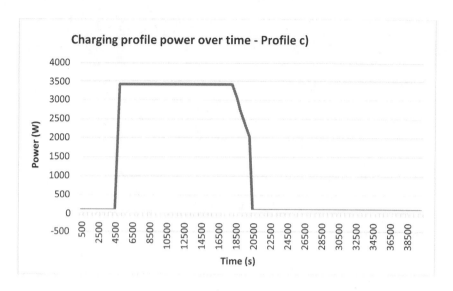

FIGURE 6.8 Profile d) part full to another higher capacity (25%–75%).

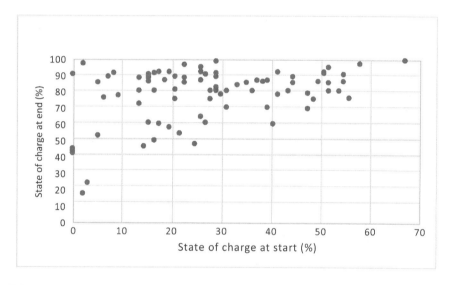

FIGURE 6.9 Public responses for usage of the PNDC rapid charger.

representative of how typical PEV drivers use public rapid chargers, in addition to the prescribed profiles found in the literature.

Training data for the NILM algorithm was logged at a sampling frequency of 100 kHz, which was exported in a technical data management system (TDMS) format, compressed, and shared on a data storage platform with the project team. In terms of file size, a TDMS data file corresponding to 1-hour monitoring equated to 10 GB of data. This learning emphasized the need for on-board edge-processing analysis via the "apps" on the ENERSYN platform, to avoid unnecessary data transmission to a server for centralized analysis.

The PEV charger data generated at the PNDC was critical to the NILM algorithm development. The PNDC is exploring further avenues of research, which could make use of this high-fidelity data. The developed NILM algorithm is now operating at a success rate of over 90% in detecting PEV charger events after being trained and tested by the data gathered at PNDC. The next stage of testing for the project will be the deployment of the ENERSYN platform on the PNDC test network and monitoring the public PEV chargers at PNDC over a longer period of time.

6.4 SYSTEM MODEL

In this section, we consider a large-scale charging station with N chargers serving PEV demand. Charging station draws grid power and employs an on-site ESS shared by all users. We denote the total charging power at time t by v_t and the ESS charge level by $i(t)$ fir $t \in \mathbb{R}^+$. It is worth noting that grid power is used to charge vehicles and storage unit whenever possible. When the total PEV demand is higher than v_t, ESS is used to support PEV demand unless it is fully empty. Customer statistics are as follows. We assume that PEVs' arrivals at the parking station is a Poisson process with rate λ. The average parking duration follows an exponential distribution with rate μ (Fan et al., 2015). Furthermore, when a vehicle is parked, its power demand follows the Poisson process with rate β. Finally, when an arriving customer finds all system resources in use, an outage event occurs. In our model, we use the outage probability as the natural performance metric.

Since PEV arrivals are independent of each other, the system state space $\{0,1,\cdots,N\}$ is represented with a birth–death process. The composite model for N slot charging station is depicted in Figure 6.10.

For the given system description, it is natural to assume that the system operates in a stable region. For this, the average demand should be strictly less than the available station capacity. To that end, we have

$$N\beta\left(\frac{\lambda}{\lambda+\mu}\right) < v_t, \tag{6.1}$$

or (6.1) can be rewritten as

$$\rho \equiv N\beta\left(\frac{\lambda}{\lambda+\mu}\right)\frac{1}{v_t} < 1, \tag{6.2}$$

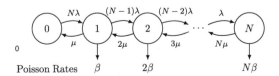

FIGURE 6.10 Birth–death process for N slot charging station.

where ρ is the utilization parameter. Furthermore, the following assumptions are made for the storage unit. Firstly, energy rating (in kWh) or the size of the energy storage is denoted by B. Secondly, energy storage efficiency (charge–discharge) is denoted by η, which takes values between 0 and 1. Note that this parameter reflects the percentage of energy transfer after losses are excluded. Thirdly, in actual ESSs, small percentage of energy is lost due to leakage. To simplify matters, dissipation losses are ignored.

During charging station operations, on-site storage unit's energy level changes in one of the following cases:

- When the storage unit is entirely discharged, i.e., $i(t) = 0$, the total demand is more than v_t. In this case, the rate of change in storage charge level would be zero.
- ESS is fully charged, i.e., $i(t) = 1$, and the total demand is less than v_t. Similar to the previous case, ESS charge level would not change.
- ESS is partially discharged, i.e., $0 < i(t) < 1$, with any level of system demand. In this case, ESS charge level would change commensurate to the difference between charging power and the system demand, i.e., $\dfrac{di(t)}{dt} = \eta\left(v_t - \sum_n L_n(t)\right)$, where $L_n(t)$ is the total demand when $n \in \{0,\dots,N\}$ chargers are on at time t.

It is noteworthy that due to the probabilistic nature of the system, by choosing storage size B, only probabilistic guarantees can be provided to system reliability. Therefore, let us define ε as the outage storage capacity, i.e., $B(\varepsilon)$, as the minimum B satisfies to serve $(100 - \varepsilon)\%$ per cent of the total load, i.e.,

$$B(\varepsilon) = \begin{cases} \min B \\ \text{subject to } P(i(t) \geq B) \leq \varepsilon \end{cases} \quad (6.3)$$

Note that our main goal is to calculate ε – outage storage capacity $B(\varepsilon)$ based on grid power, the number of PEVs, and other system parameters. To simplify mathematical notations, ESS size is scaled, and instead of B/η, we redefine B as the storage size. Furthermore, power systems planning is typically done for "peak hour" period. Therefore, in the rest of the paper, time index t is dropped and calculations for the peak statistics are made.

6.4.1 Markov-Modulated Poisson Process

Recall from the preceding discussion that at each state (see Figure 6.10), the aggregate demand generates a state-dependent Poisson process (e.g., β, 2β,). Therefore, the entire charging station can be modelled with an MMPP, and energy storage sizing option will be coupled with the computation of steady-state distribution probabilities. Let p_{in} denote the joint probability that the storage charge level is i and there are n active PEVs, i.e.,

Energy Storage Sizing for PEVs

$$p_{in} = \mathbb{P}\big(\text{ESS charge level} = i, n \text{ active customers}\big). \tag{6.4}$$

Then, the probability that ESS charge level is at level i can be written as

$$p_i = \sum_{n=0}^{N} p_{in}. \tag{6.5}$$

Given the aforementioned assumptions, the system is modelled with a two-dimensional birth–death process as depicted in Figure 6.11. Note that system states of the Markov chain are represented by a doublet (i, n), where the first dimension reflects storage charge level and varies from 0 to B, and the second dimension represents the number of PEVs in the system and varies from 0 to N. Alternatively, when customers arrive at or depart from the station, the system state moves in the horizontal direction. Similarly, when the on-site storage is charged or discharged, the system state moves in the vertical direction. Moreover, transition rates, e.g., λ, μ, are determined based

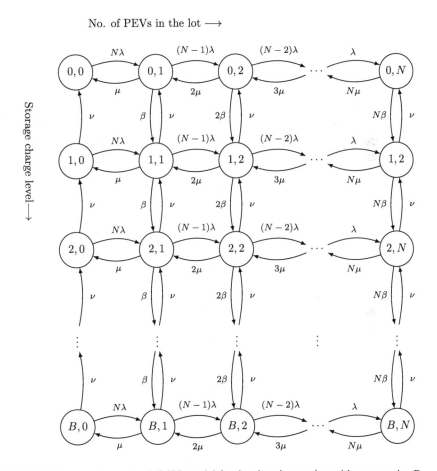

FIGURE 6.11 An illustrative MMPP model for the charging station with storage size B.

on the Poisson assumptions made earlier. It is worth emphasizing that storage sizing calculations are made based on the assumption that the storage size has an infinite capacity and the overflow probability is calculated as given in equation 6.3.

6.4.2 Matrix Geometric Approach

To compute steady-state probabilities of the MMPP model, an algorithmic solution technique called matrix geometric approach has been employed (Neuts, 1994). As a first step, balance equations for the Markov chain are written in the form of (6.4). For instance, the last raw in the Markov chain (where $i = 0$), there are three intervals: (1) $i = 0, n = 0$, (2) $i = 0$, $1 \leq n \leq N-1$, and (3) $i = 0, n = N$.

Vertical state transitions represent the energy storage charge–discharge events, while horizontal ones depict PEV arrival and departure.

For the first interval, the balance equation is

$$N\lambda p_{00} = \mu p_{01} + v p_{10}. \tag{6.6}$$

This equation can be rewritten as

$$p_{00} = (1 - N\lambda)\mu p_{01} + v p_{10}. \tag{6.7}$$

For the second interval, the balance equation can be rewritten as

$$p_{0n} = (N - (n-1))\lambda p_{0n-1} + (1 - (N-n)\lambda - n\mu - n\beta) p_{0n} + (n+1)\lambda p_{0n+1} v p_{1n} \tag{6.8}$$

The third case includes the rightmost boundary states and the balance equation can be rewritten as

$$p_{0N} = \lambda p_{0N-1} + (1 - N\lambda - N\beta) p_{0N} + v p_{1N}. \tag{6.9}$$

Balance equations for other rows. e.g., $i > 0$, can be written similar to equations 6.7–6.9 by further incorporating additional vertical state transitions. Let p_i denote $(N+1)$-element row vector consisting of probabilities defined by balance equations, i.e., $p_i \equiv [p_{i0}, p_{i1}, \ldots, p_{iN}]$. Then, balance equations, such as the ones defined in equations 6.7–6.9, can be written in a compact matrix-vector equation. For p_0,

$$p_0 = p_0 B_0 + p_1 B_1, \tag{6.10}$$

where $(N+1) \times (N+1)$ matrices B_0 and B_1 are readily given by

$$B_0 = \begin{bmatrix} (1-N\lambda) & N\lambda & 0 & \cdots & 0 \\ \mu & (1-\mu-(N-1)\lambda-\beta) & (N-1)\lambda & \cdots & 0 \\ 0 & 2\mu & (1-2\mu-(N-2)\lambda-2\beta) & \cdots & 0 \\ \vdots & \vdots & \vdots & \ddots & \vdots \\ 0 & 0 & 0 & \cdots & (1-N\mu-N\beta) \end{bmatrix}$$

$$\tag{6.11}$$

Energy Storage Sizing for PEVs

and B_1 is a diagonal matrix of the elements that contain vertical transition rate v. Remaining rows, e.g., $i > 0$, can be written similar to (6.11) by including transitions between vertically adjacent states. Hence, a complete set of balance equations for the remaining rows can be constructed from the matrix recurrences relation with three matrices such as A_0, A_1, and A_3, i.e.,

$$p_i = p_{i-1} A_0 + p_i A_1 + p_{i+1} A_2, \text{ for } i > 0 \tag{6.12}$$

Recall the assumption that the energy storage is initially assumed to have an infinite size. Then, an infinite-dimension equation p is introduced as $p = [p_0, p_1, \ldots, p_i, \ldots]$. Then from (6.12), it is easy to see that

$$p = pP, \tag{6.13}$$

where matrix P is a stochastic matrix of infinite size and called as the transition probability matrix with each row summing to one. It is trivial that matrix P is concatenated from previously constructed submatrices, namely A_0, A_1, A_2, B_0, B_1, in the following repetitive form:

$$P = \begin{bmatrix} B_0 & A_0 & 0 & 0 & \cdots \\ B_1 & A_1 & A_0 & 0 & \cdots \\ 0 & A_2 & A_1 & A_0 & \cdots \\ 0 & 0 & A_2 & A_1 & \cdots \\ 0 & 0 & 0 & A_2 & \cdots \\ \cdots & \cdots & \cdots & \cdots & \ddots \end{bmatrix} \tag{6.14}$$

In the next section, we present the solution methodology to compute minimum i that satisfies $\varepsilon = 1 - \sum_i p_i$.

6.4.3 Algorithmic Solution Technique

In this section, the algorithmic probability solution developed by Neuts (1994) is adopted. The solution to p_i is written as

$$p_{i+1} = p_i R, \ i \geq 0, \tag{6.15}$$

where R is a $(N+1) \times (N+1)$ matrix that has a non-negative solution to the following matrix equation

$$R = \sum_{k=0}^{\infty} R^k A_k. \tag{6.16}$$

Note that the calculation of p_i is equivalent to finding the minimal non-negative solution to R matrix. A recursive calculation method is used to compute the matrix R. As a first step, (6.16) is rewritten as

$$R[I - A_1] = \sum_{\substack{k=0 \\ k \neq 1}}^{\infty} R^k A_k \qquad (6.17)$$

where I is an identity matrix of $(N+1) \times (N+1)$ size. Multiplying both sides of (6.17) would yield

$$R = \sum_{\substack{k=0 \\ k \neq 1}}^{\infty} R^k A_k [I - A_1]^{-1}. \qquad (6.18)$$

R can be iteratively solved for an initial solution of $R = 0$,

$$R = \left[A_0 + R^2 A_2 \right][I - A_1]^{-1}. \qquad (6.19)$$

Once R is found, p_0 can be calculated as follows:

$$p_0 = p_0 B(R), \qquad (6.20)$$

where $B(R) = \sum_{k=0}^{\infty} R^k B_k$. Recall that for the charging station model, $B(R) = B_0 R B_1$. It is important to note that the sum of probabilities should add up to 1. Once the probabilities are found (e.g., p_0), the computed results need to be normalized by dividing each probability to the sum of all probabilities. To that end, the matrix geometric solution can be summarized as below:

1. Construct matrix R by solving the equation in (6.16) and iteration in (6.19)
2. Compute p_0 by solving the eigenvector equation in (6.20)
3. Compute p_i by solving (6.15)
4. Normalize p_0 and p_i by dividing by the sum of all probabilities
5. Calculate the minimum storage size that satisfies $\varepsilon = 1 - \sum_i p_i$.

Case studies are presented in the next section to provide more insights.

6.5 NUMERICAL EVALUATIONS

6.5.1 Computation of Station Parameters

Next, a number of case studies are presented to show how the proposed methodology can be used to size ESS sizing in a charging point. It is assumed that charging station employs typical level 2 chargers (6 kW), average parking duration is set as one hour ($\mu = 1$), and the charge request is set to $\beta = 0.9$. Three levels of customer arrival rate per charger are chosen (from $\lambda = 0.25$ to $\lambda = 0.75$) to reflect different traffic regimes, while station size is varied from $N = 50$ to $N = 150$. For system's stability (see equation (6.1)), the power drawn from the grid is chosen as

Energy Storage Sizing for PEVs

$$v = N\beta\left(\frac{\lambda}{\lambda+\mu}\right) + \Delta, \qquad (6.21)$$

where Δ is a small constant set to 0.02. Computations for the size of on-site ESS with respect to different station sizes and traffic regimes are presented in Figure 6.12. As an example, for a charging station with 150 chargers and a peak traffic regime of $\lambda = 0.50$, the size of the energy storage to provide 2% outage performances would be 128 kWh.

From the results presented in Figure 6.12, two key observations are made. Firstly, as the arrival rate increases, there is a need to a bigger energy storage to provide the same level of outage performance. Secondly, as the station size increases, the need for storage size per charger decreases due to "statistical gains". For example, consider the following two charging stations with $N_1 = 50$ and $N_2 = 250$. Both stations operate under $\lambda = 0.5N$ and employ a storage size of $B = 112$ kWh, while station 2 draws three times more power than station 1, i.e., $v_2 = 3v_1$. For these two stations, outage probabilities are calculated as $\varepsilon_1 = 0.1093$ and $\varepsilon_2 = 0.0275$. Then, it is easy to see that gains in system performance $\left(\frac{\varepsilon_1}{\varepsilon_2} = 0.0275\right)$ are higher than the corresponding capacity increase $\left(\frac{v_1}{v_2} = 3.97\right)$.

As a second evaluation, the case in which storage size and the number of chargers are known and the computation of the amount of power that is needed are investigated. In Figure 6.13, the results for arrival rate $\lambda = 0.5$ and varying station sizes $(N = 100, 200, \text{ and } 300)$ are presented. These findings help system operators to decide on appropriate amounts of power for the station. Similar to the previous case, due to statistical gains, as the station size increases, per-charger resource requirement decreases. For a target outage probability of 0.05 and on-site storage size of $B = 30$ kWh, per-charger power requirement for a $N = 100$ charging station is 4.69 kW, while this value is only 2.3 kW for a charging station with 300 slots.

6.5.2 Charging Station Economic Analysis

Recall the principal motivation to acquire on-site ESSs at charging stations to lower running cost and defer major system upgrades. A typical electricity bill of a charging station is composed of three parts, namely (1) a fixed fee, (2) energy charges (USD per kWh), and (3) demand charges (USD per kW). In this case study, actual billing tariff of a utility company in San Diego is employed with the following details. Monthly fee is $140, demand charge is $35 per highest kW, and energy tariffs are time of use based on details in Table 6.2.

It is assumed that the charging station operates between 6 am and 10 pm, and hourly traffic demand per charger is listed in Table 6.3. Moreover, the target outage probability is set to 0.005, and the maximum grid power is limited to 610 kW.

As a first step, energy storage size according to peak consumption hour is calculated using the methodology described in the previous section and found as

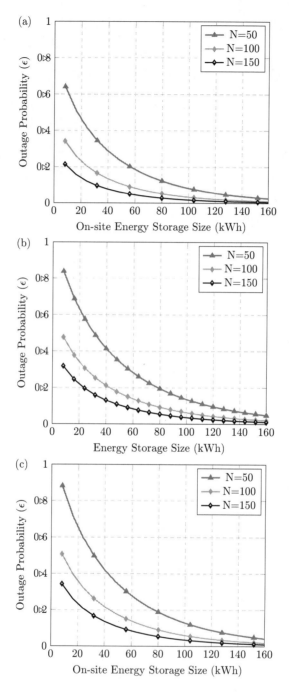

FIGURE 6.12 Computation of on-site energy storage size for different traffic regimes and station sizes.

Energy Storage Sizing for PEVs

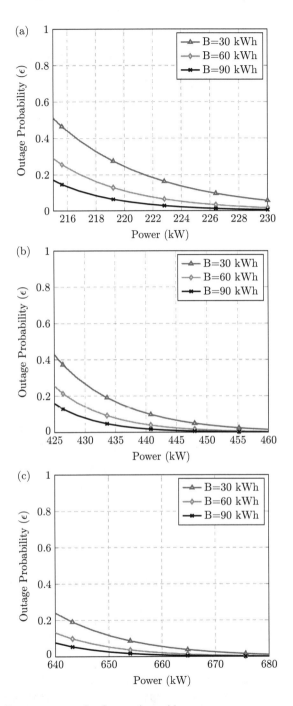

FIGURE 6.13 Energy storage size for varying grid power.

TABLE 6.2
Electric Vehicle Charging Tariffs (in US Cents per kWh) Adopted from San Diego Gas and Electric Company

Time of Day	Winter	Summer
4 pm–9 am	26	54
12 am–6 am	9	25
10 am–2 pm	9	25
Other	25	30

TABLE 6.3
Hourly PEV Demand Per Charger for the Case Study

Hour	6 am	7 am	8 am	9 am	10 am	11 am	12 pm	1 pm	2 pm
Demand	0.2	0.3	0.4	0.5	0.6	0.7	0.7	0.8	0.9

Hour	3 pm	4 pm	5 pm	6 pm	7 pm	8 pm	9 pm	10 pm
Demand	0.8	0.7	0.7	0.6	0.5	0.4	0.3	0.2

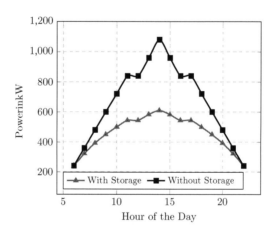

FIGURE 6.14 Hourly demand profile of the charging station.

$B = 108\,\text{kWh}$. Next, the amount of power needed to provide the target outage probability is calculated and shown in Figure 6.14. The differences between the two curves are related to peak demand reduction enabled by the employed energy storage unit. It can be observed from the presented results that station's peak demand is reduced by more than one-third.

TABLE 6.4
Comparison of Monthly Electricity Bills (in Thousands USD)

	Summer Tariff	Winter Tariff
With storage	106.64	66.03
Without storage	159	100.3

Now that necessary parameters are calculated, monthly electricity bill of charging station with and without an on-site energy storage is calculated for summer and winter tariffs.

As presented in Table 6.4, employing an on-site storage reduces typically operational cost by 34% in winter and 33% in summer. To make a fair comparison, levelized cost of electricity (LCOE), which includes acquisition, operational expenses, and financial costs of storage unit, needs to be included. According to a recent report (Henze, 2019), LCOE for lithium ion batteries has dropped to $187 per MWh. In this case study, the employed storage unit has a size of 106 kWh; hence, the LCOE would be close to $18.7. By further incorporating this cost in the presented results, it can be concluded that employing storage units makes economic sense for charging station operators.

6.6 CONCLUSIONS

In this chapter, we have presented a probabilistic capacity planning approach for PEV charging stations equipped with an on-site ESS. The system is modelled with MMPP where each system state is represented by the number of customers in the station and energy storage charge level. To solve the steady-state probability distributions, an algorithmic solution technique (matrix-geometric) was adopted. The principal goal was to compute the minimum energy storage size that can provide a good level of QoS measured by probability of outage events. In the last part, a number of case studies were presented to provide insights on how the model can be used in capacity planning. The results also showed that on-site storage systems can significantly lower station's peak demand and associated demand charges.

ACKNOWLEDGEMENT

The ENERSYN Project was funded by the European Space Agency (ESA) via the Navigation Innovation and Support Programme (NAVISP) with Programme Code NAVISP-EL2-012 (ENERSYN project and funding, n.d.).

BIBLIOGRAPHY

Aveklouris, A. N., Nakahira, Y., Vlasiou, M. and Zwart, B. (2017). Electric vehicle charging: A queueing approach. *ACM SIGMETRICS Performance Evaluation Review*, 45(2), pp. 33–35.

Bayram, I. S. and Ismail, M. (2019). A stochastic model for fast charging stations with energy storage systems. *IEEE Transportation Electrification Conference and Expo*. Novi, MI.

Bayram, I. S., Michailidis, G. and Devetsikiotis, M. (2014). Unsplittable load balancing in a network of charging stations under QoS guarantees. *IEEE Transactions on Smart Grid*, 6(3), 1292–1302.

Bayram I. S., Michailidis, G., Devetsikiotis, M. and Granelli, F. (2013). Electric power allocation in a network of fast charging stations. *IEEE Journal on Selected Areas in Communications*, 31(7), 1235–46.

Bayram, I.S. and Tajer, I. S. (2017). *Plug-in Electric Vehicle Grid Integration*. London: Artech House.

Cross, J. a. (2016). My Electric Avenue: Integrating electric vehicles into the electrical networks. *6th Hybrid and Electric Vehicles Conference*, London, UK.

ENERSYN project and funding. (n.d.). Retrieved from https://navisp.esa.int/project/details/26/show

Falvo, M. S. (2014). EV charging stations and modes: International standards. *International Symposium on Power Electronics, Electrical Drives, Automation and Motion*, pp. 1134–1139. Ischia, Italy.

Fan, P., Sainbayar, B. and Ren, S. (2015). Operation analysis of fast charging stations with energy demand control of electric vehicles. *IEEE Transactions on Smart Grid*, 6(4), 1819–1826.

Fitzgerald, G. and Nelder C. (2017). *Evgo Fleet and Tariff Analysis*. Rocky Mountain Institute.

Fotouhi, Z., Hashemi, M. R., Narimani, H. and Bayram, I. S. (2019). A general model for EV drivers' charging behavior. *IEEE Transactions on Vehicular Technology*, 68(8), 7368–7382, Boulder, Colarado.

García-Villalobos, J., Zamora, I., San Martín, J.I., Asensio, F.J. and Aperribay, V. (2014). Plug-in electric vehicles in electric distribution networks: A review of smart charging approaches. *Renew. Sustain. Energy Rev.*, 38, 717–731.

Global EV Outlook. (2019). International Energy Agency. https://www.iea.org/reports/global-ev-outlook-2019.

Han, S. H., Han, S. and Sezaki, K. (2010). Development of an optimal vehicle-to-grid aggregator for frequency regulation. *IEEE Transactions on Smart Grid*, 1(1), 65–72.

Haslett, A. (2019). *Smarter Charging-A UK Transition to Low Carbon Vehicles: Summary Report*. Loughborough: Energy Technologies Institute.

Henze, V. (2019). *Battery Power's Latest Plunge in Costs Threatens Coal, Gas*. New York: Bloomberg.

Hu, J., Morais, H., Sousa, T. and Lind, M. (2016). Electric vehicle fleet management in smart grids: A review of services, optimization and control aspects. *Renewable and Sustainable Energy Reviews*, 56, 1207–1226.

Jurgen Weiss, M. H. (2019). *Increased Transmission Investment to Support Growing Electrification Demand and Access Low-Cost Resources Can Reduce Customer Rates*. Brattle Group, San Francisco, California.

Khaligh, A. and Dusmez, S. (2012). Comprehensive topological analysis of conductive and inductive charging solutions for plug-in electric vehicles. *IEEE Transactions on Vehicular Technology*, 61(8), 3475–3489.

Krein, M. and Yilmaz, P. T. (2012). Review of battery charger topologies, charging power levels, and infrastructure for plug-in electric and hybrid vehicles. *IEEE Transactions on Power Electronics*, 28(5), 2151–2169.

Lee, Z. L., Li, T. and Low, S.H., (2019). ACN-Data: Analysis and applications of an open EV charging dataset. *Proceedings of the Tenth ACM International Conference on Future Energy Systems*, pp. 139–149.

Liang, X., Srdic, S., Won, J., Aponte, E., Booth, K. and Lukic, S. (2019). A 12.47 kv medium voltage input 350 kw ev fast charger using 10 kv sic mosfet. *IEEE Applied Power Electronics Conference and Exposition (APEC)*, Anaheim, CA.

Liu, Y., Zhu, Y. and Cui, Y. (2019). Challenges and opportunities towards fast-charging battery materials. *Nature Energy*, 4(7), 540–550.

Negarestani, S., Fotuhi-Firuzabad, M., Rastegar, M. and Rajabi-Ghahnavieh, A. (2016). Optimal sizing of storage system in a fast charging station for plug-in hybrid electric vehicles. *IEEE Transactions on Transportation Electrification*, 2(4), 443–453.

Neuts, M. F. (1994). *Matrix-Geometric Solutions in Stochastic Models: An Algorithmic Approach*. Courier Corporation, Baltimore, Maryland, USA.

Power Networks Demonstration Centre. (n.d.). Retrieved February 2020, from https://pndc.co.uk/

Ronanki, D. and Williamson, S. S. (2018). Modular multilevel converters for transportation electrification: challenges and opportunities. *IEEE Transactions on Transportation Electrification*, 4(2), 399–407.

Sarker, M. R., Pandžić, H., Sun, K. and Ortega-Vazquez, M.A. (2018). Optimal operation of aggregated electric vehicle charging stations coupled with energy storage. *IET Generation, Transmission Distribution*, 12(5), 1127–1136.

Sperling, D. (2018). Electric vehicles: Approaching the tipping point. In *Three Revolutions* (pp. 21–54). Washington DC: Island Press.

Srdic, S. and Lukic, S. (2019). Toward extreme fast charging: Challenges and opportunities in directly connecting to medium-voltage line. *IEEE Electrification Magazine*, 7(1), 22–31.

Tehrani, N. H. and Wang, P. (2015). Probabilistic estimation of plug-in electric vehicles charging load profile. *Electric Power Systems Research*, 124, 133–143.

Ucer, E., Koyuncu, I., Kisacikoglu, M.C., Yavuz, M., Meintz, A. and Rames, C. (2019). Modeling and analysis of a fast charging station and evaluation of service quality for electric vehicles. *IEEE Transactions on Transportation Electrification*, 5(1), 215–225.

Williamson, S. S., Rathore, A.K. and Musavi, F. (2015). Industrial electronics for electric transportation: Current state-of-the-art and future challenges. *IEEE Transactions on Industrial Electronics*, 62(5), 3021–3032.

Zhang, P., Zhou, C., Stewart, B.G., Hepburn, D.M., Zhou, W. and Yu, J. (2011). An improved non-intrusive load monitoring method for recognition of electric vehicle battery load. *Energy Procedia*, 12, 104–112.

Autumn P. (2018). *The 50 States of Electric Vehicles*. Raleigh, NC: NC Clean Energy Technology Center.

7 Innovative Methods for State of the Charge Estimation for EV Battery Management Systems

Zeeshan Ahmad Khan
Car.SW Org, Volkswagen AG

Franz Kreupl
Technical University Munich

CONTENTS

7.1 Introduction 143
7.2 Literature Review 145
    7.2.1 Battery Management System 146
7.3 State-of-Charge Estimation 149
    7.3.1 Kalman Filter Algorithm 153
        7.3.1.1 Extended Kalman Filter 154
        7.3.1.2 Central Difference Kalman Filter 156
        7.3.1.3 Adaptive Extended Kalman Filter 162
    7.3.2 Sliding Mode Observer 165
    7.3.3 Backpropagation Neural Network 170
        7.3.3.1 Forward Propagation 173
        7.3.3.2 Backward Propagation 174
7.4 Conclusion 176
7.5 Framework for Integrating EV Energy Storage Systems 176
Nomenclature 178
References 178

7.1 INTRODUCTION

Batteries have evolved as one of the most credible energy storage sources in the past few decades or so. A majority of the electronic devices and tools that we use daily from cell phones and laptops all the way to residential and commercial energy storage systems use batteries as the medium to store energy. In the automotive industry as well, battery-powered vehicles are being considered as an alternative for the traditional internal combustion engine (ICE) vehicles. As batteries are getting employed in more demanding applications like electric and hybrid vehicles, it becomes imperative to exactly determine the dynamics of battery during its operation. The battery forms the heart of electric powertrain in a car, and therefore, it needs a precise real-time monitoring and control system. The battery terminal voltage, current, temperature, internal resistance, state of charge (SOC) and state of health (SOH) are some of the parameters that need to be evaluated and estimated while in operation and in standby mode. The battery management system (BMS) is the part responsible for monitoring the necessary battery parameters. Along with monitoring the parameters, BMS also performs communication with other controllers; performs cell balancing; and ensures that operating limits for voltage, current and temperature both at the cell and at the pack level are not exceeded. The SOC is one of the most important parameters that always need to be known. However, it turns out that SOC is not a physically measurable quantity; rather, it is a quantity which can be estimated from other physically measurable quantities. This thesis work is dedicated to developing algorithms for SOC estimation using the lithium-ion cell modelling and comparing their performance.

In the past two decades, increasing environmental concern and heavy reliance of transportation sector on nonrenewable fossil fuels have forced the automotive industry to look for sustainable and environment-friendly alternatives. Therefore, the automotive industry has turned its attention towards electricity as the source to power the coming age vehicles. At the moment, the automotive industry is experiencing a paradigm shift from the conventional ICE vehicles to the battery electric vehicles (BEVs), hybrid electric vehicles (HEVs) and plug-in hybrid electric vehicles (PHEVs). BEVs and PHEVs have batteries as their main source of propulsion, while for HEVs lower rating batteries are present which are mostly used during vehicle acceleration and braking. Higher energy efficiency, lower maintenance of the electric motors, simple powertrain design and zero tailpipe emissions must be a few reasons to root for a future with electric vehicles (EVs) forming the heart of transportation sector.

EVs of all types require a reliable, robust and dynamic energy storage during their operation. Currently, batteries, supercapacitors, fuel cells and hydrogen-based storage are being considered for this purpose. The choice of an energy storage depends on the requirements of the application. For example, in case of HEVs, the main requirement for energy storage device is to be able to quickly provide power during vehicle acceleration and absorb power during braking. Therefore, energy storage devices which have higher power density, such as supercapacitors, lithium-polymer and nickel metal hydride (NiMH) batteries, would be more suitable for HEVs. Mercedes-Benz B250e (28-kWh battery), Toyota Prius (fourth generation) (1.31-kWh battery) and Honda Accord Hybrid (1.3-kWh battery) are some examples of HEVs. For BEVs and PHEVs, the main goal is to select a storage device that can provide a long range of operation. Therefore, lithium-ion batteries would be more suitable for operation. BMW i3 (33-kWh battery), Tesla Model S (60- to 100-kWh battery), Audi A3 Sportback e-Tron (8.3-kWh battery), Volvo XC90 T8 (9.0-kWh battery) and Nissan Leaf (30-kWh battery) are some of the BEVs and PHEVs using the lithium-ion batteries.

The lithium-based batteries have gained prominence over their lead-acid and NiMH counterparts. This could be attributed to the exceptionally high energy density, greater operating voltage range, higher capacities, possible combination with other elements to form different stable chemistries and low maintenance requirements. Additionally, what makes them attractive for EV applications is the fact that the self-discharge rate for lithium-ion battery is lower than half the self-discharge rate for lead-acid and NiMH batteries. However, there are also several challenges that are faced with lithium-ion batteries. First major issue is with protection. The batteries require a protection circuit or device to prevent the cell voltage to go beyond upper voltage limit (4.2 V in most cases) and below lower voltage limit (3.0 V in most cases and 2.4 V in some cases). A battery management circuit with a resistor for bypass can be the simplest circuit that comes in operation when limits are crossed. Ageing is another major issue with lithium batteries. Ageing can be calendric (depending on time) or cyclic (depending upon the number of charging or discharging cycles battery has undergone). Other major issues include the cost and problems in transportation. However, significant advantages offered by lithium-ion batteries outweigh the complications, and therefore, the utility and scope of lithium battery-powered EVs is on the rise and is likely to be at the centre stage in the automotive sector in the coming future.

The increasing utility of lithium-ion batteries in dynamic applications poses a challenging task of battery monitoring and control under all conditions. SOC appears to be one of the most important parameters to be determined for any battery. The concept of SOC can be simply understood by the concept of fuel gauge in IC engine car. The battery in an EV can be analogous to the fuel tank. The fuel tank requires fuel for vehicle operation, while the battery requires charge for operation. The charge is simply the product of current and instantaneous time. Higher the current applied to the battery, greater will be the charge flowing in the battery, and vice versa. The fuel level in the tank can be seen with the fuel meter on the car dashboard to have an idea of the remaining distance the car can drive. Since SOC is not a physically measurable quantity, it is estimated by algorithms that use battery voltage, current and temperature. These algorithms are implemented inside the controller in the BMS and provide estimations of SOC. In order for BMS to accurately estimate the battery SOC, BMS requires an accurate, high-fidelity, robust mathematical model of the battery. Since in an EV the battery is subjected to varying currents during operation, which causes changes in voltage and temperature, it becomes an exhaustive task to estimate SOC accurately.

7.2 LITERATURE REVIEW

This chapter performs the analysis of the advanced mathematical methods that are being widely used for SOC estimation in EVs. This chapter begins with an introduction into the BMS, which is the electronic device on which the SOC estimation algorithms are run. The BMS is the controller, which monitors the battery voltage, current and temperature, and uses these real-time values in the estimation algorithms. This is followed by the analysis of the most commonly used SOC estimation algorithms, namely the Kalman filters (KFs), sliding mode observer, and the backpropagation neural network (BPNN). The mathematical equations formulating these methods are explained in detail so that the readers understand the underlying mathematics behind these complex estimation algorithms. This understanding of the underlying mathematics will enable the readers to implement these algorithms on their own and estimate the SOC for any given battery system.

7.2.1 Battery Management System

According to Ref. [1], BMS is the central component of the battery pack and guarantees safety and efficiency of cells and the battery pack by monitoring variables such as voltage, current and temperature of each cell. Still according to Ref. [1], the BMS uses the monitored variables to estimate SOC, SOH, state of energy (SOE), and state of function (SOF) of each cell and the battery pack. In terms of safety, the BMS ensures that the cells are protected from degradation, from extremely high and low temperature conditions, from over-voltage and under-voltage conditions, from high currents, from over- and under-charging conditions, and from deep discharge condition.

The battery SOC depends on the residual capacity of cell, cell electrochemistry as a function of the anode and cathode materials, cell internal structure, charging and discharging rates, operating temperatures and the chemical reactions inside the cell. Therefore, the estimation of SOC depends on several complex factors and is a secret proprietary information of companies manufacturing and designing BMS.

Figure 7.1 illustrates a hypothetical example of the concept of battery SOC [2]. The battery SOC is a direct indication of lithium concentrations at cathode and anode. Figure 7.1a shows a charged cell, whereas Figure 7.1b shows a discharged cell. Figure 7.1c shows the variation in the open-circuit potential at the negative electrode with the quantity of lithium in Li_yC_6, where y indicates the number of lithium atoms in the chemical compound Li_yC_6. For a fully charged cell (orange colour), the concentration of lithium in Li_yC_6 (material which forms the negative electrode of cell) is very high (0.8 in the figure). Figure 7.1d shows the variation in open-circuit potential at the positive electrode with the quantity of lithium in Li_xMO_2. For a fully discharged cell (blue colour), the concentration of lithium in Li_xMO_2 (material which forms the positive electrode of cell) is very high (0.8 in the figure). Hence, we can say that the battery residual capacity at any instant depends on lithium concentrations at anode and cathode, thereby demonstrating the reliance of SOC on the internal chemical reaction of the cell. This is just one of the examples of the complex relationships of SOC with battery parameters, consequently making direct measurement of SOC a nonviable alternative. The accurate and precise knowledge of SOC is important for estimating the remaining charge in the battery and hence the remaining drivable distance; however, sampling, sensor noise and errors in parameter determination pose a significant challenge in this area [2]. V. Prajapati et al. have conducted case studies to establish examples of poor accuracy and low reliability in SOC prediction. The battery pack in an EV is composed of large number of cells, thereby increasing the complexity of the entire system and making control and management at cell level a daunting task. Therefore, this compels the use of a BMS to make sure that the utilization of each cell is maximized, and cell operates within limits. Among the multitude of tasks

EV Battery Management Systems

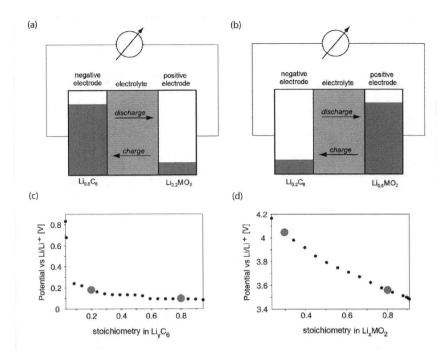

FIGURE 7.1 Schematics representing the correlation between battery SOC and residual lithium concentration in anode and cathode at [Gray portion in electrodes in Fig. a] 80% SOC and [Gray portion in electrodes in Fig. b] 20% SOC. The image is taken from Ref. [2].

performed by BMS during its operation, the most significant ones include the estimation of battery states (SOC, SOH, SOE, and SOF) at each sampling interval during its operation and measurement of voltage, current and temperature continuously at both the cell and pack levels. Other major tasks include performing cell balancing, following the specified operational limits (safety and diagnosis) and communicating back and forth with the main vehicle computer. One commonly used example of communication technology is the controller area network (CAN) bus. Figure 7.2 presents the block diagram displaying the major tasks performed by a BMS.

The BMS is a complex electronic device consisting of several components to perform the tasks listed in Figure 7.2. Figure 7.3 shows the simplified block diagram of the components of a BMS. Battery pack, as the name suggests, represents the battery under test and/or operation. Measurement unit consists of number of measurement electronic components and sensors for current measurements, voltage measurements and temperature measurements at the cell and battery pack levels. Battery controller has an algorithm implemented for estimation, which exploits a mathematical model of battery. The BMS controller adjusts the operating point based on the SOC values provided by the algorithm. Thus, the fundamental step towards developing an accurate SOC estimation is dependent on the design and selection of actual battery model implemented in algorithm. Cell balancing and equalization block comprises a comparator that compares the cell voltages and finds the difference between the

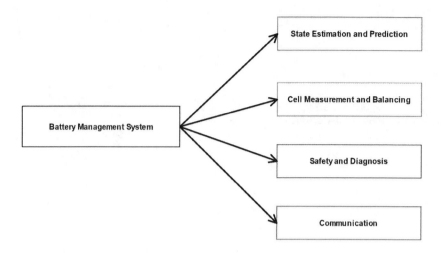

FIGURE 7.2 Block diagram listing the major tasks of a BMS. State estimation and prediction, cell measurement and balancing, safety and diagnosis, and communication are the four major tasks that are executed by a BMS when operating within a battery pack.

highest and lowest cell voltage. If this voltage difference between two cells is larger than the threshold, then the charging is stopped, and the charge is transferred from cell with the highest voltage to the cell with the lowest voltage. The cell balancing is achieved by means of charge shuttling through capacitors and MOSFETs (active balancing) or simply draining the charge via a shunt resistor (passive balancing). The CAN bus is responsible for communication between modules and the battery controller. Moreover, it is also responsible for communication between battery controller and other components of the battery pack. A battery pack during its operation is continuously receiving and sending information via CAN bus. A battery is generally housed inside a battery junction box. The battery junction box has digital communication ports that allow the transfer of information to and from the battery via the CAN bus. User interface and display is the interactive unit (mostly a computer screen), which shows the measured quantities from the BMS and permits the user to provide an input from the available selection. Based on the measurements performed by the measurement unit as shown in Figure 7.3, the battery controller estimates the SOC in real time. However, due to the complex factors involving the SOC, estimation errors can occur. Shen et al mentioned the most important sources of errors. These include capacity errors, current measurement errors, voltage measurement errors, and model-based errors. The capacity of a battery reduces as it ages; thus, error in capacity estimation can occur which can propagate to SOC estimation through ampere-hour method; these errors are referred to as capacity errors. The measurement of the electric current is performed through shunt resistors or through Hall effect sensors. Both approaches are prone to errors. These are the current measurement errors. Since the measurement of electric currents is the basis of SOC estimation, current measurement errors caused by these sensors can drastically degrade the SOC estimation. Voltage measurement errors are also possible, and in this case,

EV Battery Management Systems

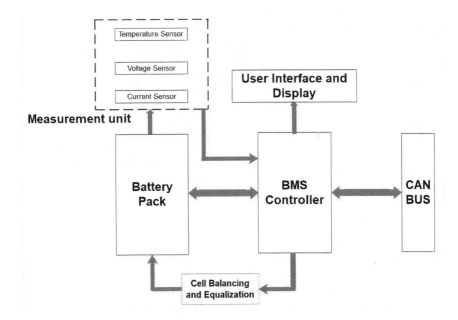

FIGURE 7.3 Simplified block diagram of a BMS. Measurement unit, battery pack, user interface and display, BMS controller, CAN bus and cell balancing, and equalization blocks are the major components of any BMS.

due to the voltage-based correction being used in the SOC estimation, the SOC estimation can be significantly degraded. Finally, model-based errors, caused by the incorrect value of parameters in a battery model, can lead to highly inaccurate SOC estimation. Generally, the models are voltage-based, and incorrect computation of terminal voltage due to an incompatible battery model can cause significant deviations in SOC from the actual value.

7.3 STATE-OF-CHARGE ESTIMATION

Historically, SOC is calculated as the ratio between the remaining capacity of cell to nominal capacity of cell. The nominal capacity can be obtained from the manufacturer's data sheet.

$$\text{SOC}(\%) = \frac{Q_a}{Q_n} * 100 \qquad (7.1)$$

where Q_a represents the instantaneous available capacity in ampere hours and Q_n outlines the nominal capacity of the battery in ampere hours specified by the manufacturer in data sheet.

Figure 7.4 provides a pictorial representation of correlation between SOC and stored energy in battery [5]. 100% SOC implies that battery has a rated capacity, which is the sum of the stored energy, energy in empty region, and energy in inactive region which will not be used under any condition. Fifty percent of SOC indicates that the battery is half empty and half with stored energy, whereas 0% SOC implies

FIGURE 7.4 Representation of SOC in battery. Light Gray indicates the stored energy in the battery, dark grey indicates the empty region, which is the energy used by the battery, and the inactive region which will not be used. The image is taken from Ref. [5].

battery is completely drained out of energy (actual capacity); however, some energy remains unused in the inactive region. The SOC estimation is one of the most important tasks to be performed by a BMS periodically during battery operation. The sampling rate of the SOC in BMS depends on the application and the type of sensors and logging devices available in the circuit. From the battery module, voltage, current, and temperature are continually measured at both the cell and pack levels. The BMS utilizes these measurements in an algorithm to provide a robust estimate for the SOC. However, the estimation of SOC is not a straightforward process; several factors influence the SOC, including applied charging/discharging rates, usable capacity, temperature, aging, internal resistance, and self-discharge. The basic method to estimate the SOC is obtained by integrating the current over the time and dividing the value with the nominal capacity of battery; this technique is known as Coulomb or Ampere-hour counting. Equation 7.2 gives the expression for SOC estimation using Coulomb counting in discrete time.

$$\text{SOC}(k+1) = \text{SOC}(k) + \frac{\eta \Delta t I(k)}{C_n} \qquad (7.2)$$

where SOC(k + 1) and SOC(k) are the SOC values at $(k+1)^{th}$ and k^{th} sampling time, respectively; η is the Coulombic efficiency assumed as 1 for charging and as 0.98 for discharging when a current pulse $I(k)$ is applied; C_n is the nominal capacity from the manufacturer's data sheet; and Δt represents the sampling interval [6].

The significant advantage of coulomb counting is that it can be applied irrespective of the cell chemistry. In addition, it requires low computing power when implemented in real time and can be implemented in conjunction with other techniques. Although coulomb counting is simple to implement, it has some serious disadvantages. As it includes the integration of current over time, the error in estimation is high due to the integral term. Also from equation 7.2, it is clear that estimation requires knowledge of previous SOC value, which makes it dependent on the accuracy on initial measurements.

Ng et al. used a smart estimation approach (termed 'enhanced coulomb counting') based on conventional coulomb counting to perform SOC and SOH estimations

EV Battery Management Systems 151

in lithium-ion batteries. The research reports that SOC estimation is accurate with conventional coulomb counting. However, as the cell ages, the error in SOH estimation increases. The research reports a reduction in estimation error of about 8% after 28th cycle (1.08% error), thus providing SOH estimation with a minimal error. Zheng et al. [8] performed coestimation of SOC, capacity, and resistance of lithium-ion batteries. Their study used proportional integral (PI) observers to realize the coestimation of battery SOC, capacity, and resistance. Further, the moving window ampere-hour counting method is used to improve the accuracy of estimation. High-fidelity electrochemical model is adopted to capture the battery dynamics. The research reports SOC root mean square error (RMSE) drops from 3.97% to 1.01%, while the maximum SOC error declines from 5.86% to 1.58% after the application of moving window ampere-hour counting method. When ageing effect is considered, SOC RMSE can be decreased to an error band of 1% of actual SOC, and the maximum SOC can be limited in an error band of 2% for all ageing levels. In order to overcome the dependency on current integration and initial SOC measurement, it is necessary that methods be used that estimate SOC independent of these quantities. KFs, observer methods, and artificial neural network can be used to estimate SOC regardless of the dependencies shown by coulomb counting. Table 7.1 shows a summary of the techniques used in SOC estimation, and lists the techniques, their possible application areas, their advantages, and their disadvantages [9–11].

TABLE 7.1
Commonly Used SOC Estimation Techniques with Their Applications, Advantages, and Drawbacks [9–11]

Techniques	Application Areas	Advantages	Drawbacks
Look-up tables	Nickel Cadmium (NiCd)	Comparison between two batteries (one with known SOC)	Offine[1], sensitive to battery and operating conditions
Current sharing method	Lithium, nickel metal hydride (NiMH), lead acid	Easy implementation, low computation time	Sensitive to battery and operating conditions
Discharge test	Used for capacity determination at the beginning of life	Easy and accurate, independent of SOH	Offine[1], time intensive, modifies the battery state, loss of energy
Physical properties of electrolyte (density, concentration, colour etc.)	Lead acid, zinc bromide, vanadium	Online[2], information about SOH	Sensitive to temperature and impurities, error due to acid stratification
Coulomb counting	All battery chemistries	Accurate if enough recalibration points are available and with good current measurements	Sensitive to parasite reactions, needs regular recalibration points
Open-circuit voltage (OCV)	Lead acid, lithium, zinc bromide	Online[2], computationally inexpensive, OCV prediction	Needs long rest phases (current = 0)

(*Continued*)

TABLE 7.1 (Continued)
Commonly Used SOC Estimation Techniques with Their Applications, Advantages, and Drawbacks [9–11]

Techniques	Application Areas	Advantages	Drawbacks
Coup de Fouet	Lead acid'	Estimating battery capacity	Sensitive to battery and operating conditions
Linear model	Lead acid, photovoltaic batteries	Online[2], easy Implementation	Needs reference data for fitting parameters
Impedance spectroscopy	All battery chemistries	Information about SOH and quality	Temperature sensitive, expensive
Internal resistance	Lead acid, lithium, nickel cadmium	Information about SOH, online[2]	Good accuracy but only for shorter intervals
Artificial neural networks	All battery chemistries	Offine[1], online[2]	Needs training data of a similar battery, backpropagation algorithms are computationally expensive and sometimes require advanced techniques for optimization
Fuzzy logic	All battery chemistries	Offine[1], online[2]	Definition of membership functions are highly subjective, does not scale to large models
Observers (sliding mode observer and Luenberger observer)	All battery chemistries	Offine[1], online[2]	Require accurate battery model in state pace format, determination of initial parameters by trial and error
Kalman filters	All battery chemistries, photovoltaic batteries, dynamic applications such as HEVs	Offine[1], online[2]	Large computing capacity, need accurate battery model, problem in the determination of initial parameters
Hybrid techniques e.g., artificial neural networks + Kalman filters, fuzzy logic + Kalman filters	All battery chemistries	Online[2], dynamic	Difficult implementation and stabilization as two separate approaches are implemented in combination

Offline[1] implies entire dataset available for algorithm, whereas
Online[2] means data is processed piece-by-piece in a sequential manner without having it available from start.

7.3.1 KALMAN FILTER ALGORITHM

KF is essentially an optimal recursive data processing algorithm. KF algorithms are among the most widely used adaptive estimation techniques for SOC estimation. Generally, KF can be applied for estimating the inner states of any dynamic system based on system model and sensor measurements. The states of the system are not directly measurable, and therefore by defining system model and obtaining the measurements, KF performs state estimation. One important point to note is that filter requires previous values of states, inputs and outputs to update the state in next iteration. KF is usually quite robust when subjected to noises and uncertainties. The filter can be applied to both discrete time and continuous time domain. However, for accurate estimations, it is necessary that the filter be supplied with the exact initial value of state and the measurements.

There exist different versions of KF. The filter can be applied to both linear and nonlinear systems, and the filter choice depends on the nature of problem. In case of SOC estimation, battery serves as the dynamic system and SOC as the inner state. Figure 7.5 provides a classification diagram for KF techniques. Based on system type, the KF is divided into two categories: linear and nonlinear Kalman filter. The nonlinear filter can be further divided into four groups: extended Kalman filter (EKF), sigma point Kalman filter (SPKF), cubature Kalman filter, and particle filters. SPKF is further divided into two categories based on the number of tuning parameters: unscented filter (2 tuning parameters) and central difference filter (1 tuning parameter). In this chapter, we discuss the mathematics behind the EKF, adaptive extended Kalman filter (AEKF), and central difference Kalman filter (CDKF).

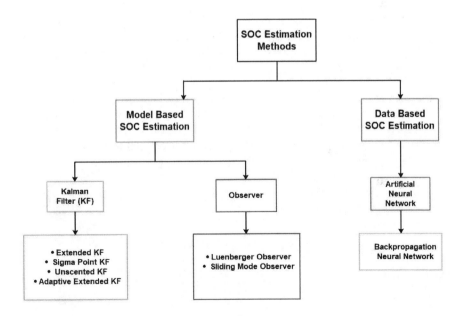

FIGURE 7.5 Kalman filter classification diagram.

7.3.1.1 Extended Kalman Filter

For nonlinear battery systems, a robust and more precise estimation can be obtained using the EKF. The EKF utilizes the same prediction and updating steps as the linear KF; however, the distinction lies in linearizing the system each point in time using the Taylor-series expansion, which renders the system as a linear time-varying system. The EKF has a very high accuracy when applied to systems with low to medium degree of nonlinearity. This can be attributed to two reasons: Firstly, when computing the output estimates for the nonlinear system, EKF assumes that the expected value of nonlinear function of unknown state is equal to the same nonlinear function evaluated at the expected value of state. That is, it approximates

$$E[\text{fn}(x)] = \text{fn}[E(x)] \tag{7.3}$$

where fn is a nonlinear function which represents the relationship between the system input, state and output. This is true only when fn(x) is linear. Therefore, the EKF works best for system with low to mild nonlinearities [6]. Secondly, when computing covariance estimates, EKF uses a truncated Taylor-series expansion to linearize the system equations around the present operating point; hence, higher-order terms are rejected [6].

Consider a nonlinear system in discrete time in state space format as below:

$$x_{k+1} = f(x_k, u_k) + w_k = A * x_k + B * u_k + w_k \tag{7.4}$$

$$y_k = g(x_k, u_k) + v_k = C * x_{k+1} + D * u_k + v_k \tag{7.5}$$

a nonlinear function representing the system output, A, B, C, D are the system matrix, input matrix, output matrix and feedforward matrix, respectively. x_k is the system state vector with a covariance P_k, u_k is the input vector, y_k is the output vector, w_k is the system noise vector, and v_k is the measurement/sensor noise vector. w_k is taken as a white noise and is assumed to be zero mean Gaussian with a covariance Q_k, and v_k is also assumed to be a zero mean Gaussian with covariance R_k.

The EKF involves mainly two steps: prediction step and update step. The prediction step is used to calculate the priori estimate of the state and the covariance of the system, while the update step provides the posteriori estimate of the same. The relevant equations describing the EKF formulation are provided below.

7.3.1.1.1 Prediction Step

State estimate time update

$$\widehat{X_{k+1|k}} = A * X_k + B * u_k \tag{7.6}$$

Error covariance time update

$$P_{k+1|k} = A * P_k * A^T + Q \tag{7.7}$$

EV Battery Management Systems

Measurement error calculation

$$e_{k+1|k} = y_{k+1|k} - y\left(\widehat{X_{k+1|k}}, u_k\right) \tag{7.8}$$

Kalman gain calculation

$$K_{k+1|k} = P_{k+1|k} * C^T * \left(C * P_{k+1|k} * C^T + R\right)^{-1}. \tag{7.9}$$

7.3.1.1.2 Measurement Update Step

State estimate measurement update

$$\widehat{X_{k+1|k+1}} = \widehat{X_{k+1|k}} + K_{k+1|k} * e_{k+1|k} \tag{7.10}$$

Error covariance measurement update

$$P_{k+1|k+1} = \left(I - \left(K_{k+1|k} * C\right)\right) * P_{k+1|k} \tag{7.11}$$

where $\hat{X}$ is the state vector of the order $n \times 1$ with n being the number of states, A is the system matrix of order $n \times n$, B is the input matrix with order $n \times 1$, C is the output matrix with order $m \times n$ with m being the number of outputs, u_k is the input matrix, P is the covariance corresponding to system state, Q is the covariance corresponding to system noise, R is the covariance corresponding to measurement noise, y is the system output, and e represents the error between the real system output and model output, respectively. Covariance P, which is a square matrix, represents the uncertainty in the system state. Depending on the dimensions of the system, this can be a scalar or a vector. As can be observed from equations 7.7 and 7.11, covariance P is updated twice during one iteration: one for a *priori* estimate at the start and the other for a *posteriori* estimate after the measurements are added to the filter. In equation 7.7, term APA^T uses the system matrix A to compute the covariance between different system states and then adds it to the covariance Q. Equation 7.7 outputs a matrix which depicts the variance of each state along its main diagonal elements and displays the correlation between the different states along its off-diagonal elements. The choice of the covariance matrix strongly affects the filter performance. A high value for covariance resembles a fast convergence of the filter to the actual state, whereas a low value of covariance indicates relatively slower convergence. The initial value of covariance P can either be selected by trial and error or be based on the sensor accuracy, if known. Covariance Q accounts for the unmodelled states and the random variations in model parameters, which contribute to system noise. Higher value of Q indicates model is accurate, while a lower value of Q indicates model is inaccurate and system heavily relies on measurements. Matrix R models the noise in sensor measurements as covariances. This matrix represents the correlation between different sensors in the system. R is a diagonal matrix as the diagonal elements represent the variance in each sensor measurement, while the off-diagonal elements represent the correlation between different sensors, which is essentially zero.

Equation 7.8 calculates the error or the residual, which is essentially obtained by subtracting the model output from the sensor output or measurement. The term $y\left(\widehat{X_{k+1|k}}, u_k\right)$ uses the predicted state to calculate the output as the state vector is transformed into the measurement through the nonlinear function that represents the output in terms of state vector and the control input. Once we know the nonlinear relation between system output and state, residual can be easily obtained. Equation 7.9 outputs the Kalman gain, which is a real number between 0 and 1. Kalman gain tells how much the estimate should change given a measurement. In equation 7.9, if $R = 0$, Kalman gain, $K = C^{-1}$. This tells us that if the measurements are very accurate, then Kalman gain is simply obtained as the inverse of C matrix. However, if $P = 0$, $K = 0$, this means that there is no change in state, so the estimates will not change at each instant and gain will be zero. Equation 10 provides the new state estimate based on the Kalman gain and the residual term. The Kalman gain has the term C^T, which when multiplied by the residual coming from the measurements provides an output in state space which is then simply added to the *priori* state estimate, $\widehat{X_{k+1|k}}$. In equation 11, I represents an identity matrix, a square matrix whose diagonal elements are 1, off-diagonal elements are 0 and determinant $= 1$. In this equation, product $K \times C$ is subtracted from I. In this product, output matrix C is being scaled by Kalman gain. Therefore, if the Kalman gain K is large, $(I - K \times C)$ is small and the resulting *posteriori* value of P will be smaller. However, if K is small, $(I - K \times C)$ is large and the resulting *posteriori* value of P will be larger. This equation gives us the ability to adjust the magnitude of P, which represents the uncertainty in state by tuning via Kalman gain, K.

The EKF is an iterative process and estimates output and state at each iteration. For the first iteration, the initial values of system state X, state covariance P, system noise covariance Q, and measurement noise covariance R are provided. The initial values of system state are either determined or taken as the first value of the experimental data. Since P and Q are related to the system state, it is difficult to provide an exact initial value. The initial values of P and Q are largely selected by trial and error. On the other hand, covariance R is related to the noise in measurement. The initial value of R can be taken as the square of the difference between the first value of measurement output and that of model output. After having determined the values of initial values of X, P, Q, R and the relevant battery parameters, these values can be utilized in EKF for SOC estimation.

7.3.1.2 Central Difference Kalman Filter

SPKF is an alternative approach to KF generalization for nonlinear systems. Due to its strong computations and resulting estimations, SPKF can be applicable to system with mild to high degree of nonlinearity. Unlike EKF, which uses Taylor-series expansion to estimate covariance matrix, SPKF performs a number of function evaluations to compute the estimated covariance matrices. This has several advantages: (1) Derivatives do not need to be computed (which is one of the most error-prone steps when implementing EKF), also implying (2) the original functions do not need to be differentiable, and (3) better covariance approximations are usually achieved, relative to EKF, allowing for better state estimation, (4) all with comparable computational complexity to EKF [12].

In SPKF, a set of special points, known as sigma points, are computed, with the condition that the mean and covariance of these points match exactly with the weighted mean and covariance of the *a priori* random variable being modelled. These points are then transformed after being passed through the nonlinear functions. A *posteriori* mean and covariance are then approximated by mean and covariance of these points. The SPKF can further be divided into two classes based on the calculation of weighting factors: unscented Kalman filter (UKF) and CDKF.

The CDKF uses the Sterling's polynomial interpolation method, avoids calculating the partial derivative, and is suitable for arbitrary functions. CDKF is capable of providing state estimates even if nonlinear function is discontinuous or has singular points, and has better accuracy than EKF [13]. Unlike UKF, which uses the three tuning parameters for the calculation of sigma points, CDKF uses only one tuning parameter 'h' for the same. The CDKF formulation and stabilization are a little bit less complex as compared to UKF; however, there is no difference in accuracy of estimation [12].

Consider a nonlinear system in discrete time as described by the following equations:

$$x_{k+1} = f(x_k, u_k) + w_k = A * x_k + B * u_k + w_k \tag{7.12}$$

$$y_k = g(x_k, u_k) + v_k = C * x_{k+1} + D * u_k + v_k \tag{7.13}$$

The description of the above equations 7.12 and 7.13 has been provided in the section EKF and will not be provided here. The CDKF also utilizes the state space matrices A, B, C, and D for estimation. The state vector x_k, control input u_k, output y_k, process noise w_k, and associated covariance Q, sensor noise v_k, and associated covariance R will all have the same definitions in the CDKF. However, some additional variables will be added and the equation for CDKF will be different from EKF. The CDKF also involves two steps: prediction step and update step. The difference lies in the fact that instead of linearizing the system through the first-order Taylor-series approximation at the mean point as in EKF, CDKF uses a set of special points called the *Sigma points* to approximate the mean and covariances between the states. These sigma points are chosen deterministically to completely reflect the dynamics of the states. They completely obtain the true mean and covariance of the state, and when propagated through a nonlinear system, they capture the posterior mean and covariance accurately to second order for a nonlinear function with any degree of nonlinearity [14]. The number of chosen sigma points generally depends on the number of states/variables to be estimated. The general rule is for n number of states, we require a minimum of $2n+1$ sigma points to completely capture the associated mean and covariance. The first sigma point is generally the mean of the state and other two sigma points are chosen to be equidistant from the mean on either side (Figure 7.6).

Behind the operation of the CDKF lies the algorithm called unscented transform. The unscented transform states that the set of chosen sigma points χ when passed through a nonlinear function yield a set of transformed sigma points in a new dimension.

$$\gamma = f(\chi) \tag{7.14}$$

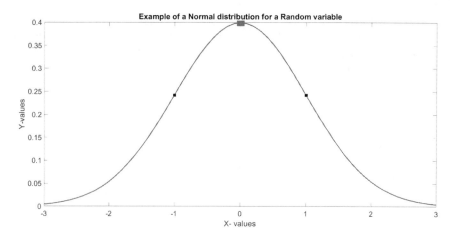

FIGURE 7.6 Example of sigma point selection for a random variable. For single variable under normal distribution, three sigma points are chosen. First sigma point is the mean (red square in the figure), while other two sigma points are chosen on either side of mean (black squares in the figure). The size of the sigma points is proportional to the associated weights, which indicates the influence of the particular sigma point over the output. In this figure, highest weight is given to the mean and equal weights are assigned to the neighbouring sigma points.

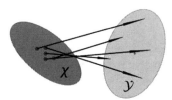

FIGURE 7.7 Pictorial representation of the sigma points. The points in dark gray ellipse denote the original sigma points, while the points in light gray ellipse are transformed sigma points after passing through the nonlinear function $f(x)$. This picture represents how the sigma points are transformed from one space to another via a nonlinear function. The picture is taken from Ref. [14].

where f is the nonlinear function, χ denotes the sigma points in the original dimension, and γ represents the set of transformed sigma points. Later, the algorithm computes the mean and covariance of the transformed sigma points. The mean and covariance become the new estimate [14]. Figure 7.7 gives a pictorial representation of sigma point taken from Ref. [14].

The formulae for sigma point calculation, weights calculation, and weighted mean and covariance calculation along with the associated constraints for sigma points are explained below.

$$\chi = \left[\bar{x}, \bar{x} + h\sqrt{P_x}, \bar{x} - h\sqrt{P_x} \right] \tag{7.15}$$

EV Battery Management Systems 159

$$w_0^m = \frac{h^2 - n}{h^2}$$

$$w_i^m = \frac{1}{2h^2} \qquad (7.16)$$

$$w_0^c = \frac{h^2 - n}{h^2}$$

$$w_i^c = \frac{1}{2h^2}$$

$$\mu = \sum_{i=0}^{2n+1} w_i^m \gamma_i \qquad (7.17)$$

$$\Sigma = \sum_{i=0}^{2n+1} w_i^c (\gamma_i - \mu)(\gamma_i - \mu)^T \qquad (7.18)$$

$$1 = \sum_{i=0}^{2n+1} w_i^m \qquad (7.19)$$

$$1 = \sum_{i=0}^{2n+1} w_i^c \qquad (7.20)$$

where $\bar{x}$ is the average of the state variable x, P_x is the associated covariance, h is the scaling factor in CDKF, μ is the weighted mean of the generated sigma points, Σ is the weighted covariance of the generated sigma points, w_i^m represents i_{th} weight corresponding to i_{th} sigma point, w_i^c represents the i_{th} covariance weight corresponding to i_{th} sigma point, γ_i is the transformed output of the i_{th} sigma point, and n is the number of states.

Equation 7.15 provides the formula for calculating the sigma points for a variable x using its mean $\bar{x}$ and covariance $P_{\{x\}}$. Equation 7.16 gives the formula for calculating the weights for sigma points, which calculate the mean in equation 7.17 and covariance in equation 7.18 of the transformed sigma points. The weighted mean and covariance provide a strong approximation of the actual mean and covariance of the system output. Each sigma point has an associated weight, and equation 7.7 provides the constraint that the sum of all the mean weights associated with the sigma points should be equal to 1. Equation 7.20 provides the same constraint on the sum of all covariances weights associated with sigma points. The weights for mean and covariances can have different values or even the same value, which are calculated empirically using a set of equations given in equation 7.16. Now with this background knowledge, the CDKF can be developed. The prediction step provides the *priori* estimate of the states and covariance, while the measurement update step provides the *posteriori* estimate of the same. These estimates are calculated by incorporating the concept of sigma points, which form the backbone of the CDKF algorithm. The relevant equations associated with CDKF in discrete time are provided below.

Nonlinear state space model

$$x_{k+1} = f_k(x_k, u_k, w_k)$$
$$y_{k+1} = g_k(x_{k+1}, u_{k+1}, v_{k+1}) \tag{7.21}$$

Variable definition

$$x_k^a = \begin{bmatrix} x_k^T, w_k^T, v_k^T \end{bmatrix}^T$$
$$\chi_k^a = \begin{bmatrix} (\chi_k^x)^T, (\chi_k^w)^T, (\chi_k^v)^T \end{bmatrix}^T \tag{7.22}$$

Initial values ($k=0$)

$$\widehat{x_0^{a,+}} = \begin{bmatrix} (\widehat{x_0^+})^T, \overline{w}, \overline{v} \end{bmatrix}^T$$

$$P_{x,0}^{a,+} = \text{diag}\begin{bmatrix} P_{x,0}^+, Q, R \end{bmatrix} \tag{7.23}$$

7.3.1.2.1 Prediction Step
State estimate time update

$$\chi_{k-1}^{a,+} = \widehat{x_{k-1}^{a,+}}, \widehat{x_{k-1}^{a,+}} + h\sqrt{P_{x,k-1}^{a,+}}, \widehat{x_{k-1}^{a,+}} - h\sqrt{P_{x,k-1}^{a,+}}$$
$$\chi_{k,i}^{x,-} = f_{k-1}\left(\chi_{k-1}^{x,+}, u_{k-1}, \chi_{k-1}^{w,+}\right) \tag{7.24}$$

$$\widehat{x_k^-} = \sum_{i=0}^{2n+1} w_i^m * \chi_{k,i}^{x,-}$$

Error covariance update

$$P_{x,k}^- = \sum_{i=0}^{2n+1} w_i^c * \left(\chi_{k,i}^{x,-} - \widehat{x_k^-}\right)\left(\chi_{k,i}^{x,-} - \widehat{x_k^-}\right)^T \tag{7.25}$$

Output estimate

$$Y_{k,i} = g_k\left(\chi_{k,i}^{x,-}, u_k, \chi_{k-1}^{v,+}\right)$$

$$\widehat{y_k} = \sum_{i=0}^{2n+1} w_i^m * Y_{k,i} \tag{7.26}$$

EV Battery Management Systems

Output covariance

$$\sigma_{y,k} = \sum_{i=0}^{2n+1} w_i^c * (Y_{k,i} - \widehat{y_k})(Y_{k,i} - \widehat{y_k})^T \tag{7.27}$$

Cross-covariance

$$\sigma_{\hat{x}\hat{y},k} = \sum_{i=0}^{2n+1} w_i^c * (\chi_{k,i}^{x,-} - \widehat{x_k^-})(Y_{k,i} - \widehat{y_k})^T \tag{7.28}$$

Kalman gain

$$K_k = \sigma_{\hat{x}\hat{y},k} * (\sigma_{y,k})^{-1} \tag{7.29}$$

7.3.1.2.2 Measurement Update Step

State estimate measurement update

$$\widehat{x_k^+} = \widehat{x_k^-} + K_k * (y_k - \widehat{y_k}) \tag{7.30}$$

Error covariance measurement update

$$P_{x,k}^+ = P_{x,k}^- + K_k * \sigma_{y,k} * K_k^T \tag{7.31}$$

where k is the notation for discrete time, x_{k+1} is the state vector at instant k+1, $f\_\{k\}$ is the nonlinear input function, and g_k is the nonlinear output function forming equation 7.21. In equation 7.22, augmented state vector x_k^a and augmented sigma points vector χ_k^a are defined. x_k^a consists of state vector x_k, process noise w_k, and measurement noise v_k. χ_k^a is the vector for augmented sigma points consisting of sigma points x_k^x related to state variables, sigma points χ_k^w related to process noise, and sigma points χ_k^v related to measurement noise. Since CDKF is an iterative algorithm, after defining the variables it is necessary to provide initial values at $k = 0$ for the state and covariance vector. Equation 7.23 provides an expression for initial values of state vector $\widehat{x_0^{a,+}}$ and covariance $P_{x,0}^{a,+}$. $\widehat{x_0^{a,+}}$ consists of initial values of state variables in mean value of $\widehat{x_0^+}$, process noise $\bar{w}$, and mean value of measurement noise $\bar{v}$. Since both the noises are white noises (zero mean and unit standard deviation), both the terms $\bar{w}$ and $\bar{v}$ are taken as zero in expression. The second expression in equation 7.23 computes the covariance matrix for the system consisting of covariance corresponding to state variables $P_{x,0}^+$, covariance for process noise Q and covariance for measurement noise R. This is a diagonal matrix with the different elements placed along the main diagonal.

After the variables are defined and initial values have been provided, the iterative filter can be started. At first, we compute the *priori* state estimate in equation 7.24. This equation consists of three expressions. In the first expression, we calculate the sigma points for the estimate. Recall equation 7.15 provides the expression for calculating sigma points, this expression is essentially like equation 7.15. This expression uses the *posteriori* state estimate $\widehat{x_0^{a,+}}$, *posteriori* covariance estimate $P_{x,0}^{a,+}$ from the previous time instant and scaling factor for CDKF h for calculating augmented sigma points $\chi_{k-1}^{a,+}$. In the second expression, the sigma points $\chi_{k-1}^{x,+}$ and $\chi_{k-1}^{w,+}$ are passed through the nonlinear input function f_{k-1} to provide the *priori* sigma points $\chi_{k,i}^{x,-}$. Finally, these sigma points are multiplied by the corresponding mean weights w_i^m and summed together to generate *priori* state estimate $\widehat{x_k^-}$. Equation 7.25 computes the *priori* covariance estimate $P_{x,k}^-$ on the transformed sigma points using the covariance weights w_i^c and difference between sigma points $\chi_{k,i}^{x,-}$ and state estimate $\widehat{x_k^-}$. This is calculated for each sigma point, multiplied with the respective covariance weight and summed together.

After the *priori* estimates have been calculated using the estimates from previous iteration, it is now the time to correct them using the available measurements. In equation 7.26, output estimate $\widehat{y_k}$ is obtained by first calculating the sigma points for output $Y_{k,i}$ using the transformed sigma point for state $\chi_{k,i}^{x,-}$, sigma points related to measurement noise $\chi_{k-1}^{v,+}$ and control input u_k. These values are passed through the nonlinear output function g_k to yield the sigma points in output space. These sigma points are then multiplied with corresponding mean weights to provide the output estimate $\widehat{y_k}$. This output estimate will serve as a feedback in updating the state at the second step of this algorithm after obtaining Kalman gain. The relevant output covariance $\sigma_{y,k}$ is provided by equation 7.27 using the covariance weights w_i^c, multiplying them with difference between output sigma points $Y_{k,i}$ and output estimate $\widehat{y_k}$. A peculiar feature of SPKF algorithms is the calculation of cross-covariance. The output of equation 7.28 is a cross-covariance matrix $\sigma_{\hat{x}\hat{y},k}$, in which each matrix element shows the correlation between the state and measurements at each time instant. This matrix shows how the output gets affected as the internal state variables change over the different time instants.

The output covariance $\sigma_{y,k}$ and cross-covariance $\sigma_{\hat{x}\hat{y},k}$ are used to calculate the Kalman gain K_k in equation 7.29. This Kalman gain serves as the scaling factor when providing the *posteriori* state estimate $\widehat{x_k^+}$ and *posteriori* covariance estimate $P_{x,k}^+$ in equations 7.30 and 7.31. respectively. These posteriori estimates are again used in the next iteration, and the algorithm runs continuously to provide the desired state and output estimates.

7.3.1.3 Adaptive Extended Kalman Filter

AEKF is an extension of the EKF algorithm with the incorporation of an updating step for process noise covariance matrix (Q) and/or measurement noise covariance matrix (R). The matrix Q represents the covariance of the system noise, and it is very difficult to determine it precisely. The matrix R is the measurement noise covariance and can be determined if the accuracy of the measuring instrument is known. One big issue in designing a KF is the selection of the values of matrices Q and R. Improper choice of Q and R can considerably deteriorate the performance of KF algorithm and in some extreme cases, can even cause it to diverge. Presently, a majority of research works adopt a makeshift approach for determining the values of Q and

EV Battery Management Systems

R matrices. In this method, these matrices are kept constant throughout the iterative state estimation procedure and are manually adjusted by trial-and-error approach outside the loop to obtain a suitable value [15]. Mehra in 1970 [16] published a paper suggesting the suitability of Bayesian approach, correlation matrix approach, covariance matching approach, and/r maximum-likelihood approach for adaptive estimation. Any of these approaches can be used for performing the adaptive Q and/or R estimation. Akhlaghi et al. [15] use the covariance matching approach to include an update step for Q and R matrices in the recursive estimation loop. They propose an *innovation-based approach to adjust Q matrix* and a *residual-based approach to iteratively adjust R matrix*. The algorithm is similar in functioning to EKF along with two additional steps to update Q and R. The innovation is defined as the difference between the measured value and the predicted value before the state is updated, whereas the residual is defined as the difference between the measured value and the predicted value after the state has been updated inside the loop. With the help of an example, they find that AEKF is more accurate than conventional EKF and simultaneously more robust to improper initial values of Q and R. Since this method recursively updates Q and R in a loop, the designer must no longer worry about using trial and error to select an appropriate initial value for Q and R.

In this chapter, the explained AEKF uses only the adaptive estimation of Q matrix. The matrix R is taken as a constant throughout the run time of the loop. The study has implemented only the adaptive estimation of Q because of the filter stability. During the implementation of the adaptive steps of Q and R simultaneously, they had problems trying to stabilize the filter and therefore opted to implement only the adaptive step for process noise covariance matrix, Q, in this study for battery SOC estimation. The adaptive step for Q is implemented using the maximum-likelihood estimation. This approach defines some adaptive parameters that are used with the conventional EKF estimation [17]. The filter stores the last N values of Q and processes the data at each time step $k \leq N$. Furthermore, the Q matrix is considered completely stationary over any given interval of N sample periods, and tuning is required for finding the suitable size of the moving window to achieve closed loop convergence [18]. The relevant equations describing the AEKF in discrete time are provided below.

7.3.1.3.1 Prediction Step

State estimate time update

$$\widehat{X_{k+1|k}} = A * X_k + B * u_k \qquad (7.32)$$

Error covariance time update

$$P_{k+1|k} = A * P_{k|k} * A^T + Q \qquad (7.33)$$

Measurement error calculation

$$e_{k+1|k} = y_{k+1|k} - y\left(\widehat{X_{k+1|k}}, u_k\right) \qquad (7.34)$$

Kalman gain

$$K_{k+1|k} = P_{k+1|k} * C^T * (C * P_{k+1|k} * C^T + R)^{-1} \quad (7.35)$$

7.3.1.3.2 Measurement Update Step
State estimate measurement update

$$\widehat{X_{k+1|k+1}} = \widehat{X_{k+1|k}} + K_{k+1|k} * e_{k+1|k} \quad (7.36)$$

Error covariance measurement update

$$P_{k+1|k+1} = (I - (K_{k+1|k} * C)) * P_{k+1|k} \quad (7.37)$$

Residual calculation

$$\text{Res}_{k+1|k+1} = y_{k+1|k+1} - y(\widehat{X_{k+1|k+1}}, u_k) \quad (7.38)$$

Moving window estimation

$$H_{k+1|k+1} = \frac{\text{Res}_{k+1|k+1} * \text{Res } e_{k+1|k+1}}{N} \quad (7.39)$$

Process noise covariance update

$$Q_{k+1|k+1} = K_{k+1|k} * H_{k+1|k+1} * K_{k+1|k} \quad (7.40)$$

where $\hat{X}$ is the state vector of the order $n \times 1$ with n being the number of states, u_k is the input matrix, P is the covariance corresponding to system state, Q is the covariance corresponding to system noise, R is the covariance corresponding to measurement noise, y is the system output, and e represents the error between the real system output and model output, respectively. Covariance P, which is a square matrix, represents the uncertainty in the system state. Depending on the dimensions of the system, this can be a scalar or a vector. N is the size of the moving window, Res is the residual which is the difference between the real system output and the model output after the state has been updated through Kalman gain, and H is the output of the moving window estimation. Equations 7.32–7.37 are the same as the equations 7.6–7.11 used in EKF. The description of these equations has been provided in the section of EKF and therefore, will not be provided here again. The additional equations from equation 7.38 to 7.40 describe the implementation of the adaptive step for process noise covariance Q. Equation 7.38 calculates the residual, which gives the difference between the real measurement and the estimate value from the filter at that particular time step. The residual helps to analyse the offset in the filter output at the given time step and is used to update Q later. In equation 7.39, the moving average filter is applied to the stored values of Q to obtain a smoothed output. The moving average filter provides an average single value over the N stored samples. Equation 7.40 provides the updated value of Q using the output of the moving window and the Kalman gain. The multiplication of Kalman gain with the moving window output scales the value of Q to bring it closer to its theoretical value, thus increasing the accuracy of the estimator.

7.3.2 Sliding Mode Observer

The structure of the SMO is like any other state observer; however, in its estimations, it is highly robust against errors induced in the system during modelling and the system parameter variations. The robustness of the SMO can be attributed to the fact that it generates a sliding motion on the error between the measured plant output and the output of the observer, ensuring that the produced state estimates are precisely commensurate with the real plant output [19]. This sliding motion is obtained by inserting a nonlinear discontinuous term (switching term) into the observing system, depending upon the output error. This discontinuous injection is designed in such a way that the trajectories of the system are forced to remain on some sliding surface in the error space. This resulting motion is referred to as sliding mode [20]. The presence of the discontinuous term allows the system to reject disturbances, and the classes of mismatch between the actual system and the model can be used for design [20]. However, this discontinuous term that produces the sliding mode can cause fast oscillations in the modelled system dynamics. These fast oscillations are termed as 'chattering effect' [21]. The chattering effect is observed in the estimated states as the fast oscillations that cause the estimations to deviate from the actual values. Higher the magnitude of the switching term, greater are the oscillations in the modelled output. Therefore, the methods of chattering suppression are developed so that the magnitude is decreased properly holding the establishment of sliding mode. First option is to decrease the magnitude along with the system states [22]. The second one implies that the magnitude is the function of an equivalent control derived by a low-pass filter. The method can be applied for the plants subjected to unknown disturbances [22].

Figure 7.8 represents the structure of a SMO in discrete time. The upper part of the diagram is the discrete time state space representation of the real system. The lower part of the diagram represents the modelled system in real time. The gain of the observer, H, is obtained using the difference between the real system output and the modelled system output. Furthermore, the block representing the term 'sliding motion' represents the discontinuous switching term that provides the sliding motion for the model. This switching term is a combination of the switching gain multiplied by the uncertainty input matrix and the Lyapunov function. These terms in combination with the system matrix, A, are used to obtain the state estimate, $xhat(k)$. The state estimate is then applied to the output matrix, C, to get the estimated output, $yhat(k)$. The difference between the real and estimated output is calculated and the algorithm operates iteratively.

The switching term which introduces a sliding motion is basically a signum function, which is discontinuous in nature and switches in zero time (infinite frequency). In case of presence of noise in signals, the sliding mode observer can reduce the output estimation error to zero in finite time. This robustness of the observer is due to the strong mathematical design, which allows it to perform strongly even in the presence

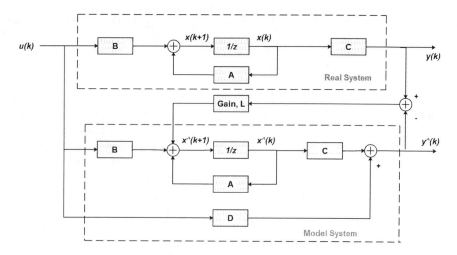

FIGURE 7.8 Sliding mode observer implementation structure in discrete time. The model system (system at the bottom) and the real system (system at the top) are shown in parallel. The observer design requires an accurately developed and parametrized state model.

of any disturbances. Furthermore, the strict condition of observability is not mandatory for the SMO. The discontinuous switching function is defined in equation 7.41.

$$v = \rho \text{sign}(e_y) \qquad (7.41)$$

where v is the discontinuous switching term, ρ is a positive scalar, and e_y is the error between the observer output and the real system output. The signum function takes in e_y as its input argument and depending on the sign of e_y gives the output. When e_y is positive, output is 1; output is zero when e_y is zero; and output is -1 when e_y is negative (Figure 7.9).

The signum function introduces a discontinuity in the system, which leads to a phenomenon known as chattering. The problem of chattering is quite common in SMO, and it appears as noise in the estimated states. In practical control systems, this induced noise can come across as a serious problem that could undermine the accuracy of the designed controller or observer. The solution for this problem can be provided by a sigmoid function. The switching term v is now represented by equation 7.42.

$$v = \rho \frac{e_y}{|e_y| + \delta} \qquad (7.42)$$

where δ is a small positive scalar. The switching term in equation 7.42 when used in the SMO can solve the problem of unwanted noise; however, the performance of the system is unaffected by this modification to the switching term. The SMO gets its name from the sliding motion that occurs on the sliding surface when the error in the output estimation e_y goes to zero. The e_y is referred to as the sliding variable. When

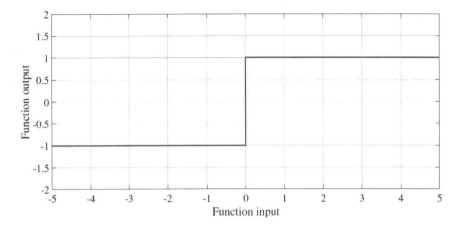

FIGURE 7.9 Example showing the output of a signum function. In this example, the value of input varies between −5 and +5. When the input is −5, the output is 1. If the input is 0, the output is 0; when the input is +5, the output is 1.

the value of e_y becomes zero, the sliding motion begins. The error e_y slides along the surface, which is defined by the equation of e_y. When the sliding motion occurs, the error e_y continues its movement on the surface and stays there. The switching term v oscillates along the vertical line defined by the signum function. With the error in output estimation is reduced to zero, the observer output converges to the real system output and the error in state estimation also reaches a very small value. The size of the sliding surface can be adjusted by the value of ρ. Higher the value of ρ, larger the area of sliding surface. However, this also increases the problem of chattering. Therefore, a trade-off between the accuracy of observer and the severity of chattering needs to be made when selecting the value of ρ.

Equations 7.41 and 7.42 will be used in the observer that has been designed for performing SOC estimation iteratively. Additionally, it is required that starting values of the states and control input should be provided during the observer initialization. These starting values will be utilized in the first iteration of SOC estimation. The relevant equations in discrete time that are used to design SMO for SOC estimation are described below.

Nonlinear state space model

$$x_{k+1} = f_k(x_k, u_k, w_k)$$

$$y_{k+1} = g_k(x_{k+1}, u_{k+1}, v_{k+1}) \tag{7.43}$$

Initial values ($k = 0$)

$$\widehat{x_0} = \left[\left(\widehat{x_0} \right)^T \right] \tag{7.44}$$

Parameters for solving Riccati equation

$$R = 1$$

$$Q = \begin{bmatrix} 1 & 0 & 0 \\ 0 & 1 & 0 \\ 0 & 0 & 1 \end{bmatrix} \quad (7.45)$$

Parameters for solving Lyapunov equation

$$W = 1$$

$$Q_f = \begin{bmatrix} 1 & 0 & 0 \\ 0 & 1 & 0 \\ 0 & 0 & 1 \end{bmatrix} \quad (7.46)$$

Observer output

$$\widehat{y_{k+1}} = g_k(x_k, u_k) \quad (7.47)$$

Measurement error calculation

$$e_{k+1} = y_{k+1} - \widehat{y_{k+1}} \quad (7.48)$$

Solution to Riccati equation

$$P_k = -Q/(A + A^T - C^T * R^{-1} * C) \quad (7.49)$$

Observer gain matrix

$$H_k = (R^{-1} * C * P_k)^T \quad (7.50)$$

Solution to Lyapunov stability equation

$$P_{f,k} = -Q_f * ((A - H*C) + (A - H*C)^T) \quad (7.51)$$

Model uncertainty factor

$$\gamma_k = (W * C * P_f^{-1})^T \quad (7.52)$$

State estimate measurement update

$$\widehat{x_{k+1}} = A * \widehat{x_k} + B * \widehat{u_k} - H_k * (y_{k+1} - \widehat{y_{k+1}}) + \rho * \gamma * \text{sign}(y_{k+1} - \widehat{y_{k+1}}) \quad (7.53)$$

Equation 7.43 describes the structure of the nonlinear system in discrete time. Equation 7.44 provides the initial values of the states for the first iteration.

EV Battery Management Systems

These values should be known and provided accurately as a model-based estimation process is affected strongly by the starting values. In order to obtain the gain matrices for the SMO state estimate equation in 7.53, we use the Riccati equation. The constants for Riccati equation are R and Q that are provided default values in equation 7.45. The asymptotic stability of the designed SMO in state estimation is proved by Lyapunov stability. Therefore, it is necessary to define the constant for solving the Lyapunov equation after a suitable candidate function has been selected. The constants W and Q_f for Lyapunov stability equation are provided in equation 7.46. Equations 7.44–7.46 are defined outside the iterative loop for SOC estimation. The first step inside the recursive loop is the calculation of the observer output. The observer output is dependent on the control input and the present value of the system states. These are passed through the nonlinear function g_k to get the observer output for the current iteration given by equation 7.47. After observer output has been calculated, the error in output estimation needs to be found. Equation 7.48 gives the formula for output estimation error, which will be fed back to the observer at the end of each iteration. Equation 7.49 provides the solution to the Riccati equation. The solution involves the use of system matrix A and the output matrix C. The solution of Riccati equation is a $n \times m$ matrix, where n is the number of rows of A and m is the number of columns of C. If the denominator is large, then matrix P has a small determinant, and when the denominator is small, the matrix P has a large determinant. The value strongly affects the value of observer gain matrix H in equation 7.50, and when P is small, the resulting gain is small. However, with larger values of P, observer gain matrix H has a higher value. After having calculated the observer gain matrix H, it is important to check the stability of the designed observer by the Lyapunov stability equation. The selected candidate Lyapunov function is provided in equation 7.54.

$$V = e^T * P * e \qquad (7.54)$$

This candidate function is differentiated with respect to time to find its derivative and check whether the observer error e_y slides along the surface. If the condition is satisfied, then the sliding motion occurs when e_y goes to zero; however, when this is not satisfied, other candidate functions can be chosen to make e_y slide along the surface. After having checked for the sliding condition, the solution to the Lyapunov stability equation can be found using equation 7.51. This solution is used to consider the uncertainties associated with the model using the matrix γ calculated in equation 7.52. Since the observer considers all these uncertainties within the model itself, it is more robust to the noise/disturbance present within the system states and the control input. After having calculated both the gain matrices, H and γ, to account for observer gain and model uncertainty, these can now be used to provide the state estimate measurement update in equation 7.53. The third term in equation 7.53 scales the error with the observer gain and provides a linear effect on the system state. However, the fourth term multiplies the uncertainty with the discontinuous switching function to show the sliding effect when the error converges to zero. After the state has been updated, the recursive loop starts all over again from equation 7.47 and continues until the maximum number of iterations are reached.

7.3.3 Backpropagation Neural Network

Backpropagation Neural Networks (BPNN) are one of the most widely used neural network architectures; with the intuition that the error between actual output and network output is propagated backwards from output towards input. BPNNs are capable of accurately mapping any nonlinear input–output relationship and can be used for both regressions and classification problems. BPNNs use the learning principle of error backpropagation.

Figure 7.10 shows the signal propagation in a BPNN. The forward signal (blue) refers to the signal that propagates in the forward direction from input layer and generates an output by passing thorough neurons and undergoing transformation by means of weights and activation functions. The error signal (red) represents the offset between the real output and the network output. Since we have the knowledge of real output and not the intermediate output from different layers in a BPNN, the error needs to be propagated in a backward direction in order to adjust the weights and biases. The weights and biases are adjusted by means of a learning technique called gradient descent. The network can be designed to operate for offline as well as online applications.

BPNN presents a novel data-based approach to SOC estimation problem. The aforementioned methods require the preparation of a model in state space in order to perform SOC estimation. However, the BPNN does not require any model and uses the measurement data for SOC estimation. The data from the measurement is filtered and normalized before it is fed into the BPNN. This normalization is necessary as different inputs to the network have different range of values. These different ranges

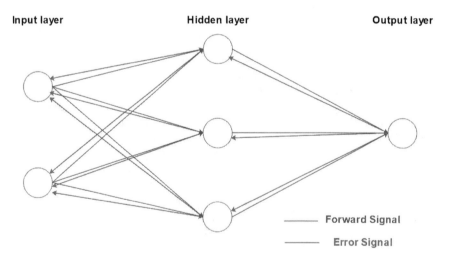

FIGURE 7.10 Forward signal and error signal flow in a BPNN. The circles represent the neurons in different layers. Arrows in forward direction indicate the movement of the forward signal from the input layer to the output layer, while arrows in backward direction indicate the movement of error signal from the output back towards the input. The movement of error signal from output towards input is termed as 'backpropagation'.

mean that inputs have different scales. Due to different scales of inputs, the network performance will be slow and will take a longer time to converge. After normalization has been done, all the different inputs to the network lie within the range −1 to +1 for the whole dataset. The normalization is performed outside of the iterative estimation loop. In addition to normalization, the network configuration needs to be defined before the training begins. The number of layers in the network, the number of neurons in each layer, the activation function for each neuron in these layers, the learning rate for the network, and the error threshold for training need to be predefined. This network configuration is stored and will be subsequently used when we test the network in the absence of a real system output.

The training is performed using an optimization algorithm, which allows the network to reach a global minimum solution. This optimization algorithm corrects the weights and output associated with each neuron through a set of mathematical equations. The most commonly used optimization algorithms include the gradient descent, scaled conjugate gradient descent, and Levenberg–Marquardt, among others. In this study, stochastic gradient descent (SGD) algorithm is implemented for training the network. Gradient descent aims to achieve the global minimum value of the cost function by iteratively updating its parameters and checking the function value at those parameters. The cost function in neural networks is given by calculating the squared difference between the real value and the predicted network output over the total number of used samples. Gradient descent is the algorithm that is applied to this cost function in order to minimize it by obtaining the optimal values of the variables that constitute the expression of this cost function. The gradient descent operates by calculating the partial derivative (gradient) of the cost function with respect to each of the parameters. If the gradient is negative implying that the cost function is being minimized, then we continue in this same direction to arrive at optimal parameter values. The partial derivative is calculated iteratively for each sample, and after reaching an error threshold defined by the user, the algorithm stops and does not minimize any further. This point will be the global minimum of the function, and the optimal parameter values will be obtained. The three forms of gradient descent algorithm are batch gradient descent, mini-batch gradient descent, and SGD. The fundamental equations are the same for each of three methods; however, the main difference lies in the quantity of data samples used to compute the gradients in each step. Here, only the description of SGD will be provided. In SGD, the gradient is calculated, and the parameters are updated for each sample. Therefore, the training of weights and neurons happen at each iteration in the network. The samples for training are usually selected at random from the whole dataset and fed into the network. The cost function is calculated for each sample in every iteration; however, under such conditions, the time to converge to a global minimum can be longer and the path can contain more deviations before reaching the minima.

The rate at which the parameters are updated in each iteration of gradient descent algorithm is defined by the learning rate. In this study, the learning rate for the network is defined as a constant parameter with a value between 0 and 1. The learning rate updates the weights of the network in each iteration so as to achieve the optimal value of weights that minimize the loss function. Smaller values of learning rates provide small updates to the weights, leading to slower decay of the loss function towards the global minimum. This implies that a network with a very small learning rate will require more updates to the weights during training before weights reach an optimal value; on the other hand, when the learning rate is too high, the updates to the weights are quite large. However, under these conditions, it could happen that the weights might skip the optimal value and the loss network instead of decaying starts to rise. Therefore, an optimal value of learning rate needs to be chosen when performing training of the network. This learning rate and the network weights are saved so that they can be used when the network is being tested against an unknown input.

The BPNN gets its name from the implemented backpropagation algorithm, which propagates the error in backward direction from the output layer towards the input layer. After the network output has been obtained the difference between the network output and the real output (reference), the error is calculated. This error is propagated backwards to update the weights and output of neurons. The weights of the final layer are updated first, and the weights of the first layer are updated last. In this network, a concept of backpropagation is implemented with the SGD algorithm. This means that the gradients corresponding to the weights and neurons in the last layer will be calculated first, and then, this will be propagated towards the first layer. The gradients of one layer are used in computing the gradients of the previous layer. This connected approach in gradient computation allows the efficient exchange of information between the different layers of the network. The set of relevant equations used to implement BPNN in this study are provided below.

Network parameter initialization

$$I = \text{Number of inputs to the network}$$

$$O = \text{Number of outputs from the network}$$

$$HN = \text{Number of hidden layer neurons}$$

$$\eta = \text{Learning rate}$$

$$\lambda = \text{Regularization parameter}$$

$$\text{Error threshold} = \text{User defined}$$

$$\text{Total layers} = \begin{bmatrix} I\,HN\,O \end{bmatrix}$$

Network weights and biases initialization

$$WH_1 = 0.01 * \text{randn}(HN, I)$$

$$W_O = 0.01 * \text{randn}(O, HN)$$

$$B_1 = 0.01 * \text{randn}(HN, 1)$$

$$B_O = 0.01 * \text{randn}(O, 1) \tag{7.55}$$

EV Battery Management Systems

$$\text{Delta } WH_1 = 0.01 * \text{randn}(HN, I)$$

$$\text{Delta } W_O = 0.01 * \text{randn}(O, HN)$$

$$\text{Delta } B_1 = 0.01 * \text{randn}(HN, 1)$$

$$\text{Delta } B_O = 0.01 * \text{randn}(O, 1)$$

where WH_1 is the matrix of weights for hidden-layer neurons, W_O is the matrix of weights for output-layer neurons, B_1 is the matrix of biases for the hidden-layer neurons, B_O is the matrix of bias for the output-layer neuron, Delta WH_1 is the matrix of regularization coefficients for hidden-layer neurons, Delta W_O is the matrix of regularization coefficients for weights of the output-layer neurons, Delta B_1 is the matrix of regularization coefficients for the hidden-layer neurons, and Delta B_O is the matrix of regularization coefficients for the output-layer neurons. The values initialized in equation 7.55 will be used outside the recursive loop to calculate the gradients for the first sample. *randn* is the command in MATLAB to generate random numbers between 0 and 1. This is multiplied by 0.01 to make the coefficients even smaller as higher initial values may affect the values, which will be calculated subsequently inside the loop.

7.3.3.1 Forward Propagation

Output of each neuron of the hidden-layer unit

$$z_{HN,1} = \sum_{i=1}^{HN} WH1_{i,I} * a_I + B_{i,1} * 1.0$$

$$y_{HN,1} = -1 + 2/(1 + \exp(-2 * z_{HN,1})) \tag{7.56}$$

Network output

$$z_o = \sum_{i=1}^{O} WH1_{i,I} * a_{HN} + B_{i,1} * 1.0$$

$$y_o = -1 + 2/(1 + \exp(-2 * z_o)) \tag{7.57}$$

Estimation error calculation

$$E = y_m - y_o \tag{7.58}$$

Cost function calculation

$$J = 1/M \sum_{k=1}^{M} (y_m - y_o)^2 \tag{7.59}$$

7.3.3.2 Backward Propagation

Regularization for weights of hidden and output layers

$$RegWH_1 = \lambda*DeltaWH_1$$

$$RegW_O = \lambda*DeltaW_O \qquad (7.60)$$

Output-layer error calculation

$$\delta_{out} = \frac{\partial J}{\partial y_o}*(1+y_o)*(1-y_o) \qquad (7.61)$$

Hidden-layer error calculation for each neuron

$$\delta_H = W_{o,i}^T*\delta_{out}*(1+y_{HN,i})*(1-y_{HN,i}) \qquad (7.62)$$

Gradient of a single output-layer weight

$$Grad_o = y_{HN,i}*\delta_{out} \qquad (7.63)$$

Gradient of a single hidden-layer weight

$$Grad_H = a_{I,i}*\delta_H \qquad (7.64)$$

Gradient of output-layer bias

$$Grad_{B,O} = 1.0*\delta_{out} \qquad (7.65)$$

Gradient of hidden-layer bias

$$Grad_{B,H} = 1.0*\delta_H \qquad (7.66)$$

Updating weights and bias matrices

$$W_O = W_O - eta*Grad_o - RegW_O$$

$$WH_1 = WH_1 - eta*Grad_H - RegWH_1$$

$$B_1 = B_1 - eta*Grad_{B,H} \qquad (7.67)$$

$$B_O = B_O - eta*Grad_{B,O}$$

Equations 7.56–7.67 represent the set of equations used to implement the BPNN. The first step is the forward propagation, which is then followed by the backward propagation. Equation 7.56 consists of two equations. The first equation is simply the sum of the product of the neuron output and the weight that originates from that neuron. This equation gives the sum of all the weights that are connected to the first neuron in the hidden layer from each neuron in the first layer. A bias term $B_{i,1}$ is also added as each layer has an associated bias neuron except the first layer. This bias has a constant value of 1 and has an associated weight. After getting the sum of weights and biases, the output of the first equation is fed into the second equation. The second equation is the activation part of a neuron where the summation is fed into an activation function (the activation function used here is the hyperbolic tangent sigmoid function). This step gives the output of each neuron from the hidden layer. This small loop is run for each hidden-layer neuron, and the results are stored in an array. Equation 7.57 is similar to the previous equation. This equation gives the predicted output from the neural network. After we have an output from neural network y_o, it is then subtracted from the reference value from the measurement dataset y_m. Furthermore, in equation 7.59, we calculate the cost function of this problem, which needs to be optimized. After calculating the cost function, the forward propagation step is completed. Now we start with backpropagation. In this step, there are fundamentally four equations. The first equation is for calculating the error associated with each neuron output, which is essentially the deviation of a neuron output from the actual value that it should have. This is first done for the output layer and then subsequently backpropagated for the remaining layers except the input layer. The second equation is the calculation of the gradients with respect to weights of the network. Gradient is the rate of change of cost function with respect to a weight in the network. Third equation is the gradient for the biases in the network. Lastly, after calculating the gradients, the existing weights and biases are updated and algorithm runs all over again with these updated values in the next iteration. Equation 7.60 calculates the matrices for applying regularization to the updates of weights and biases. Regularization is performed to avoid overfitting of the network to the output. Regularization reduces the values in the weight matrix, which cause a reduced summation part in the neuron output, leading to a reduction in the impact of the activation function on the neuron output in the forward propagation step. This will tackle the problem of overfitting. The regularization is controlled by the parameter λ, which is constant and tunable. Higher values of λ imply a stronger effect of regulation, while smaller values of λ correspond to a smaller regularization effect. Equation 7.61 calculates the error δ_{out} associated with the output neuron. The first term in the equation shows the change of the cost function with respect to the output neuron. If the cost function does not strongly depend on the output neuron, then the resulting δ_{out} will be smaller. The second term is simply the derivative of the activation function at the neuron output. Equation 7.62 shows the implementation of the backpropagation principle. The δ_{out} from the output neuron is used to calculate the error associated with each neuron in hidden layer. The formulae for calculating the error associated with neurons in the output layer and hidden layer are different. In this step, the error is propagated backwards in the network, which gives us the intuition about error in the output of neurons in a particular layer. After the errors associated with neurons have been calculated, we move towards calculating the gradients for weights in the network. Equation 7.63 shows the formula for calculating gradient for one weight of the output layer. This is simply the product of the output of the neuron in the previous layer from which the weight originates and the delta of the neuron to which it connects. This formula is repeated to obtain the gradient corresponding to all the weights connected to the output-layer neuron. If the neuron output is small, then the gradient will also be small. Smaller gradient implies that the weights learn slowly and do not change too much in one step. Equation 7.64 shows the gradient for the one weight of hidden layer. Similarly, equations 7.65 and 7.66 provide the formulae for calculating gradients associated with biases in the network. After the gradients have been calculated, the weights and biases are updated using the gradients multiplied by the learning rate and regularization term. These two terms are subtracted from the weights to bring the weights closer to the optimal value with each iteration. These values are then used as the initial values in the next iteration, and the network operates until a global minimum solution is reached.

7.4 CONCLUSION

This chapter explains the underlying mathematics behind the five most widely used methods for battery SOC estimation. The relevant mathematical equations for each model are explained in detail and serve as a base for developing these algorithms. These methods after being developed can be implemented on a BMS with ease to perform the desired estimation. All these methods are quite robust in their design and have high accuracies. However, when implementing on a BMS, the computation time, complexity, and memory requirements must also be kept in mind. The implementation of a BMS would require that the code of the selected algorithm based on user requirements be written in C language. This can be done in two ways: One would be to rewrite the whole code in C and the other would be to write the code in MATLAB and then use the inbuilt MATLAB C-coder. Both methods would work fine and will give the desired results when implemented correctly.

One possibility that arises from this work is the estimation of further states in the battery. With the results of SOC estimation and a chosen algorithm, techniques to evaluate SOH, state of power (SOP), and SOE can be developed. The algorithms to further the problem of state estimation in a battery can also be implemented in real time.

Another possibility would be to further analyse the implemented algorithms, and attempts should be made to dissect their structure. The in-depth analysis of complex mathematical equations forming the algorithm could be done to understand the effect of altering a single step or a single variable on the result. This would be an interesting research topic for enthusiasts, engineers, and researchers with a background in mathematics and electrical engineering.

7.5 FRAMEWORK FOR INTEGRATING EV ENERGY STORAGE SYSTEMS

The energy storage systems form an integral part of the electrification process. The widespread adoption of EVs to replace the ICE vehicles depends to a large extent on the availability of the electrical energy storage systems. The electrical energy storage systems in the EVs make it possible to transfer electrical energy bidirectionally between vehicle and grid, vehicle and home etc. The available energy in any electrical energy storage system is generally estimated by means of some mathematical algorithms. In case of batteries, the present energy in battery is interpreted in terms of the remaining charge. The SOC of the battery is estimated in the background by means of different algorithms whose complexity depends on the level of precision demanded and the feasibility of implementation in the BMS. The BMS has its own engine control unit (ECU) inside the vehicle where it communicates back and forth with the different ECUs over the CAN bus.

The algorithms for SOC estimation have evolved over the period. Earlier simpler methods like the internal resistance method were used to get an idea of the SOC, where the value of internal resistance was measured and SOC was estimated based on it. Nowadays, a wide range of mathematical algorithms similar and even more advanced to the ones mentioned in this chapter are being implemented as well. This can also be attributed to the increased capabilities of the processors and controllers on which these complex algorithms are implemented. This would also imply increased costs of the whole estimation process. Although the algorithms are costly and complex, the accuracy and precision of estimation make them worthy of implementation. Researchers all over the world are also looking to implement hybrid methods wherein they combine two different methods and try to make them work together. This approach is even more complex but is not applicable in all the cases, as the increment in accuracy of estimation is not very much when either of the two methods is implemented independently. Therefore, for the industry standard, only one mathematical method is developed in software and then implemented on the BMS.

Along with the implementation of SOC algorithm, BMS also monitors the temperature, voltage level, and current level across each cell and the whole module as well. Since a battery is made up of several different modules, each module has its own BMS and these BMSs exchange information with a central BMS in a battery pack to monitor the essential variables at all levels. Since a battery pack consists of thousands of cells arranged in series and/or parallel combinations, the task of these controllers becomes more complex. There have been incidents when the cells got overheated and went into thermal runaway state causing the whole battery to catch fire. Incidents have also occurred where the battery caught fire due to faulty wirings.

Since in an EV the battery forms the heart of electrical energy transfer, a lot of precautions need to be taken while designing the battery module and integrating it into the whole vehicle. Measures also need to be taken to ensure that there is no fault on the software level and that the implemented software is working as per the requirements. Also, as the battery is integrated into the EV, test and trial runs need to be conducted thoroughly taking into account every possible test case to avoid these incidents from happening in the near future.

As the EVs become more and more popular, the whole infrastructure surrounding the EVs also needs to be developed. Proper quality and safety assurances need to be met according to the specified standards to ensure a safe and reliable operation of the whole EV ecosystem. In the near future, EVs would see the implementation of more sophisticated and advanced technologies, which would raise the safety and quality standards to make the whole process more secure and steadfast.

NOMENCLATURE

AEKF:	Adaptive extended Kalman filter
Ah:	Ampere hour
ANN:	Artificial neural network
BEVs:	Battery electric vehicles
BPNN:	Backpropagation neural network
CAN:	Controller area network
CDKF:	Central difference Kalman filter
EKF:	Extended Kalman filter
EVs:	Electric vehicles
HEVs:	Hybrid electric vehicles
ICE:	Internal combustion engine
LKF:	Linear Kalman filter
LO:	Luenberger observer
MOSFET:	Metal oxide semiconductor field effect transistor
OCV:	Open-circuit voltage
PHEVs:	Plug-in hybrid electric vehicles
SMO:	Sliding mode observer
SOC:	State of charge
SOE:	State of energy
SOF:	State of function
SOH:	State of health
SOP:	State of power
SPKF:	Sigma point Kalman filter
UKF:	Unscented Kalman filter

REFERENCES

1. S. Dhameja. Electric vehicle battery systems. In *Electric Vehicle Battery Systems*, Newnes, Woburn, 2002.
2. Marie-Therese von Srbik, *Advanced Lithium-Ion Battery Modelling for Automotive Applications*, PhD thesis, Imperial College London, 2015.
3. V. Prajapati, H. Hess, E. J. William, V. Gupta, M. Hu, M. Manic, F. Rufus, A. Thakker, and J. Govar, A literature review of state of-charge estimation techniques applicable to lithium poly-carbon monofluoride (li/cfx) battery, In *India International Conference on Power Electronics 2010* (IICPE2010), pp. 1–8, Jan 2011.
4. P. Shen, M. Ouyang, X. Han, X. Feng, L. Lu, and J. Li, Error analysis of the model-based state-of-charge observer for lithium-ion batteries, *IEEE Transactions on Vehicular Technology*, 67(9), 8055–8064, Sep. 2018.
5. P. Shrivastava, T. K. Soon, M. Y. I. B. Idris, and S. Mekhilef, Overview of model-based online state-of-charge estimation using Kalman Filter family for lithium-ion batteries, *Renewable and Sustainable Energy Reviews*, 113, 109233, 2019. https://doi.org/10.1016/j.rser.2019.06.040.
6. G. L. Plett, Extended Kalman filtering for battery management systems of LiPB-based HEV battery packs part 1. Background, In *Extended Kalman Filtering for Battery Management Systems of LiPB-Based HEV Battery Packs Part 1*, 134, 252–261, 2004. https://doi.org/10.1016/j.jpowsour.2004.02.031.

7. K. S. Ng, C.S. Moo, Y. p. Chen, and Y.-C. Hsieh, Enhanced coulomb counting method for estimating state-of-charge and state-of-health of lithium-ion batteries, In *Enhanced Coulomb Counting Method*, 86, 1506–1511, 2009. https://doi.org/10.1016/j.apenergy.2008.11.021.
8. L. Zheng, L. Zhang, J. Zhu, G. Wang, and J. Jiang, Co- estimation of state-of-charge, capacity and resistance for lithium-ion batteries based on a high-fidelity electrochemical model, In *Co-Estimation of State-of-Charge*, 180, 424–434, 2016. https://doi.org/10.1016/j.apenergy.2016.08.016.
9. B. R. Pattipati, C. Sankavaram, and K. R. Pattipati, System identification and estimation framework for pivotal automotive battery management system characteristics, *IEEE Transactions on Systems, Man, and Cybernetics, Part C (Applications and Reviews)*, 41, pp. 869–884, 2011.
10. S. C. Piller, M. Perrin, and A. Jossen, Methods for state-of-charge determination and their applications, In *Methods for State-of-Charge Determination and Apps*, 96, 113–120, 2001. https://doi.org/10.1016/S0378-7753(01)00560-2.
11. V. Pop, H. J. Bergveld, P. H. Notten, and P. P. Regtien, State-of-the-art of battery state-of-charge determination, In *State-of-the-Art of battery*, 16(12), R93–R110, 2005. https://doi.org/10.1088/0957-0233/16/12/R01.
12. G. L. Plett, Sigma-point Kalman filtering for battery management systems of LiPB-based HEV battery packs: Part 1: Introduction and state estimation, In *Sigma Point Kalman Filtering for Battery Management Systems of LiPB-Based HEV Battery Packs Part 1*, 161, 1356–1368, 2006. https://doi.org/10.1016/j.jpowsour.2006.06.003.
13. E. A. Wan, Sigma-point Fillters: An overview with applications to integrated navigation and vision assisted control, *2006 IEEE Nonlinear Statistical Signal Processing Workshop*, pp. 201–202, 2006, Cambridge, UK. doi: 10.1109/NSSPW.2006.4378854.
14. R. van der Merwe, E. A. Wan, and S. I. Julier, Sigma-point Kalman filters for nonlinear estimation and sensor-fusion: Applications to integrated navigation, In *Sigma-Point Kalman Filters for Nonlinear Estimation and Sensor Fusion*, 2004. AIAA 2004-5120. AIAA Guidance, Navigation, and Control Conference and Exhibit.
15. S. Akhlaghi, N. Zhou, and Z. Huang, Adaptive adjustment of noise covariance in Kalman filter for dynamic state estimation, In *2017 IEEE Power Energy Society General Meeting*, pp. 1–5, July 2017, Chicago, USA. doi: 10.1109/PESGM.2017.8273755.
16. R. Mehra, On the identification of variances and adaptive Kalman filtering, *IEEE Transactions on Automatic Control*, 15(2), pp. 175–184, April 1970.
17. C. Taborelli and S. Onori, State of charge estimation using extended Kalman filters for battery management system, In *2014 IEEE International Electric Vehicle Conference (IEVC)*, pp. 1–8, Dec 2014, Florence, Italy. doi: 10.1109/IEVC.2014.705612.
18. J. G. Silva, J. O. d. A. Limaverde Filho, and E. L. F. Fortaleza, Adaptive extended Kalman filter using exponential moving average, In *Adaptive Extended Kalman Filter using Exponential*, 51(25), 208–211, 2018. https://doi.org/10.1016/j.ifacol.2018.11.106.
19. S. K. Spurgeon, Sliding mode observers: A survey, *International Journal of Systems Science*, 39: 751–764, 2008.
20. J. A. Moreno and M. Osorio, A lyapunov approach to second order sliding mode controllers and observers, *2008 47th IEEE Conference on Decision and Control*, pp. 2856–2861, 2008, Cancun, Mexico. doi: 10.1109/CDC.2008.4739356.
21. I. Boiko, L. M. Fridman, and M. I. Castellanos, Analysis of second - order sliding-mode algorithms in the frequency domain, *IEEE Transactions on Automatic Control*, 49: 946–950, 2004.
22. V. I. Utkin and H. Lee, Chattering problem in sliding mode control systems, *International Workshop on Variable Structure Systems*, 2006. VSS'06, pp. 346–350, 2006, Alghero, Italy. doi: 10.1109/VSS.2006.1644542.

8 High-Voltage Battery Life Cycle Analysis with Repurposing in Energy Storage Systems (ESS) for Electric Vehicles

Mamdouh Ahmed Ezzeldin, Ahmed Alaa-eldin Hafez, Mohamed Adel Kohif, Marim Salah Faroun, Hossam Hassan Ammar
Nile University

CONTENTS

8.1 Introduction 182
8.2 Literature Review 182
    8.2.1 Conventional Cars and Electrical Vehicles 183
    8.2.2 Life Cycle 184
    8.2.3 Manufacture 184
        8.2.3.1 Battery Cell 185
        8.2.3.2 Packaging 187
        8.2.3.3 Battery Management System 187
        8.2.3.4 Battery Pack Assembly 188
        8.2.3.5 Solutions to Minimize the Impact Due to Manufacturing..... 188
    8.2.4 Battery Life cycle Analysis with Repurposing in Energy Storage Systems (ESS) 189
        8.2.4.1 First Use in Electric Vehicles 189
        8.2.4.2 Second Use in Energy Storage Systems 191
    8.2.5 Environmental Approaches for Battery Disposal 192
        8.2.5.1 Introduction and Background Information 192
        8.2.5.2 Currently Applied Recycling Techniques 192
8.3 Methodology 195
    8.3.1 Power Peak Shaving 195
        8.3.1.1 A Sample of Current Simple Comparative Algorithms 196
    8.3.2 The Proposed Simple Comparative Algorithm 197
        8.3.2.1 Definition of Variables 197
        8.3.2.2 Solution Flow 198

8.4	The Methodology Study Design	200
8.5	Factors of the Methodology's Ideal Environment	201
	8.5.1 Drivers	201
	8.5.2 Barriers	203
8.6	Case Study	205
	8.6.1 System Briefing	205
	8.6.2 System Parameters & Assumptions	205
8.7	Results	207
8.8	Conclusion	209
Acknowledgements		209
Bibliography		209

8.1 INTRODUCTION

Since the world has discovered electricity as an energy form, to be used in a massive and exponentially growing rate, it has been recognized as a clean energy source and form, until it was discovered that the damage that could be evolved by using electricity as a clean form of energy was not prevented, but the methods used to generate electricity are the problem. Burning of coal and oil causes a damage to the environment in every aspect – one of the causes is emission process that occur during the burning. However, there are efforts to implement other cleaner ways of generating electricity such as solar panels and wind turbines to meet the increasing need for electricity. To narrow down our vision upon the expected definite environmental disasters, we have to analyse the pattern of the market needs and its consequences on the short term to verify the expected negative impacts that we assume on the long term, so by looking at the late twenty years, it could be obvious to see how the rate of disposed lithium-ion batteries (LIBS) has increased since the demand on electrical energy storage is escalating. It should be elaborated that the chemicals of LIBS have their own consequences on the environment such as damaging the soil and water supply to kill the biological form of any exposed area. In addition, as the industry of transportation is witnessing a leap and transformation into electrical vehicles with a slow development rate in the field of recycling LIBS into the bargain. Expectations should be confirmed that if there is neither a solution to fully restore heavy metals from LIBS with a commercial applicability, nor a new technology for electricity storage, there will be fields of LIBS disposals and closed loop of environmental disaster. In this chapter, we discussed how the currently conducted methods are not efficient on the long term, and suggested a strategy to hold down the speed of reaching peak point as it cannot be ended or stopped its consequences, by studying the need for LIBS in our life and repositioning it to serve longer before disposed.

8.2 LITERATURE REVIEW

The first section compares the differences between the electrical vehicles and the conventional cars, while the second section talks about the life cycle of the battery and its components.

8.2.1 CONVENTIONAL CARS AND ELECTRICAL VEHICLES

Since the invention of electrical vehicles, they are always in comparison with the conventional cars, and nowadays there is a great controversy between both conventional cars and electrical vehicles, which leads to some confusions of which one is the perfect to use. In order to have a considerable judgement between the two vehicles, some points should be clarified. Firstly, the conventional vehicles are those that depend on the internal combustion for propulsion without any electric motors or mechanism. An internal combustion engine exploits the fuel/air mixture's repeated explosion to move a collection of pistons down a corresponding set of cylinders (Granovskii et al. 2006). Those engines, in effect, provide a mechanical rotational energy via an attached crankshaft. It is observed throughout the years that the conventional cars have better power, their maintenance cost is low, and they are at economically feasible price. Secondly, the electrical vehicles (EVs) are the vehicles that run with one or more electric motors which are responsible for propulsion. They consist of four essential elements: an on-board converter that converts the household AC energy to DC, batteries that stock charged electricity, an inverter that regulates the electricity flow from the batteries to the motor, and motors that convert the electricity to propulsion power (Mebarki 2017). Although conventional vehicles have better power and are of lower cost than the electrical ones, they have a huge negative impact on the environment. They deploy harmful CO_2 emissions, which affect the global warming (Granovskii et al. 2006). They are also responsible for tangential environmental impacts such as human toxicity potential, terrestrial toxicity potential, freshwater toxicity potential, mineral depletion potential, and fossil fuel depletion potential (Mebarki 2017). In the last decade, electrical vehicles have witnessed an increase in research and development. As more technological enhancements are processed to extend EV battery life and improve its performance, a great demand for EV has risen. Electrical vehicles have lots of advantages over the conventional vehicles due to their low environmental impacts since no carbon dioxide-containing exhaust gases are emitted and 97% reduction in CO_2 emissions are observed in contradistinction to conventional vehicles; in addition, they have lower charging cost as compared to the fuel cost (Mebarki 2017). But contrarily, the lack of the infrastructure of the charging stations and the charging time needed to recharge the battery are considered as the drawbacks of the EVs in addition to the secondary environmental impacts caused by the manufacturing of the batteries and their disposal (Li et al. 2014). However, there are some methods to make these batteries environmentally friendly, which will be discussed later.

Although conventional car batteries are similar to the electrical vehicle's ones, there are major differences between both types of batteries. As the batteries used in the conventional cars can't replace the ones used in the EV, and vice versa. Each battery has its capabilities and abilities to withstand the conditions of the car in which it is implemented, such as the heat and the power. Moreover, the batteries in the EV are used to move the whole car, while in the conventional ones, they are used just to start the car.

The battery of the EV is made up of two subpacks which are connected in parallel. Each subpack consists of 6 battery modules and 12 totals; each module is made up of equal number of battery cells, which are different in each type of car (Granovskii et al. 2006). The weight of the battery is almost 60% of the total weight as it can weigh 253 kg and can have an energy capacity of 26.6 kWh. In addition, the batteries in the EV are made of thick lead plates and can be recharged without causing any damage (Ellingsen et al. 2014). Generally, the most commonly used batteries are the LIBS. On the other side, conventionally, a standard battery is made up of six cells that are connected in series and provide two volts, thus constituting a total of twelve volts. The weight of the battery (18 kg) – which is almost 1% of the total mass of a car – compared to the whole mass of the conventional cars is negligible. Moreover, the most commonly used batteries are the lead acid batteries that are made of thin lead plates which allow the car to start fast, and these batteries are available for single use only as they can't be recharged.

LIBS are the mostly used type of batteries in the EV because they are of low weight and have a great energy density. EV may seem that it has no environmental impact, as it is going to reduce the greenhouse gas (GHG) emissions in 2030 compared to those emissions in 2005, but studies showed the opposite as only the manufacturing of the cells includes toxic chemicals, which increase the global warming potential (GWP) (Ellingsen et al. 2014). Although manufacturing of the LIBS has a lesser impact on the environment than manufacturing traditional ones, it still significantly affects the environment and its effects on the environment will increase as the estimated number of EV in 2050 is predicted to reach almost 2 billion and more. With this large demand, the impact on the environment (GHG and the climate) will dramatically increase (Ellingsen et al. 2014).

8.2.2 Life Cycle

The life cycle of a battery has a lifespan, and after a certain period of time, it needs to be replaced with a new one. Several studies showed that a 26.6-kWh battery cell can perform in an efficient way till reaching 80% of its initial capacity under the normal use. And after reaching 80% of its initial capacity, the battery isn't recommended to be used in the car as it will drain fast. On the other hand, it can be reused in other applications till attaining 65% of the initial condition, and then it can be recycled or disposed away. Studies showed that the GWP for a driven distance is calculated by dividing the impact of the total battery manufacture over the total distance covered by the battery's life while operating in the vehicle. So by assuming the battery cycle number to be 3,000 cycles and supplying 0.5 MJ km^{-1} power, the assumption will result in the emission of 11 gCo_2 – eq kg^1 (Ellingsen et al. 2014).

The life cycle of the battery depends on several stages such as material extraction, processing of the material, manufacturing of the components, usage in car, battery reuse, and end of life. The whole process, including the manufacturing, usage, recycling, and disposal of the materials, shapes the life cycle of the battery, as shown in Figure 8.1.

8.2.3 Manufacture

EVs are called the green vehicles as they have a zero emission impact on the environment, but studies showed the opposite (Ellingsen et al. 2014). Several studies showed

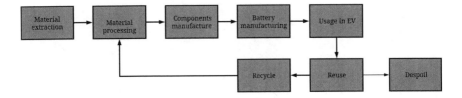

FIGURE 8.1 The life cycle of the battery.

that the batteries used in the manufacturing of EV have minor effects on the environment; on the contrary, the process of producing the battery itself is considered the main burden to the environment and has a large impact on it (Notter et al. 2010). An average of 4.6 tons of carbon dioxide is emitted into the environment, which is caused only by the manufacturing of 26.6-kWh, 253-kg battery pack. Other studies showed that not only the manufacturing of the battery but also the mining and refining processes of the material used are also the reasons for its impact on environment (Romare & Dahll¨of 2017a).

Lithium ion is the mostly used in manufacturing of battery in the EV cars. Several types of LIBS are used, such as lithium manganese oxide (LMO), lithium iron phosphate (LFP), and lithium manganese cobalt oxide (NMC). The different types of batteries are attributable to the different materials of the cathode while using the graphite as an anode material (Notter et al. 2010).

The manufacturing of the cell pack is made by assembling four components: cooling system, battery cell, packaging, and battery management system (BMS). Each component of the cell pack has several subcomponents, as shown in Figure 8.2.

8.2.3.1 Battery Cell

The manufacturing of a battery cell consists of two stages. The first is mining and refining of the materials, and the second is assembling the cells altogether.

8.2.3.1.1 Mining of the Material

Mining of the materials used in the manufacturing of the battery cell is the first stage of its life cycle since several materials are extracted from the ground. Various details regarding the impact of these materials on the environment are found at different databases. In this chapter, three databases are compared together: GaBi (Romare & Dahll¨of 2017b), Ecoinvent (Dones et al. 2003), and GREET (Li et al. 2017). The primary cobalt emits 90 and 8.8 kg CO_2 – eq kg^{-1} in GaBi and Ecoinvent equivalent. Moreover, the primary nickel emits 19.1 and 11.2 kg CO_2 – eq kg^{-1} in GaBi and Ecoinvent equivalent. The primary copper emits 2.7, 3–5, and 3–4 CO_2 – eq kg^{-1} in GREET, Ecoinvent, and GaBi.

8.2.3.1.2 Assembling the Battery Cell

While dissociating the battery cell, five subcomponents are found: anode, cathode, separator, cell container, and electrolyte. The anode is made from a mixture of copper current collector and a layer of negative electrode paste, which mainly consists of graphite and a small amount of binders. The cathode is made from a mixture of aluminium current and a layer of positive electrode paste (PEP), which mainly

FIGURE 8.2 Manufacturing and assembling a battery pack.

consists of positive material such as lithium manganese or NMC (Li et al. 2014). The electrolyte is a mixture of solvents and a salt, mainly lithium hexafluorophosphate (LiPF6). The separator, which acts as a filter, is made of porous polyolefin film to achieve its function (Ellingsen et al. 2014). The container is made of many tabs and pouch which surround the cell component. After adding all the components, the welding of both copper tab and aluminium tab to the negative and positive current collector, respectively, occurs.

Studies showed an industrial process of manufacturing a LMO which is used in several cars such as Chevrolet Volt and Nissan Leaf. The study showed the manufacturing of a 24-kWh battery pack that consists of 192 prismatic cells, and the manufacturing processes include coating, baking, welding, mixing, cutting, and finally assembling. A total of 88.6 GJ energy is used in the manufacturing process: 29.9 GJ is consumed by the material, and 58.7 GJ is utilized in the fabrication of battery (Yuan et al. 2017).

Another study showed the percentage of the components used in the generation of electricity that is needed for the manufacture of a battery cell: 46% coal, 33% nuclear, 15% gas, 4.4% oil, and the remaining 1.6% (which is gained from renewable resources such as wind, hydraulic, and solar energy) (Notter et al. 2010). The manufacturing of the cell, PEPs, and negative current collector (NCC) has the most significant effect on the environment with 56%–87% of the total effect caused by the batteries. For the manufacturing of the battery cells, coal, natural gas, and uranium are used to supply the energy needed, which is found to cause 51% of the battery's total GWP impact, 32% of the battery's total fossil deficiency positional (FDP), and 31% of the battery's total ozone depletion potential (ODP). After some analyses have been conducted, the study showed the total impact of the manufacturing on 13 different impact categories and the percentage of production of the cell (MOC), PEPs, and NCC in each of these impact categories. All values are at the lower boundary value (LBV) of the battery and listed in Table 6.1 (Ellingsen et al. 2014).

Some materials used in the manufacturing of the battery components are responsible for most of the negative environmental impacts, such as magnesium and copper, which are used in manufacturing of the PEP and NCC battery, respectively. The copper, for example, results in the disposal of sulfidic tailings, which is the reason behind the pollution of the most waterbodies, such as freshwater eutrophication potential (FEP), freshwater ecotoxicity potential (FETP), and marine ecotoxicity potential (METP) with a percentage of 62%, 65%, and 54%, respectively (Ellingsen et al. 2014).

Another study showed three different scenarios of generating energy for manufacturing the cells of the battery. In the first scenario, coal is used to generate electricity for the manufacture of the battery, which increases the GWP of the battery by exceeding 40%. In the second scenario, natural gas is used to generate electricity, which has found to have corresponding carbon intensity as the current uses mixed electricity. In the third scenario, hydroelectric power is used to manufacture the cell. In comparison with the current GWP, the usage of the hydroelectric power decreases the GWP by 60% and more, as shown in Table 8.1 (Notter et al. 2010).

8.2.3.2 Packaging

The process of packaging is divided into three processes. The first process is the module packaging composed of inner and outer frames, connectors, and a cover. The inner and outer frames are made of nylon cassettes surrounding the cells, which are used to protect and support the battery cell. Besides every frame, an aluminium heat radiator is found to make sure that the optimal conditions are achieved for the cells. The battery retention system is used to keep the modules in place using foams, constrains, and leach to keep it inside the battery tray. After making all the modules, the components are placed inside the tray made of steal and closed with a lid (Ellingsen et al. 2014).

8.2.3.3 Battery Management System

The BMS has five subcomponents: battery module boards (BMBs), integrated battery interface system (IBIS), fasteners, high-voltage (HV) system, and low-voltage system. For every module in the battery, one BMB is found and is used for monitoring the voltage and the temperature of the battery cells. While the BMBs monitor the cells, the IBIS controls the BMBs and the percentage of the HV. The HV system contains cables, fuse, aluminium lid, and connectors. The low-voltage system contains clips and straps (Romare & Dahllöf 2017a).

TABLE 8.1
Total Impact on 13 Different Impact Categories

		Impact		Percentage Impact	
Impact	Units	LBV	Impact	PEP	NCC
GWP	kg CO_2-eq	4,580	62	9	2
FDP	kg oil-eq	1,320	59.5	9	2
ODP	kg CFC-11-eq	2.8E-04	43	12	3
POFD	kg NMVOC	18	37	21	16
PMFP	kg PM10-eq	16	22	29	29
TAP	kg SO_2-eq	51	22	36	26
FEP	kg P-eq	8	13	7	64
MEP	kg N-eq	6.4	7	42.5	6.5
FETP	kg 1,4-DCB-eq	256	6.5	10.5	67.5
METP	kg 1,4-DCB-eq	276	6	10	69
TETP	kg 1,4-DCB-eq	1.3	6	25	50
HTP	kg 1,4-DCB-eq	15,900	4	7	76
MDP	kg Fe-eq	4,100	1	40	43

Impact categories: GWP, global warming potential; FDP, fossil depletion potential; ODP, ozone depletion potential; POFP, photooxidation formation potential; PMFP, particulate matter formation potential; TAP, terrestrial acidification potential; FEP, freshwater eutrophication potential; MEP, marine eutrophication potential; FETP, freshwater toxicity potential; METP, marine toxicity potential; TETP, terrestrial eutrophication potential; HTP, human toxicity potential; and MDP, metal depletion potential.

8.2.3.4 Battery Pack Assembly

By adding all the systems and the modules together, the process of assembling the pack is done. Besides all the energy needed in the manufacturing of the battery cells, the systems, and the components inside it, additional energy is consumed in the assembly process itself. Studies show that 0.3 GJ energy is used for assembling the battery pack (Yuan et al. 2017).

8.2.3.5 Solutions to Minimize the Impact Due to Manufacturing

The studies showed that the manufacturing of the batteries consumes a lots of energy, which is gained from burning the fossil fuel, which in turn affects the environment. In contrast, if the manufacturing of the batteries is done with clean energy, such as the hydroelectric, the impact of the manufacturing on the environment will reduce. Another study showed that the impact done by manufacture can be reduced by more than 60% if the energy needed is gained by the hydroelectric power instead of the current electricity mix. Besides changing the way of energy used to reduce the GWP, reducing the energy needed for manufacturing the cell, recycling the materials, and increasing the lifetime of the battery are other solutions to decrease the GWP. Impact can be reduced by these suggestions till it reaches a point where EVs are very cleaner than the internal combustion engine vehicles (ICEVs) (Ellingsen et al. 2014).

8.2.4 Battery Life cycle Analysis with Repurposing in Energy Storage Systems (ESS)

In this section, the life cycle analysis of the LIBS will be given. This will be done taking into consideration the use of the batteries firstly in mobility solutions such as their use in electric vehicles (EVs) and secondly in the smart grid solutions represented here as their use in the energy storage system (ESS).

8.2.4.1 First Use in Electric Vehicles

The first stage of the scenario proposed for the life cycle analysis will be discussed as the mobility application of the batteries is meant for use in EVs after the manufacturing process.

8.2.4.1.1 Background

As people shift from petroleum-derivative vehicles towards EVs, they demand EVs due to their cost-effectiveness, cleaner ways of generating electricity, and inexhaustible nature, and also sustainable power sources can be met. Be that as it may, the existence of environmental impact cycle profile shifts from use-related emission weights as the case is with internal combustion engines, to material mining, manufacturing, and assembling in EVs progressively. In this light, commercial transportation and personal vehicles account for around 10% of worldwide energy use and – consequently – leading to harming emission discharges. However, as far as the full cycle of supply and demand is concerned, the natural assets depletion and waste disposal into the environment reach their utmost contribution at the production stage of the EVs. This can be – to a great extent – the counterbalance contrast with fossil fuel-powered internal combustion powertrain systems at the utilization stage (Granovskii et al. 2006). However, the main disadvantages of the batteries are their high initial costs and high material depletion potential considering the chemical and physical compositions of all the metals used in the manufacturing of batteries. On top of that, batteries fail to hold their efficiencies at an operable level for the EVs beyond the 8-year mark. The efficacy of batteries drops to be as far as 80% of their original effectiveness when they were primarily manufactured during those first 8 years of their life cycle. That is why, those batteries with 80% efficiency should be reused in a less-demanding environment than those used in mobility services. Here, the focus will be on repurposing in energy storage utilities such as a stationary ESS. With this in hand, spent EV batteries can be repurposed for reusing them. Although repurposing demands a limited attempt for dismantlement; analysis for deterioration and malfunction; wrapping the batteries for a reuse; and attaching control systems, safety systems, and electrical equipment to the repurposed packs, it is proven that this will make the investment of energy and money in batteries of a better and greener return of investment (ROI) (Ahmadi et al. 2017a, 2014).

8.2.4.1.2 Types of Capacities

Since batteries are the main source of power in EV given the electric power train, the batteries need to be of matching capacity to the operational loads imposed on the vehicle to be able to meet the mobility needs of the operators. This means that for the diverse mobility solutions offered in the markets, there is a meeting diversity in terms of the types and capacities of the ESSs here referred to as batteries (Kushnir 2015). As heterogeneous as this market is, there is something in common with the majority of those types and capacities. This common facet between most of the battery types is their reliance on (Li-ion) in the chemistry of battery. In this light, the previously mentioned batteries can be manufactured using diverse amalgamations of materials comprising both the anode and the cathode. Nevertheless, as far as it is a mobility and transportation-related application in which the batteries will be used, the batteries are evaluated according to certain criteria, the most important of which is power and energy per kg (Kushnir 2015). This is why, some of those materials are preferred by battery manufacturers more than others, as shown in Table 8.2.

On the other hand, anode materials are not as diverse as those of the cathode materials. Most of those cathodes rely on graphite as their anode material.

Graphite is preferred as an anode material as it is cheap and relatively easy to manufacture. Moreover, some synthetic compounds of graphite can be used as well. Overall, graphite has a decent lifetime in the battery and is considered to have a great value for its cost of excavating and manufacturing. Contrarily, lithium titanium oxide is an alternative material for anode manufacturing that is found on a much smaller scale than graphite. This alternative offers excellent power and life cycle. Nonetheless, it operates at a lower voltage, and thus, it has lower energy content and is more expensive than graphite indeed (Kushnir 2015).

TABLE 8.2
Overview of the Most Common Battery Chemistries Used in Manufacturing Cathodes Along with their Advantages and Disadvantages (Kushnir 2015)

Cathode material	Abbr.	Use	Advantages	Disadvantages
$LiCoO_2$ Lithium cobalt oxide	LCO	Mainly in small-scale electronics	Performance, well understood	Safety, uses nickel and cobalt
$LiNi0.33\ Mn0.33\ Co0.33\ O_2$ Lithium manganese cobalt oxide	NMC (333)	Common in EVs	Better safety and performance than LCO	Cost, nickel and cobalt
$LiFePO_4$ Lithium iron phosphate	LFP	High-power option, potential choice for EVs	Excellent power, lifetime and safety, abundant materials	Low energy density
$LiMn_2O_4$ Lithium manganese oxide	LMO	Historically used in EVS, now less common	Cheap, abundant, high power	Lifetime, low capacity means low energy density
$LiNi0.8\ Co0.15\ Al0.05\ O_2$ Lithium nickel cobalt aluminium oxide	NCA	Used in some EVs	High capacity and voltage, high power	Safety, cost, uses nickel and cobalt

8.2.4.2 Second Use in Energy Storage Systems

For achieving the high efficiency, the batteries should be retained after being used in EVs; research shows the implication for a second use of the batteries in order to increase their utilization factor and account for the emissions and materials used in their manufacturing. This repurposing use in ESS is proposed in the scenario of the life cycle analysis.

8.2.4.2.1 Definition of ESS

ESSs can be defined as a complementary system added to a power generation source (mainly intermittent) to serve as a substitute power supplier when the source fails to meet the demands or loads imposed on it by the grid connected to it. Since there are several ways to store this energy, ESS can be divided according to the form of the stored energy inside the system: mechanical, electrical, and chemical (Ahmadi et al. 2017a, Aghajani & Ghadimi 2018). Here, the discussion will be about the repurposing of the spent LIBS as an electrochemical ESS. The main idea of electrochemical ESSs involves converting electrical energy into chemical energy stored in the bonds between reactants and the products formed with respect to the cathode, anode, and electrolyte of the batteries. Thus, the chemical part of the system requires the reaction of the – at least two – reactants existing in the system; then, the output of this chemical reaction is electrons that are obligated to flow through a circuitry forming a potential difference between the poles of the battery, which is referred to as voltage difference or simply voltage (Kushnir 2015).

8.2.4.2.2 Power Peak Sheaving

Battery charge/discharge profiles are greatly connected with what is happening on the subject of the primary power sources that in times when the supply is surpassing the demand, the batteries are toggled to start charging off the excessive power found on the system. On the other side, the batteries serve to discharge when the primary power sources are outcompeted by the loads imposed on the system. In the scope of the studies proposed, the depth of discharge of the batteries, which is the amount of the total charge contained in the battery when it is fully charged with the amount of current charge subtracted from it, is 75% (Ahmadi et al. 2017a, 2014). Power peak shaving is the process of meeting the peak hour demands on the grid by supplying energy through the ESSs. This can come in handy because of the price variations imposed by electricity companies during the peak demand hours. This extra pricing enables the utility company to increase the power capacity needed to supply for the peak demand. Without the usage of ESS, this extra supply is generated using older, more expensive, and less environmentally friendly power generation equipment. The peak pricing also encourages customers to reduce demand in order to cut down electricity costs. According to economists, the strategy of pricing is what gives peak shaving an attractive outlook for organizations with large electrical consumption during peak times (Olsson et al. 2018). In the scenario when this equipment is replaced by battery-powered ESSs, the consumption of those organizations will be moderated as they paid for power during low-utility pricing yet used it during the peak pricing times. This can save a lot of money for the third-party organizations and also save the environment from the emissions produced by the obsolete equipment used to meet the peak demands (Ahmadi et al. 2014, 2017b).

8.2.4.2.3 Use with Intermittent Power Sources and in Smart Grids

In today's world, the reliance on intermittent power sources of renewable energy such as solar and wind is on a hype. This can be contributed to many reasons; among them are the current environmental status and the pressing demands for clean energy. However, the reliability of those sources always comes to question as they have predetermined power cycles related to occasional phenomena such as the winds or the sunshine. That is why, a complementary battery energy storage systems will increase the reliability of those sources. In the same vein, an increased interest arises in the use of microgrid systems. In the microgrids, a small network of fossil and renewable energy sources, battery ESSs, and loads can be connected. As mobile as this is, microgrids can be utilized in inaccessible locations or when the pricing of the utility provider is not feasible (Aghajani & Ghadimi 2018, Azaza & Wallin 2017).

8.2.5 Environmental Approaches for Battery Disposal

In this section, the dimension and the disposal and recycling of the LIBS will be discussed.

8.2.5.1 Introduction and Background Information

Based on the aforementioned environmental impacts of LIB disposal, it is a necessity to search for an appropriate method for disposal. Spent LIBS contents of heavy metals and toxic electrolyte exceed 50%, which varies from one manufacturer to another; hence, land filling or burning is not considered as safe disposal method since toxic substances find their way to underground waterbodies and atmosphere, respectively. Environmental pollution as well as the loss of such significant amounts of active metals is the trigger behind recycling research. 4,000 tons of spent LIBS contains more than 1,100 tons of heavy metals. Thus, research has been conducted to find the optimum recycling of spent LIBS. Some techniques showed a great potential to be effective, while some other techniques have already made their way to the execution phase.

8.2.5.2 Currently Applied Recycling Techniques

The main purpose of the recycling processes is the extraction of heavy metals from spent LIBS since they hold a great economic value. Procedures are mainly categorized into three stages: pretreatment processes, metal extraction process, and product preparation process. Pretreatment processes are mainly conducted to avoid short circuiting or spontaneous combustion. Meanwhile, the extraction process is considered to be the most fundamental stage where metals are separated to ease their recovery in the next step. After acid leaching, the obtained leachate contains large amounts of metal ions, so solvent extraction, crystallization, and chemical precipitation are conducted in order to recover these metals (Zheng et al. 2018).

8.2.5.2.1 Pretreatment Processes

The target of pretreatment processes is not only avoidance of short circuits but also to reduce waste: the better the separation, the more optimum the extraction will be. At this phase, either mechanical separation or manual dismantling is used to disassemble the spent battery components after it is immersed in a salt solution although mechanical methods are more common. Firstly, the plastic shell is removed so that harmful substances are deactivated when the battery is immersed in liquid nitrogen. Then, the end of the battery shell is removed, and anode, cathode, and separator are separated. At the next step, they are dried in an oven, but further separation is required before metal extraction; thus, several methods were developed for this advanced separation, which are as follows:

A. Solvent Dissolution Method

In this method, an organic solvent is used to reduce adhesion between the cathode material and the aluminium foil. Multiple organic solvents can be used, but for each set of circumstances, there is a specific optimum organic solvent; these circumstances include the cathode material, the binder type, the temperature, and the electrodes using the rolling method (Zheng et al. 2018).

B. NaOH Dissolution Method

Depending on the amphoteric property of aluminium, the cathode material is leached with NaOH solution. In this way, both the aluminium foil and the protecting layer are dissolved, but recovery of this dissolved aluminium is complicated. It is a simple efficient method; however, waste NaOH is the opposite to environmentally friendly nature (Zheng et al. 2018).

C. Ultrasonic-Assisted Separation

This method was found to be the most effective when both ultrasonic treatment and mechanical agitation are applied simultaneously. Ultrasonic treatment causes cavitation whose effect is enhanced when combined with agitation effect (Zheng et al. 2018).

D. Thermal Treatment Method

Thermal treatment is used to weaken the bonding between the cathode material and the binder material to ease separation. Temperature varies with cathode and binder materials. This technique was even developed to vacuum pyrolysis where heat was not only used to ease separation but also used to vaporize or decompose the binder material (Zheng et al. 2018).

E. Mechanical Method

The mechanical method is considered to be the simplest one. However, the main disadvantage of the mechanical method is that organic materials penetrate through metals within the battery structure, which makes mechanical separation inefficient. However, sometimes, the extent of separation that can be achieved by using this method is satisfactory since it is only considered as a pretreatment procedure (Ordoñez et al. 2016).

8.2.5.2.2 Extraction Processes

Metal extraction is the main process in battery recycling. Metals are transformed to their metal alloys; the solution is formed as a step towards their recovery in the final step. Several methods are presented to conduct this separation, but the three main methods are found to be the most common and effective, which are as follows:

A. Hydrometallurgy

Hydrometallurgical methods depend on leaching for separation and recovery by dissolving. Multiple leaching agents are in use, such as organic acids, inorganic acids, and ammonia-ammonium salts. Inorganic acid leaching is considered to be the most effective, but it produces acidic waste that is harmful to the environment, which makes organic acids more favourable. Several conditions are there to affect the efficiency of the leaching process and the selection of optimum agent to use, such as temperature, reaction time, concentration of reducing agents, and concentration of leaching agents (Zheng et al. 2018).

B. Pyrometallurgy

Pyrometallurgical methods mainly depend on heat for separation. With high-temperature smelting, heavy metals are reduced to their metal alloys for recovery. Pyrometallurgical methods and hydrometallurgical methods are mostly combined for powerful separation. Several hybrid techniques have been mainly developed to enhance separation; specifically, the main struggle in the separation process arises from lithium separation. The basic technique had the battery components inside a furnace directly, which reduced them to their metal alloys. Later, the battery components are purified with sulphuric acid to produce cobalt oxide and nickel hydroxide. This basic technique has the disadvantage such as a complete lithium loss, which triggers further development. A second reduction smelting method was conducted, which could recover valuable metal alloys (iron, cobalt, nickel, and manganese) and turn lithium into dust, which is purified with sulphuric acid leaching. A third method uses high temperature, vacuum evaporation, and a particular carrier gas evaporation to evaporate lithium; this technique consumes huge amounts of energy, which is a significant drawback (Zheng et al. 2018).

C. Biometallurgy

Biometallurgical methods may not be as common as pyrometallurgical and hydrometallurgical methods; they are only conducted on a laboratory scale so far, but they are so promising that they are expected to significantly replace other methods.

8.2.5.2.3 Product Preparation Processes

A. Recovery of Metals from Leachate

This stage starts with a variety of metal ions all dissolved in solution. Several methods are there with the purpose of recovery and separation of these metals, such as chemical precipitation, crystallization, and solvent extraction; usually, a combination of these methods is used in this process. The main drawback of this stage is the large amount of reagents used. This loss is considered to be unavoidable since impurities in the reagents significantly affect the product purity (Zheng et al. 2018).

B. Preparation of Cathode Material

This stage may not sound very useful after the final formation of metals – the last stage – could reach, but the similar nature of some metals such as cobalt, nickel, and manganese is the main trigger behind this step. Separation is difficult because of this mutual behaviour among metals. Thus, an intermediate material is introduced to ease separation by adjusting the leaching solution composition. Later, either of two methods (coprecipitation and sol-gel) is used for the regeneration of final cathode material. Coprecipitation is more commonly used despite its appeal to produce impurities since many factors can influence the precipitation process; it is used because it requires a simple equipment. On the other hand, sol-gel requires shorter time for the reaction to occur and lower temperature but is accompanied with poor productivity, which limits its application (Zheng et al. 2018).

8.3 METHODOLOGY

In this section, real-life examples reflecting the importance of the usage of ESSs are given in continuation to what was mentioned before. The methodology will be divided into two sections: the use in power peak shaving applications and the usage in developing smart grids.

8.3.1 Power Peak Shaving

Due to the variant consumer behaviour imposed by both individuals and entities regarding the electrical power consumption, the rise in demand and fluctuations in consumption caused a noteworthy variation in power peaks and valleys as seen from the substation side. Factually, this has induced difficulties in the operation of the grid, and this causes problems in controlling the stability and reliability of the grid. Various techniques have been suggested to determine charging and discharging quantities and their respective time periods in the quest to remove the peaks and valleys from the load configuration. These techniques depend on analysis methods such as swarm particle optimization, dynamic programming, and nonlinear programming (Rahimi et al. 2013). On the other hand, some of these techniques rely on resizing of the BESS to deliver optimal efficiency and power for boosting the financial advantage by cutting the energy bill. In this section, a simple comparative approach is proposed as an algorithm for power peak load shaving. The algorithm compares the combined load profile with the mean loading profile over a specific time span; the next step is then triggering a charge or discharge response from the batteries according to the corresponding accompanying load profile. For the algorithm to be functional, several parameters are needed to be defined, among which are "energy bars" and "weighting factors".

8.3.1.1 A Sample of Current Simple Comparative Algorithms

Although the literature is rich with many algorithms solving the optimization problems related to the power peak shaving using BESS, this algorithm proposed in Dones et al. (2003) was chosen for its simplicity and straightforwardness. This peak shaving algorithm is independent of both the system topology and the imposed impedances of its wiring. Here, an assumption of a variable time window is taken into consideration. Through a sequence of intervals of (τ) for their following utilization periods (U_p), the method iterates the calculation of the combined load profile (P_av). Consequently, the contribution of each battery module in both charging and discharging processes will depend on their individual capacity compared to the total capacity of the system. In this light, the charging and discharging are estimated for the next suitable time period (τ), which is calculated by the utilization factor (U_f). Utilization factor ($0U_f1$) can be defined as the area utilized by energy bars (the targeted area to be shaved) over the total area above (or below) the average load line, as shown in the figure. The variable window approach mentioned earlier enables the system to act in an up-winding approach for load peak shaving. This means that the periodic calculation of (P_av) minimizes the effect of inaccurate load forecasting on load peak shaving. In Figure 8.3, a schematic illustration of the method is pictured, in which the usage time span (U_p), energy bars, and time intervals of (τ) are depicted. In the figure, (P) is the combined load drew from the power station. In the time when ($U_f = 1$), the batteries will be toggled to charge or discharge. Contrarily when ($U_f = 0$), the batteries will not be used at all. Meanwhile, ($0U_f 1$), the energy bars represented will be the only times when charge or discharge takes place. The amount of charged (discharged) energy from the arbitrary battery (m) can be calculated by:

$$\left| E_{bat,m} \right| = \left(SOC_{max,m} - SOC_{min,m} \right) \times \frac{\left| E_1 \right|}{\sum_{i \in S} \left| E_i \right|}, \tag{8.1}$$

where (S) is the set of all energy bars determined by (U_f).

From 8.1, the absolute charged or discharged power from the arbitrary battery (m) at the grid side is as follows:

$$\left| P_{bat,m} \right| = \eta_m \cdot \frac{\left| E_{bat,m} \right|}{\tau}. \tag{8.2}$$

As previously stated, the power charged/discharged from (m) is calculated in equation 8.2. However, the power needs to be confined regarding two conditions to accurately represent the real operating conditions. In equation 8.3, the first condition, where the charged/discharged power of the BESS is limited to the maximum rate of charge/discharge, is explained.

$$\left| P_{bat,m} \right| \geq \frac{\left| E_{bat,m} \right|}{\tau}. \tag{8.3}$$

The second case, as shown in equation 8.4, confines the state of charge of the battery (m) to its minimum (during discharge) or maximum (during charge) intervals, respectively.

High Voltage Battery Life Cycle Analysis

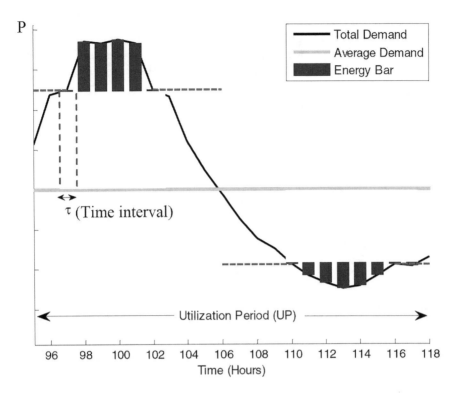

FIGURE 8.3 Sample power consumption (load) distribution over time for the proposed algorithm (Ahmadi et al. 2017a, 2014).

$$SOC_{max,m} \geq SOC_m \geq SOC_{min,m}. \tag{8.4}$$

Figure 8.3 depicts an immediate t of the intersection of the dotted line with the load profile in order to calculate the energy bars used for calculations in equation 8.1. The dotted line is formalized according to the utilization factor (U_f).

8.3.2 The Proposed Simple Comparative Algorithm

Following the previous section, here the researchers will propose a novel comparative algorithm for the determination of charge–discharge cycles of the batteries in the ESS. An arbitrary power–load curve is going to be used to clarify the analysis and determine the parametric equations of the important parameters.

8.3.2.1 Definition of Variables

where $M(t)$ is the characteristic load–time curve.

(P_H): The limit at which the load levels reach the limit that the obsolete equipment pieces – of low efficiencies – are used. Thus, the cost of the energy unit – usually KWH – increases after the (P_H) mark.

(P_L): The lower limit of the lowest valley in the load curve indicating the best time to charge the batteries.

$C(t)$: The cumulative batteries charge corresponding to time (t).
(δ): The rush–peak time over charge factor.
(α_{gen}): The GHG emission per unit energy generated from a specific power source.
($\beta_G$hg): The values of the GHG emissions released into the atmosphere on generating power from a specific source.
(η_P): Energy generation efficiency in the peak.
(η_v): Energy generation efficiency in the valley.
(η_{Ess}): The efficiency of the batteries.

We know from Ahmadi et al. (2017a, 2014) that the efficiency of the batteries through their repurposing life 10 years in stationary ESSs ranges from:

$$65\% \leq \eta_{Ess} \leq 80\% \tag{8.5}$$

Time marks: (T_1) start time of the peak, (T_2) ultimate load time, (T_3) end time of the peak, and (T_4) weighted average load correspondence time to the actual load.

$$\therefore M(T_4) = P_{avg} \tag{8.6}$$

Then (T_4) can be determined by

$$T_4 = \frac{\int_{T_1}^{T_3} M(t) \cdot dt \times T_2 + \int_{T_5}^{T_7} M(t) \cdot dt \times T_6}{\int_{T_1}^{T_7} M(t) \cdot dt} \tag{8.7}$$

(T_5) start time of the valley, (T_6) minimum utilization time, and z(T_7) end time of the valley are shown in Figure 8.4.

8.3.2.2 Solution Flow

Although the previously stated algorithm is one of the easiest battery ESS's control algorithms in the literature, the algorithm nomenclature is targeting the research segment of the audience, which may make it hard to be understandable by the operators or the general audience (Figure 8.5). This is why, a new simple comparative control algorithm will be proposed here to further simplify the analysis and to highlight the value proposition of BESS to the public. Inspired from Rahimi et al. (2013), (τ) will be the proposed time step in order to iterate the calculations of the curve points to be defined later.

Consequently, this will decrease the error due to inaccurate load forecasting. Moreover, the algorithm works to be in a state of continuous comparison of parameters that the iterations will continuously check the value of the battery charge percentage and correspondingly determine the state of charge or discharge. This will be further illustrated in the flowchart of the algorithm. Finally, the amount of money and emissions saved by the utilization of the BESS will be calculated.

Algorithm since peak area to be shaved is defined by (T_1), (T_3),

High Voltage Battery Life Cycle Analysis

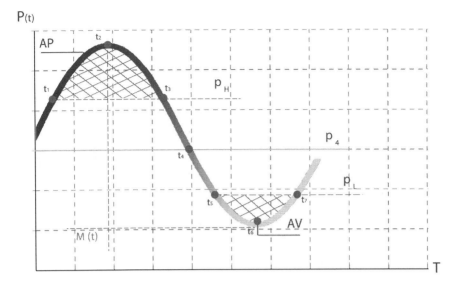

FIGURE 8.4 Arbitrary load–time curve.

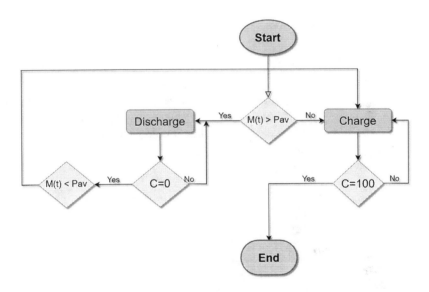

FIGURE 8.5 Flowchart for the control algorithm.

$$\therefore A_p = \int_{T_1}^{T_3} M(t) \cdot dt. \tag{8.8}$$

To make up for the excessive consumption at this peak, the batteries should charge utilizing the area of the valley,

$$\therefore A_v = \int_{T_5}^{T_7} M(t) \cdot dt, \qquad (8.9)$$

where

$$\int_{T_1}^{T_3} M(t) \cdot dt = \eta_{Ess} \times \int_{T_5}^{T_7} M(t) \cdot dt. \qquad (8.10)$$

To calculate the environmental impact (GHGs) of generation of each energy area, Peak area,

$$\left(\beta_{Ghg}\right)_p = \alpha_{gen} \times \frac{A_p}{\eta_P} \qquad (8.11)$$

Valley area,

$$\left(\beta_{Ghg}\right)_v = \alpha_{gen} \times \frac{A_v}{\eta_v} \qquad (8.12)$$

$$\therefore \text{Environmental benefit} = \left(\beta_{Ghg}\right)_p - \left(\beta_{Ghg}\right)_v. \qquad (8.13)$$

8.4 THE METHODOLOGY STUDY DESIGN

It is against the flow to have such a plan and aim to change the current system direction so, by having a good strategy and tools, it is achievable. Ahlborg and Hammar (2014) and Painuly Painuly (2001) proposed a qualitative study design to collect data in Tanzania and Mozambique, and made a literature review, interviews with power sector representatives, and at-site monitoring. The literature review consists of six categories negotiating the rural electrification (RE): institutions and representative's work; economy and finance; social dimensions; technical system and its management; technology diffusion and adaption; and rural infrastructure. The representatives interviewing part covered seven subjects: technical, economic, and social aspects of RE, off-grid, and renewable energy technologies (RETs): current state of the electricity infrastructure in rural areas; RE strategies (including capacities of both own and other organizations), roles, and relations between actors; institutional, social, and economic drivers and barriers to RE; productive uses of electricity; potential for off-grid and renewable energy technologies; domestic involvement in electrification processes; and effect of electricity on people's lives (Painuly 2001).

Seventeen interviews were followed through with government officials, international funding agencies, expert advisers, and the civil society organization (TaTEDO). Depending on their impact in and expertise of RE processes. The accomplished technique relied on the open-ended questions, without neglecting the professional experience while constructing the questions. Site visiting was applied to the operating off-grid systems using solar and diesel generators, and pico-hydro was in the favour of discussions. The limited data was not included in the analysis. There were some weak spots in this study design, which are needed to be considered while it is applied to other countries, such as the limited scope of analysis, the number of interviewees, and time assigned to each interview.

8.5 FACTORS OF THE METHODOLOGY'S IDEAL ENVIRONMENT

The previously mentioned methodology has its own ideal conditions to be fully beneficial and customized, which is explained in the study design. However, there are no neglectable drivers, which form a base of support for both the study design and methodology, and barriers that stand opposing the flow of the methodology sequence, which plays a major factor in this whole process.

8.5.1 Drivers

The power sector is mainly centralized and controlled by the government; the power sector structure is expecting new actors and a change in the sector dynamics and structure. The market actors are playing a major role in strongly affecting the drivers and barriers in both of their organizational capacity. Describing the contribution of the actors of power sector to RE development. Local initiatives, local demand, and other drivers are followed by the policy at national level compiled, as shown in Figures 8.6 and 8.7.

"According to interviews, the core drivers category for RE in both Tanzania and Mozambique is political priorities" (Ahlborg and Hammar 2014). When it comes to RE, it is an expected result since the government has a significant role in that field in most of the countries. Also, the politics has its priority over many other things on the frequently mentioned list of important aims such as better healthcare and education. Electrification for some actors might seem to hold back the urbanization and decrease the rate of birth in the rural areas. Regarding the use of financial aids and the pro-poor policies, the connection rate in the rural areas increases by the drivers, according to Akesson and Nhate (2006). As well, Akesson mentioned the affordability for the targeted customers as it is under the required tariffs in the case of no subsidized privet actors' off-grid systems. As the focus in both countries was directed towards grid extension, Mozambican donor and consultant B is claiming that the political interests are inflating in the fields of RETs and off-grid electrification. It is considered as a driver when the donor (international and local donors) support for RE is directed. However, a major part of the financial support in the energy budget is provided by the international donors aiming for reforms. It is noteworthy to mention that influential politicians advocate electrification for their districts during the elections time. The study was divided into categories under the drivers and barriers sections DS & BS. As Figures 8.6 and 8.7 are negotiating seven major aspects and themes, from how currently the electricity infrastructure state is through RE strategies, organizational, social and economic electrical power drivers and barriers in rural areas, passing by the beneficial uses of electricity, the future strength of technology regarding off-grid systems and renewable energy, how deep is the local role in the electrification process and finally the alter of electricity on people.

The first component of the DS is policy and poverty mitigation ambitions drivers. Figure 8.6 shows the part which the interviewed actors picked from there point of view the effective actors are related to the government and policies drivers and the mean of the answers points at the governmental policies as the main affecting driver. In the second part of the DS, local initiatives drivers are being illustrated in Figure 8.6 as the interviewees showed generally an equal importance to all the listed drivers. Except for the Churches installing or promoting RETs. That was not as expected as the churches are a major implementor of RE from there projects to support rural hospitals and communities. Following up with the DS, Figure 8.6 shows the local demand DS as the collected data from the participants shows how effectively the increasing demand drives the RE developments. The last part in Figure 8.6 demonstrates other drivers which are not under a specific category: high cost of grid extension drives off-grid and required rising in sustainability in grid.

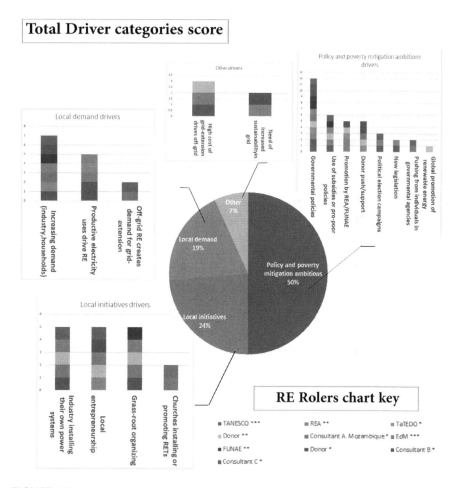

FIGURE 8.6 Mega diagram for the RE drivers.

8.5.2 Barriers

Tanzania and Mozambique encountered similar obstacles towards RE, which occurred because of many factors such as geographical, political, financial, and social aspects. Figure 8.7 shows the barriers to RE. The "economic issues, funder

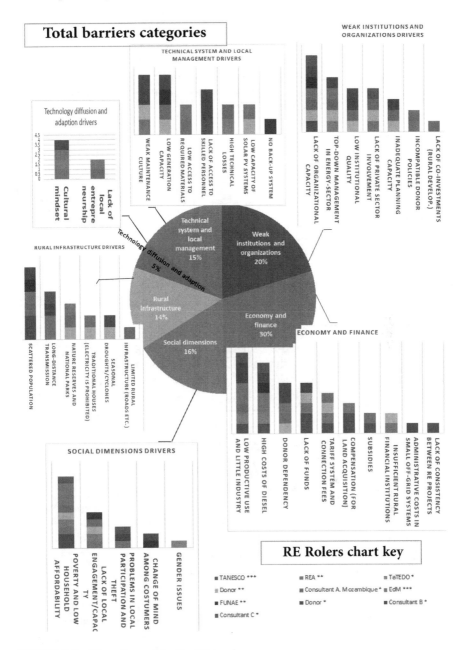

FIGURE 8.7 Mega diagram for the RE barriers.

subordination, lack of private sector involvement and institutional and organizational barriers" with a combination of other factors make up the barriers to grid extension, which is the reason why RE is difficult for both countries. Electrifying rural areas is considered one of the sustainability plans as it has a beneficial impact on the national economy, which is restricted by the grid extension of the RE in most of the cases, according to Bugaje (2006). RE is dependent on international financing due to the shortage of capital. Actors in Tanzania and Mozambique preview RE as being held back due to poverty because it is donor dependent and there is a shortage of fund as well as being rural poverty. The grid can be extended; however, this means the payback interval will increase, while the population density in other progressing countries within the vicinity is low and settlements are dispersed. Many users have reported the conveyance over long distances, which has resulted in high cost and has become a geographical barrier. Households are not able to afford the cost of transmission as it equals to several months of income due to the undeveloped economy of the rural area – this barrier is stated by Barnes (2010) and Haanyika (2008). Also, a problem considering the traditional building methods raises as these houses are unfit for electrification. Residences made out of clay and grass in Tanzania are not recognized for linkage. Therefore, only 10% of the rural population can pay for the connections and install electricity in their houses.

Customers with low income face problems in covering fees for connection, which is an important barrier as stated by Kankam (2009) and Murphy (2001). As a result, the government and donor projects provide help by using subsides and tariffs for the certain users so as to increase the rate of connections. Although the tariffs are low, a majority of people are not able to pay it. Also, rural users that are connected to the grid can pay for more expensive commercial tariffs; on the other hand, Hindman Persson asserts that the main trouble is rather the flow of cash, which needs to be solved using the applicable payment strategy. The financial outlook for the power sector can be developed by steadily increasing the tariffs; the only obstacle is that increasing tariffs is a tricky political constrain. Furthermore, there are specimens of grid and off-grid projects where users pay the fees without financial aid, which is said by Akesson (2006) and Ilskog et al. (2005). Another disadvantage for using variant tariffs while reducing the fees for low-income groups is that the threshold value remains the expenditure rate at the lowest margin. Power generation becomes less appealing to private properties when subsidized tariffs are used. In Tanzania, the cost of grid electricity generation has increased to double the present tariff; thereafter, its circulation has become a continuous loss, which caused TANESCO to not be able to afford its operation costs. As a result, financial aid works as a barrier and a driver as stated by Kankam (2009) and Thomalla et al. (2006). The costs of the supported linkage have been donor-driven projects intrusive to projects paid by the government where people are demanded to pay full cost of the connection fees. In addition, the objections were raised due to the low payoff for property loss (such as farmland) alongside power lines, since donor-driven projects dispense greater payoffs than governmental RE projects.

The upcoming diagram is handling the BS to RE by grid extension. Firstly, it discusses the barriers under the weak institutions and organizations category, and the interviewees stated that the lake of organizational capacity has the main role among the rest of the barriers; however, they did not neglect the management techniques that have been used in the energy sector, since it came in the second place. Economy could be one of the highest affecting barriers if it is not the centre of them.

As it is known, economic interests can drive or stop a development process in most of the fields, by putting the financials in the priority. As an output for these practices, the rollers of the RE are witnessing many barriers under this category. These barriers lead to the low productive and little industry coefficient, followed by the high cost of diesel as Figure 8.7 shows. The study did show an equal importance to the social dimension's barriers as shown in Figure 8.7, as it is a cornerstone in the study, and they picked the poverty and low household affordability as major drivers in the drivers' category. The category in Figure 8.7 has consensus on low maintenance culture and the truncate generation volume as the highest affecting drivers. The rest are the average priority from there point of view.

The research did include the spread of the technology in the energy industry, and how it is getting usual for the new adapting communities, and the results revealed the cultural mindset as the major barrier, but with fifty percent of the results for the lake of entrepreneurial mindset as Figure 8.7 shows.

If it is not the reason, which can be related to cultural mindset driver? At the end, the investment in rural areas infrastructure has been known for its lake of priority from the governmental developments and investors. However, they cannot take the blame for the scattered population as it will not serve as much people as the investment can cover, so it came at the first place to concord the results, as Figure 8.7 demonstrates.

8.6 CASE STUDY

8.6.1 System Briefing

After developing the operational methodology for the smart grid in general, the BMS is specific, and the next step is to test the system. To test the hypothesis and try to quantify the efficiency of the system, a model scenario of a future city's smart grid during its daily cycle was studied. In this scenario, the residential and industrial power demands are met through various renewable and nonrenewable power sources. In addition, mobility load was taken into consideration in order to model various mobility loads as well. This was done by forecasting the behaviour of different age groups and customer segment, which was in turn modelled by battery charging loads deployed at different times of the day to match their respective scenarios. In future, the reliance on electric cars is projected to increase. Consequently, this added load on the power grid can be used to alter the behaviour of the load curves as a form of "demand-side management" (DSM) depending on the forecasted scenarios of the users' usage cycles.

8.6.2 System Parameters & Assumptions

In this exemplary grid, the subsystems are modelled to be either active (generating power) or passive (grid power loads). However, the problem here is to optimize the operation of the grid to meet the power needs in the greenest way relying on the renewable energy sources rather than the base load plant that runs on diesel.

A. Power Loads
- The residential metropolitan loads are modelled as simple resistive loads for the household supplies and lighting.
- On the other hand, the industrial loads are modelled as electric motors, which portray the driving force for the industrial machinery.
- Vehicle-to-grid systems for charging mobility solutions (V2G) is comprised of 125 vehicles each of (40 kW) rated power and a rated capacity of (85 kWh) (Figure 8.8). The overall system efficiency is set to be 90%. This mobility consumption is encompassed of 5 different user profiles. For each user profile, there are a certain amount of vehicles assigned. This is what will be used later to manage the demand side and help make the overall consumption more dependent on renewable sources rather than diesel fuels. The profiles are divided as follows:
  1. Working professionals commuting to work and charging their cars on the premises of their workplace. (35 cars)
  2. Working professionals commuting to work but for longer distances and charging their cars on the premises of their workplace. This means a longer delay before recharging in the morning. (25 cars)
  3. Working professionals commuting to work but with no access to charging stations on the premises of their workplace. (25 cars)
  4. Working professionals working from home, i.e., no morning commutes. (20 cars)
  5. Working professionals working on night shift. Morning commutes are replaced by night commutes. (20 cars)
- Overall, the power loads sum up to a 15 MW representing the loads of a medium consumption day during spring or fall seasons (Figure 8.9).

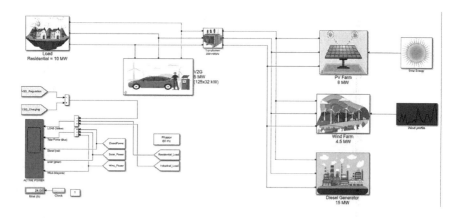

FIGURE 8.8 Exemplary smart grid with V2G capacity.

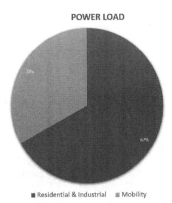

FIGURE 8.9 Power loads.

B. Power Generation (Maximum Capacity)
- The base load of the grid is met by a typical steam power plant that runs on diesel to produce a nominal power of 15 MW. However, the goal of the grid management algorithm and the DSM of the mobility loads is to minimize the reliance on the fossil fuel in order to produce less emissions and greener electricity.
- In addition, a photovoltaic farm with an area of 8,000 m^2 and an efficiency of 10% is capable of delivering 8 MW of nominal power in daytime. The dependency of the solar power production on the exemplary solar irradiance and cloud shading among other properties were taken into consideration while modelling the real conditions of an arbitrary PV farm.
- Finally, a wind farm represents the other part of the renewable energy sources. The wind turbine is capable of producing a nominal power of 4.5 MW. The wind profile is modelled by varying data points throughout the day ranging from a maximum of 15 m/sec and a nominal value of 13.5 m\sec.
- Collectively, the power output is modelled to be stepped down using a transformer station 25 kV/600 V before it is stepped down once more before getting to the usable form 220 V/60 Hz at the respective consumption areas (Figure 8.10).

8.7 RESULTS

After simulating the performance of the arbitrary grid, comparison curves between various power generation sources' performance were plotted. As shown in Figure 8.11, the load fluctuates according to the variation in the activities of the users during the day. Collectively, those varying demands of varying peaks and valleys (at their corresponding times of the day) were met using all the power generation stations and farms, respectively. The demand management system behaviour was apparent in peak shaving using the renewable sources. This was done by maximizing the utilization of

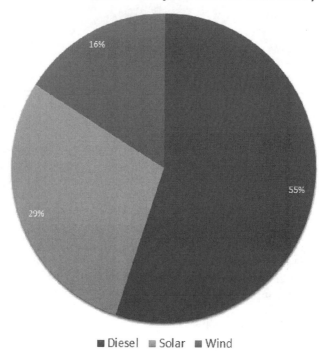

FIGURE 8.10 Power generation (maximum capacity).

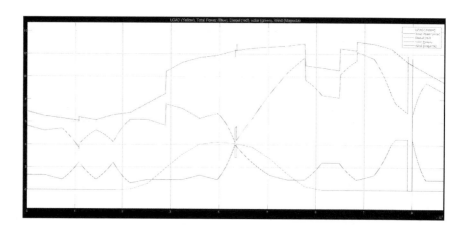

FIGURE 8.11 Load curve vs production curves through 24-hour duration.

wind and solar power (when available) to decrease the reliance on the diesel power plant in order to decrease emissions and lessen the environmental impact of charging the batteries used for the mobility solutions. Consequently, it was proven that the system can be utilized to decrease the overall environmental footprint of reliance on EVs for mobility rather than relying on fuel-powered vehicles. Although the analysis was not discussed quantitatively – as a result for the arbitrary assumptions that will render the results useless in specific real-life scenarios – the trends of the graphs simulated should be reassuring enough to promote the conversion to electric mobility solutions.

8.8 CONCLUSION

In this chapter, the environmental impacts of the EV batteries' repurposing and disposal were meant to be examined. Firstly, this was done by framing the differences between the conventional cars and EVs in terms of the main energy source and propulsion mechanisms. In addition, the current environmental status was discussed to compare between the impacts of the different mobility solutions. Secondly, a life cycle analysis of the batteries was made referring to research from the literature about the environmental impact metrics and their quantification for the EV batteries' product life cycle. This evaluation was conducted on all processes from the mining for the material till the disposal of the spent batteries in addition to their complementary parts as the BMSs. Afterwards, repurposing applications for the Li-ion-spent batteries were discussed mentioning their use in power peak shaving and smart grids. Moreover, a conclusive overview of the recycling process of the batteries was given, passing through all the steps and procedures involved. To wrap up, operational methodologies of the energy management systems were discussed reflecting on models from the literature and proposing a novel one. Finally, a study was conducted on the drivers and barriers to RE in the sub-Saharan region using smart grids. This study design can be transformed into a general formula to be applied on other cases taking most of the aspects which are similar to many cases without even editing, and in other cases, it will have to adapt as it is ready for any market or society or country to be applied depending on percentages and ratios between the usage of different energy sources.

ACKNOWLEDGEMENTS

We would like to thank Prof. Dr. Ahmed G. Radwan (Vice President for Research, Nile University, Egypt) for his helpful efforts and continued supervision throughout this chapter. Also, we express our sincere gratitude to Nile University for providing us with the all the facilities and equipment used in the case study.

BIBLIOGRAPHY

Aghajani, G. & Ghadimi, N. (2018), 'Multi-objective energy management in a micro-grid', *Energy Reports* **4**, 218–225.

Ahlborg, H. & Hammar, L. (2014), 'Drivers and barriers to rural electrification in Tanzania and Mozambique–grid-extension, off-grid, and renewable energy technologies', *Renewable Energy* **61**, 117–124.

Ahmadi, L., Fowler, M., Young, S. B., Fraser, R. A., Gaffney, B. & Walker, S. B. (2014), 'Energy efficiency of li-ion battery packs re-used in stationary power applications', *Sustainable Energy Technologies and Assessments* **8**, 9–17.

Ahmadi, L., Young, S. B., Fowler, M., Fraser, R. A. & Achachlouei, M. A. (2017a), 'A cascaded life cycle: reuse of electric vehicle lithium-ion battery packs in energy storage systems', *The International Journal of Life Cycle Assessment* **22**(1), 111–124.

Ahmadi, L., Young, S. B., Fowler, M., Fraser, R. A. & Achachlouei, M. A. (2017b), 'A cascaded life cycle: reuse of electric vehicle lithium-ion battery packs in energy storage systems', *The International Journal of Life Cycle Assessment* **22**(1), 111–124.

Akesson, G. & Nhate, V. (2006), 'Study on the socio-economic and poverty impact of the rural electrification project rib´au`e/iapala, nampula, mozambique', *Maputo: Swedish Embassy in Mozambique*.

Azaza, M. & Wallin, F. (2017), 'Multi objective particle swarm optimization of hybrid microgrid system: a case study in Sweden', *Energy* **123**, 108–118.

Barnes, D. F. (2010), *The Challenge of Rural Electrification: Strategies for Developing Countries*, Earthscan.

Bugaje, I. M. (2006), 'Renewable energy for sustainable development in Africa: a review', *Renewable and Sustainable Energy Reviews* **10**(6), 603–612.

Dones, R., Bauer, C., Bolliger, R., Burger, B., Faist Emmenegger, M., Frischknecht, R., Heck, T., Jungbluth, N., R¨oder, A. & Tuchschmid, M. (2003), 'Sachbilanzen von energiesystemen', *Final Report Ecoinvent 2000* **6**.

Ellingsen, L. A.-W., Majeau-Bettez, G., Singh, B., Srivastava, A. K., Valøen, L. O. & Strømman, A. H. (2014), 'Life cycle assessment of a lithium-ion battery vehicle pack', *Journal of Industrial Ecology* **18**(1), 113–124.

Granovskii, M., Dincer, I. & Rosen, M. A. (2006), 'Economic and environmental comparison of conventional, hybrid, electric and hydrogen fuel cell vehicles', *Journal of Power Sources* **159**(2), 1186–1193.

Haanyika, C. M. (2008), 'Rural electrification in Zambia: a policy and institutional analysis', *Energy Policy* **36**(3), 1044–1058.

Ilskog, E., Kjellstr¨om, B., Gullberg, M., Katyega, M. & Chambala, W. (2005), 'Electrification co-operatives bring new light to rural Tanzania', *Energy Policy* **33**(10), 1299–1307.

Kankam, S. & Boon, E. K. (2009), 'Energy delivery and utilization for rural development: lessons from northern Ghana', *Energy for Sustainable Development* **13**(3), 212–218.

Kushnir, D. (2015), 'lithium ion battery recycling technology 2015', *Current State and Future Prospects; ESAReport* **18**.

Li, B., Gao, X., Li, J. & Yuan, C. (2014), 'Life cycle environmental impact of high-capacity lithium ion battery with silicon nanowires anode for electric vehicles', *Environmental Science & Technology* **48**(5), 3047–3055.

Li, Q., Cai, H. & Han, J. (2017), *Life-Cycle Analysis of Greenhouse Gas Emissions and Water Consumption–Effects of Coal and Biomass Conversion to Liquid Fuels as Analyzed with the Greet Model*, Technical report, Argonne National Lab.(ANL), Argonne, IL.

Mebarki, A. (2017), 'Civil engineering and urban planning'.

Murphy, J. T. (2001), 'Making the energy transition in rural east Africa: is leapfrogging an alternative?' *Technological Forecasting and Social Change* **68**(2), 173–193.

Notter, D. A., Gauch, M., Widmer, R., Wager, P., Stamp, A., Zah, R. & Althaus, H.-J. (2010), 'Contribution of li-ion batteries to the environmental impact of electric vehicles'.

Olsson, L., Fallahi, S., Schnurr, M., Diener, D. & Van Loon, P. (2018), 'Circular business models for extended EV battery life', *Batteries* **4**(4), 57.

Ordon˜ez, J., Gago, E. & Girard, A. (2016), 'Processes and technologies for the recycling and recovery of spent lithium-ion batteries', *Renewable and Sustainable Energy Reviews* **60**, 195–205.

Painuly, J. P. (2001), 'Barriers to renewable energy penetration; a framework for analysis', *Renewable Energy* **24**(1), 73–89.

Rahimi, A., Zarghami, M., Vaziri, M. & Vadhva, S. (2013), A simple and effective approach for peak load shaving using battery storage systems, in *2013 North American Power Symposium (NAPS)*, IEEE, pp. 1–5.

Romare, M. & Dahll¨of, L. (2017a), 'The life cycle energy consumption and greenhouse gas emissions from lithium-ion batteries', *Stockholm. Zugriff am* **23**, 2017.

Romare, M. & Dahll¨of, L. (2017b), 'The life cycle energy consumption and greenhouse gas emissions from lithium-ion batteries', *Stockholm. Zugriff am* **23**, 2017.

Thomalla, F., Downing, T., Spanger-Siegfried, E., Han, G. & Rockstr¨om, J. (2006), 'Reducing hazard vulnerability: towards a common approach between disaster risk reduction and climate adaptation', *Disasters* **30**(1), 39–48.

Yuan, C., Deng, Y., Li, T. & Yang, F. (2017), 'Manufacturing energy analysis of lithium ion battery pack for electric vehicles', *CIRP Annals* **66**(1), 53–56.

Zheng, X., Zhu, Z., Lin, X., Zhang, Y., He, Y., Cao, H. & Sun, Z. (2018), 'A mini-review on metal recycling from spent lithium ion batteries', *Engineering* **4**(3), 361–370.

9 Charging Infrastructure for Electric Taxi Fleets

Chandana Sasidharan
Alliance for an Energy Efficient Economy (AEEE)

Anirudh Ray
School of Planning and Architecture (SPA)

Shyamasis Das
Alliance for an Energy Efficient Economy (AEEE)

CONTENTS

9.1 Introduction: Background and Driving Forces 214
9.2 Commercial Electric Taxi Fleets 216
    9.2.1 Case Study: Uber Electric Vehicle Trial in London 216
    9.2.2 Case Study: Ola Electric Mobility Pilot 217
    9.2.3 Key Findings from the Study of Fleet Operations 217
9.3 Important Charging-Related Aspects of Electric Cars 218
    9.3.1 Battery Capacity and Range 218
    9.3.2 Charger Capacity and Charging Time 218
    9.3.3 Factors Affecting Charging Time 220
9.4 Charging Technologies for Electric Cars 222
    9.4.1 EV Charging Standards 223
    9.4.2 Charger Classifications Worldwide 225
    9.4.3 Charging Technologies 226
        9.4.3.1 AC Charging 227
        9.4.3.2 DC Charging 228
        9.4.3.3 Wireless Charging 229
        9.4.3.4 Battery Swapping 230
    9.4.4 Charging Technology Trends 231
        9.4.4.1 Mobile EV Charging 231
        9.4.4.2 Solar EV Charging 231
    9.4.5 Charging Station Safety 232
9.5 Categories of Commercial Four-Wheeler Passenger Fleet 232
    9.5.1 Ride-Hailing Fleet 232
    9.5.2 Corporate Fleet 233

9.6 Plausible Locations for Charging Electric Taxi Fleet....................................233
    9.6.1 Charging Facilities for Taxi Fleet..234
        9.6.1.1 Public Charging Hubs for En route Charging....................234
        9.6.1.2 Charging Facility at Public Parking Spaces 234
        9.6.1.3 Captive Charging Facilities..234
    9.6.2 Critical Factors for Siting Charging Facilities....................................235
9.7 Techniques for Locating Charging Facilities 235
    9.7.1 Analytic Hierarchy Process (AHP) 235
    9.7.2 Technique for Order of Preference by Similarity to Ideal
Solution (TOPSIS) 237
    9.7.3 Rationale for Using AHP and TOPSIS..238
9.8 Configuring a Charging Facility for an Electric Taxi Fleet 238
    9.8.1 Selection of Charging Technology 238
        9.8.1.1 Identification of Charging Technologies for Evaluation 238
        9.8.1.2 Selection of Parameters for Decision-Making...................238
        9.8.1.3 Deciding Relative Weights of Parameters 238
        9.8.1.4 Ranking of Parameters 239
        9.8.1.5 Preparing Decision Matrix 239
9.9 Recommendations for Fleet Charging...240
    9.9.1 Public Chargers Are Required to Support Fleet................................240
    9.9.2 Role of AC Charging for Fleet..240
9.10 Grid Interaction and Integration of Fleet..242
Nomenclature...243
Works Cited...243

9.1 INTRODUCTION: BACKGROUND AND DRIVING FORCES

Electrification of commercial ride-hailing operations is identified as an effective way to realize sustainable regional transportation (Rokhadiya, et al., 2019). Many cities, including London, California, Boston, New York, San Francisco, and Shenzhen, have rolled out policies to support the replacement of internal combustion engine (ICE) taxi vehicles (Nicholas, Slovik & Lutsey, 2020). The prime drivers for the adoption of electric vehicles (EVs) are mitigation of greenhouse gas emissions and decreased local air pollution. A study in New York found that the replacement of one-third of city taxi fleets would lead to 18% reduction in CO_2 emissions (Kettles, 2016). In Beijing, electric taxi fleets were introduced in order to reduce vehicular emissions (Merkisz-Guranowska & Maciejewski, 2015).

    The transition to electric mobility is also an economical option for fleet operators as it can lower their operating costs. The accrued savings becomes an important factor to support commercial EV adoption, as the average travel distances of commercial fleets are higher than those of private vehicles. A study on ride-hailing fleets in the United States showed that the payback period for commercial EVs is half of that of private vehicles (Pavlenko, Slowik & Lutsey, 2019). This fact was also corroborated by an electric taxi pilot programme in India (Arora & Raman, 2019). A case study on electric taxi operation in Stockholm showed that electrification led to lower total cost of ownership (TCO) and slightly higher profitability than the investigated ICE taxis (Hagman & Langbroek, 2019).

Charging infrastructure for electric taxi fleets is an area that requires special attention to planning and strategy. As the average annual mileage of a commercial EV is higher than that of a private vehicle, the requirement for charging would be higher in the case of the former (Clairand et al., 2019). Jager et al. (2017) found that 250-km range is ideal for electric taxi operation in Munich and that an average of 22% of the time is used for charging. But there has not been a deep investigation on the range of existing EV models and their suitability for electric taxi fleets.

Research has highlighted the importance of adequate charging infrastructure to minimize the number of detours needed for charging and revenue loss to drivers (Hagman & Langbroek, 2019). Wang and Cheu (2013) studied the Singapore taxi problem and proved that there is a direct correlation between the decline in visits to charging facilities and increase in revenue per taxi. Wang, Cheu, and Balal (2016) continued the research and found that though increasing charging facilities was beneficial, the utilization of charging facilities was not the same. Their study found that utilization of a charging station depends on location and geographical distribution of customer demand. Hence, it is important to select appropriate locations for fleet charging. Jia et al. (2017) transformed the charging station allocation problem to a location problem and proposed a data-driven model for the allocation of charging stations, which was tested against the data of commercial taxis in Beijing. As data asymmetry is a practical challenge, there is a need to develop simpler yet robust tools for the selection of locations.

Moreover, taxi fleets can be divided into two segments: ride-hailing fleets and corporate fleets. Demand for clean employee transport is leading to the electrification of corporate fleets, but there is hardly any research on the charging needs of these fleets (WBCSD, 2019). The operations of these two types of electric taxi fleets differ based on the purpose of use/ mobility, origin and destination points of the trips, trip attraction/ generation models, service/business catchment area, etc. As the characteristic features of both the fleets are different, there is a need to assess the requirements for corporate fleet electrification. Both electric taxi fleets have increased public charging infrastructure utilization in some markets (Jenn, 2019). The charging needs will be dependent on their operational patterns, and consequently, charging preferences of fleets is a key factor while planning of charging infrastructure (Das, Sasidharan & Ray, 2020).

The multiple dimensions make fleet charging a complex riddle to solve. The existing research on fleet charging options has primarily focused on fleet optimization and allocation of charging stations. The need of the hour is a comprehensive understanding of the charging-related aspects of EVs, and linking them with the operational characteristics of fleets. This chapter is dedicated to answering the key questions related to the shift to electric technology, particularly related to charging.

1. Where are EVs currently being used as taxis? What are the learnings from fleet operation with EVs?
2. What are the important charging-related aspects of EVs? What are the factors that affect the charging time?
3. What are the charging options available for electric cars? Which of the following options are applicable to fleet charging?
4. What are the typical commercial passenger fleets and their operational aspects?

5. Where can the electric taxis be recharged? Is it possible to identify locations in the cities to set up a new set of chargers for electric car charging?

Though there is no dearth of literature on EV charging and taxi fleets, there are no studies that can comprehensively answer all of the above questions. Most of the existing works are focused on solving different aspects of the fleet charging conundrum in isolation. This chapter presents a holistic overview of the multiple dimensions of the charging puzzle, including transport characteristics, charging technologies, and practical tools that can be effectively used. Firstly, a few instances of commercial electrical fleet operations are studied.

9.2 COMMERCIAL ELECTRIC TAXI FLEETS

Europe has been the focal point for the introduction of many taxi fleets. The common denominator for taxi fleet electrification in all places has been the establishment of charging facilities. One of the first taxi electric fleets for Europe was set up in Amsterdam in 2011, supported by the subsidy for the procurement of EVs. Special taxi charging facilities were set up for the taxis apart from the normal public charging points. Subsequent to that, the city of Rotterdam introduced electric taxis fleets along with 500 charging facilities. In 2014, electric taxis fleets were introduced in Zurich along with a network of fast charging stations (Merkisz-Guranowska & Maciejewski, 2015; Nicholas, Slovik & Lutsey, 2020).

London has launched a charging facility plan to support the fleet electrification. The city has adopted a two-pronged approach with dedicated taxi charging. It has already dedicated 25% of existing public charging facilities for taxi fleets and is keen on installing more on-street fast charging infrastructure in priority locations. The locations are identified in commercial hubs based on a mapping of taxi driving patterns and the capacity of the electricity grid. London is also planning to launch ultrafast charging stations for electric fleets (Nicholas, Slovik & Lutsey, 2020; Hall & Lutsey, 2020).

Many of the Chinese cities are currently the front runners in electric taxi fleets. Electric taxis was initially introduced in Beijing and Shenzhen, but it has spread over to multiple cities across the nation. Most of the Chinese fleets comprise home-grown electric BYD e6 models. In China, the EVs in taxi fleets are charged twice a day and travel 360–500 km daily. Pilot projects are launched in many North American and South American cities, including New York, Mexico, Rio de Janeiro, and Sao Paulo (Hirschfeld, 2019; Merkisz-Guranowska & Maciejewski, 2015; Teixeira & Sodre, 2017).

The two case studies on the EV taxi fleet in the UK and India, and their findings are presented below.

9.2.1 Case Study: Uber Electric Vehicle Trial in London

Uber conducted an electric taxi trial in London with 50 EVs between August 2016 and January 2017. Three different car models, namely Nissan Leaf, Tesla Model S, and BYD E6, were used in the trial, and typically, drivers would recharge with 10%–40% remaining charge. The major challenges throughout the trial were range anxiety and lack of adequate fast charging points. The drivers found the

Electric Taxi Fleets

charging network inadequate with respect to the numbers and its distribution. They found that the time taken for waiting at charging stations had a compounding effect on charging time. Ninety-nine percent of drivers were dependent on public charging facilities for charging, whereas home charging accounted for only 26%. Still, one of the important aspects of the trial highlighted is that home charging was a supportive factor for taxi fleet adoption. The ability to charge at home allowed 40% of drivers to charge at domestic electricity rates, thus minimizing the cost of charging. Most of the charging happened at on-street parking as the majority of partner drivers lacked access to off-street parking. In summary, drivers preferred to use public fast charging facilities supported with slow charging at their homes (Lewis-Jones & Roberts, 2017).

9.2.2 Case Study: Ola Electric Mobility Pilot

In May 2017, Ola Electric started an EV taxi trial in the city of Nagpur, India. The EV used in the trial was Mahindra e2o with 15-kWh battery and a travelling distance range of 100 km. The average daily travel distance for the cars was 142 km, and typically, the cars underwent a fast charge and slow charge cycle in a day. They found that EVs usually have 20%–25% less running time compared to ICE vehicles because of the time taken for charging. In the hot summer season in India, the situation worsened as the time taken for fast charging was impacted by ambient temperature. When temperatures increased over 45°C, the charging time increased from 1.5 to 3 hours. Ola established a network of charging stations at times partnering with conventional fuel pump operators. The power demand at charging stations peaked twice a day, which was coincident with a lean traffic demand period. Between the faster and slow charging options offered in public charging facilities, drivers preferred the fast charging option. Similar to the London trial, home charging was also a strategy used in this trial. Ola assisted in setting up of charging points at driver partner residences for overnight charging (Arora & Raman, 2019).

9.2.3 Key Findings from the Study of Fleet Operations

The major observations from studying the fleet operations are as follows:

- In all cases of electrification, public charging infrastructure is a key component to successful fleet operations.
- For fleet operation, drivers preferred the faster charging options at public charging facilities.
- Dedicating a percentage of charging facilities for taxi fleet is a good tactic.
- The best strategy for fleet operations includes the provision of home charging facilities at driver homes.
- Charging demand might peak during a lean transport demand time and could lead to an increase in waiting time at charging facilities.
- The choice of EV has an impact on the range and the charging time. The driving range of EV and the travel distance determine the charging patterns. In the next section, this interrelationship is explored.

9.3 IMPORTANT CHARGING-RELATED ASPECTS OF ELECTRIC CARS

The choice of an ideal electric car for a taxi fleet will depend on the techno-economic feasibility of using an EV. While operating an electric fleet, charging time of the electric car represents a loss of revenue for the taxi fleet operator (Arora & Raman, 2019). Hence, the charging-related aspects are of prime importance for choosing a suitable electric car. In this section, a brief synopsis of the factors that impact the charging-related aspects of the electric car models in the market is presented. The battery capacity and energy efficiency together determine the range and charging need of an electric car. To plan charging, other parameters, including the battery chemistry, voltage, and maximum charging rate, are also relevant. Additionally, the factors including the capacity of the on-board chargers are an essential consideration to plan charging. Mapping of the important charging-related aspects of major electric car models available in different geographies is undertaken. The section summarizes the charging-related aspects by studying the specifications of over 45 electric car models available in major geographies of the world. It should be noted that there are disparities in the global adoption of EVs, and not all EV models are available in all geographies.

9.3.1 Battery Capacity and Range

One of the key parameters considered to select an electric car for a taxi fleet is the driving range of the vehicle. The range of an EV represents the distance it can travel in a single charge. The range of a car is dependent on the battery capacity. The range of electric cars available in the world is anywhere between 75 and 525 km, and battery capacities range between 11 and 100 kWh, as shown in Figure 9.1. The ratio between battery capacity and range is a proxy indicator of the energy efficiency for the electric car. For the cars studied in this research, the energy efficiency is estimated to be between 14.7 and 25.7 kWh/100 km. It should be noted that the actual range of an EV will depend on the operating conditions such as ambient temperature and traffic conditions. A study on EV adoption showed that the EVs with higher battery capacity (80–100 kWh) are preferred for taxis in markets where these models are available (Clairand et al., 2019). In cost-sensitive and nascent markets like India, the choice of EVs is limited, and the selected models have a battery capacity under 20 kWh (Das, Sasidharan & Ray, 2020).

9.3.2 Charger Capacity and Charging Time

Charging time is an important consideration with respect to operation of an electric taxi fleet. Almost all electric cars have an on-board charger that facilitates the basic charging requirement of the vehicles. However, the power of these chargers is typically very low, and this on-board charging power corresponds to the maximum charging time reported by the manufacturers. Manufacturers also specify the maximum charging power for the EV models, which corresponds to the minimum charging time. Maximum charging power associated with the on-board charger of EVs studied is shown in Figure 9.2.

Electric Taxi Fleets

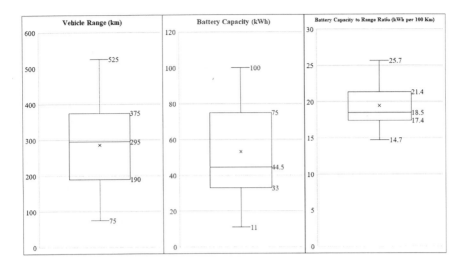

FIGURE 9.1 Mapping of battery capacity, efficiency, and range of electric cars.

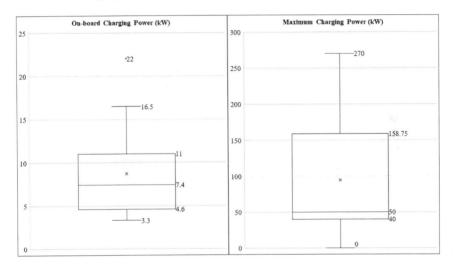

FIGURE 9.2 Mapping of charging capacity of electric cars.

It is observed that as the size of the battery pack in an EV increases, both on-board charger and maximum charging power increase as seen in Figures 9.3 and 9.4. Most of the on-board chargers have capacities under 11 kW. In the case of the maximum charging power, a key observation is that there are a few car models currently with a charging power over 150 kW. Typically, the maximum charging power needed to charge the existing electric cars is less than 100 kW. With the advancement in technology, this trend may change.

The minimum and maximum charging times of the cars provided by manufacturers are presented in Figure 9.5. The maximum charging time can be correlated with the on-board charger output. It can be seen that 75% of cars take 3–9 hours for a full

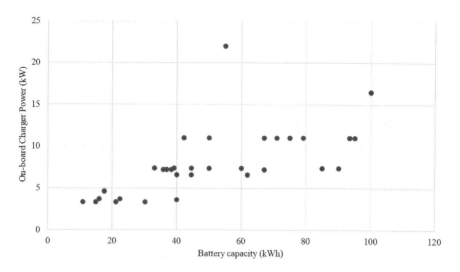

FIGURE 9.3 Scatter plot of battery capacity against on-board charging power.

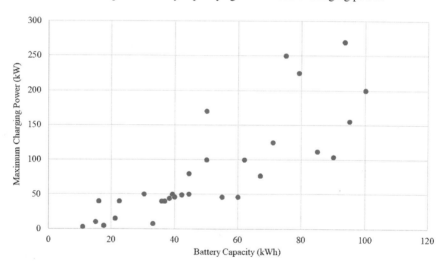

FIGURE 9.4 Scatter plot of battery capacity against the maximum charging power.

charge using on-board chargers. The minimum charging time for 80% of electric cars is less than 60 minutes. This implies that a driver might need to wait for an hour to charge an electric taxi. As charging time is a vital consideration, the factors impacting the charging time are investigated in the next section.

9.3.3 Factors Affecting Charging Time

Estimation of charging time of EV batteries is a challenging exercise as the charging time depends on the charger output power as well as the battery characteristics. Charging is often described as 'slow' or 'fast'. Typically, the fast chargers have

Electric Taxi Fleets

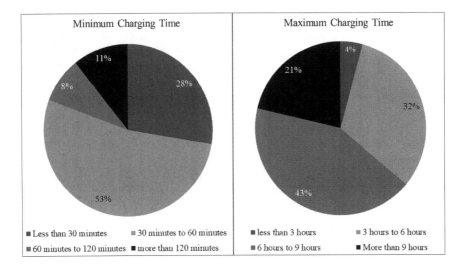

FIGURE 9.5 Snapshot of charging time of electric cars.

higher power output rating and are capable of charging EVs under an hour. On the other hand, slow chargers with lower power output often take six hours or more to charge an EV. Still, it is not possible to demarcate fast and slow chargers based on the level of output power of chargers alone. A 50-kW rated charger could be a fast charger for an electric car with a battery of 40 kWh as it would be able to charge the car within an hour. At the same time, the same charger would take more than two hours to charge an electric car with a 100-kWh battery. Hence, it is better to understand charging from the perspective of the battery. Batteries themselves are complex energy storage devices wherein the available energy, the chemistry, configuration within the battery pack, and the safe operating limits would determine the charging rate and time (Sasidharan, Ray & Das, 2019).

   a. *Battery chemistry.* Electric car batteries can have different material chemistries even within lithium-ion type of battery. The common battery chemistries that are found in electric four-wheeler batteries are lithium nickel manganese cobalt oxide (NMC), lithium nickel cobalt aluminium oxide (NCA), lithium manganese oxide (LMO), and lithium iron phosphate (LFP) (Coffin & Horowitz, 2018). Each battery chemistry has its own set of advantages when it comes to charging.
   b. *Battery pack configuration.* The three most common design variants of lithium-ion batteries are cylindrical, pouch, and prismatic cells. Each design has an implication on the thermal management of the battery. The electric car batteries are generally designed as high-voltage packs, with the voltage rating above 110 V. However, in India, there are electric cars in the market, which have low-voltage battery packs below 110 V. The battery voltage determines the voltage of the battery charger and hence is an important aspect for planning charging.

c. *C-rate.* Batteries of every capacity have an associated maximum limit for the charging and discharging. The C-rate of the battery is the measure of the rate at which a battery is discharged or charged relative to its maximum capacity. For example, a rate of 3C means the battery discharges in one-third of an hour, and a rate of 0.3C means battery discharges in 3 hours. The charging power requirement of a battery can be correlated with the C-rates of charging. A 30-kWh battery might require a 90-kW charger to charge at 3C rate, whereas at 0.3C rate, the charger power output needed is only 10 kW. However, the actual charging power requirement would be lower as batteries are never fully discharged.

d. *State of charge.* Batteries are not charged fully all the time. This means that the charging energy requirement of any battery is not the same as its capacity all the time. Batteries always have some residual energy, which is expressed in terms of state of charge (SoC) of the battery. The SoC describes the existing condition of a battery and presents the battery capacity as a percentage of the maximum capacity of the battery (MIT, 2008). The maximum depth of discharge (DoD) of a battery is the threshold beyond which battery is not discharged. Generally, the percentage of the battery capacity that has been discharged is expressed as a percentage of maximum capacity for DoD. An SoC of 40% implies a DoD of 60%. The maximum charging energy requirement happens at the maximum DoD of the battery (Sasidharan, Ray & Das, 2019).

To summarize the discussion, the factors affecting charging time are battery chemistry, configuration, C-rate, and state of charge. The next piece of the puzzle is identifying the charging technologies for electric taxi fleet, which is dealt with in the next section.

9.4 CHARGING TECHNOLOGIES FOR ELECTRIC CARS

One of the most important questions that need answering is about the most ideal charging technology for the fleet. There is a need to study charging technologies in detail. EV chargers currently deployed worldwide for charging electric cars are quite diverse in their method of electricity transfer, power output levels, control and communication capabilities, etc. (IEA, 2018). A mapping of the important parameters of commonly available chargers for over 40 EVs in the market is presented in Figure 9.6. It is evident from the analysis that most of the chargers available in the market are designed considering the maximum charging rate possible for the battery packs used in the available electric car models (Refer to Section 9.3.2). Most of the chargers for EVs have an output power between 3.3 and 200 kW. Typically, the output current for most of the chargers in the market is between 32 and 200 A, though the maximum current is 500 A. Most of the chargers are designed to have low-voltage (<1,000 V) alternating current (AC) inputs and provide either direct current (DC) or AC output. In the case of AC output chargers, there is not much difference between input and output voltage levels. However, the most DC output chargers offer output voltages between 130 and 720 V DC. In order to make an objective assessment of charging technologies, the existing standardization practices for charger needs are examined in the next section (Sasidharan, Ray & Das, 2019).

Electric Taxi Fleets

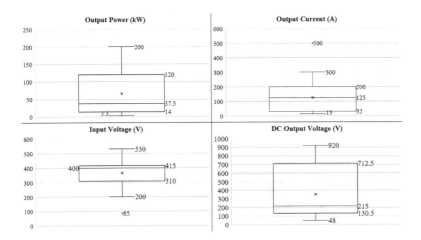

FIGURE 9.6 Mapping of major parameters of electric car chargers in market.

9.4.1 EV Charging Standards

There is no universally accepted standard for EV charging – the reason for which could be attributed to the global variations in EV and electricity grid design. One can observe standardization in prominent EV markets, including the United States, Europe, China, and Japan. Multiple international organizations are invested in the development of standards for EV charging these markets. One of the premier institutions associated with standards development is the International Electrotechnical Commission (IEC). The IEC 61851 standard is a popular standard associated with the conductive charging of EVs. IEC is also in the process of making standards for battery swapping and wireless charging.

China is one of the major EV markets, which has developed its own standards for EV charging, which are commonly called GB/T. The country also has a pivotal role in many of the recently announced EV standards for wireless charging (CEC, 2020) and battery swapping (Bloomberg, 2020). China also plays a key role in the development of a high-power DC charging standard dubbed 'ChaoJi', which is an updated version of the popular DC charging protocol CHAdeMO (2020). The CHAdeMO protocol developed in Japan, the first fast charging standard, is already codified as part of two international standards: IEC 61851 and IEEE Standard 2030 (CHAdeMO, 2015). There are also standards covering other important aspects for EV charging, including communication, security, and interoperability. For example, ISO 15118 is an international standard that is associated with the digital communication protocol important for vehicle grid integration.

The Society of Automotive Engineers (SAE) is the major EV charging standards organization in North America. All major vehicle and charging system manufacturers in the United States follow their standards. SAE J1772 and J1373 standards are associated with conductive and inductive charging, respectively. SAE J2931 covers and establishes the security requirements for digital communication between EV, EVSE (electric vehicle supply equipment), and the metering equipment. Details of the common standards associated with EV charging are given in Table 9.1.

TABLE 9.1
Summary of EV Charging Standards

Standard	Description
IEC 61851	Electric vehicle conductive charging system
	Part 1: General requirements
	Part 21: Electric vehicle requirements for conductive connection to an AC/DC supply
	Part 22: AC electric vehicle charging station
	Part 23: DC electric vehicle charging station
	Part 24: Digital communication between a DC EV charging station and an electric vehicle for control of DC charging
IEC 61980	Electric vehicle wireless power transfer (WPT) systems
	Part 1: General requirements
	Part 2: Specific requirements for communication between electric road vehicle (EV) and infrastructure
	Part 3: Specific requirements for the magnetic field wireless power transfer systems
IEC 62196	Plugs, socket outlets, vehicle connectors, and vehicle inlets – conductive charging of electric vehicles
IEC 62840	Electric vehicle battery swap system
	Part 1: General and guidance
	Part 2: Safety requirements
GB/T 20234	Conductive charging of electric vehicles
	Part 1: General requirements
	Part 2: AC charging coupler
	Part 3: DC charging coupler
GB/T 38775	Wireless charging system for electric vehicles
	Part 1: General requirements
	Part 2: Communication between on-board chargers and charging equipment
	Part 3: Special requirements
	Part 4: Electromagnetic environment limits and test methods
GB/T 29317	Terminology of electric vehicle charging/battery swap infrastructure
GB/T 27930	Communication protocols between off-board conductive charger and battery management system for electric vehicle
SAE J1772	Electric vehicle and plug-in hybrid electric vehicle conductive charge coupler
SAE J1773	Electric vehicle inductively coupled charging
SAE J2847	Communication for smart charging of plug-in electric vehicles
SAE J2836	Wireless charging communication for plug-in electric vehicles
SAE J2954	Wireless power transfer for light-duty plug-in/electric vehicles and alignment methodology
SAE J2931	Security for plug-in electric vehicle communications
SAE J3068	Electric vehicle power transfer system using a three-phase capable coupler
SAE J3105	Electric vehicle power transfer system using conductive automated connection devices
ISO 15118	Road vehicles – vehicle-to-grid communication interface
ISO 19363	Electrically propelled road vehicles – magnetic field wireless power transfer – safety and interoperability requirements

9.4.2 Charger Classifications Worldwide

One of the commonly recognized characterizations worldwide for EVSEs is categorized into levels, models, and types.

  a. *Levels.* EV chargers are classified based on charging power into 'levels'[1] as early as 1996 (Morrow, Karner, & Frankfort, 2008). This level classification was extended subsequently with the technology advancements. Currently, there are many EV charging levels as shown in Level 6, which is expected to be new entrant into the classification with the recent launch of 'ChaoJi' standard. The level classification of EV is important as with every ascending level, the complexity of charging increases (Lee & Clark, 2018; CHAdeMO, 2020).
  b. *Table 9.2.* Level 1 charging and 2 charging are associated with AC charging and are defined according to the single-phase voltage available in residential and commercial buildings. Charging at higher voltages, via DC, is classified into Level 3 and above. Level 6 is expected to be new entrant into the classification with the recent launch of 'ChaoJi' standard. The level classification of EV is important as with every ascending level, the complexity of charging increases (Lee & Clark, 2018; CHAdeMO, 2020).
  c. *Modes.* In Europe, charging is classified into four modes[2] (refer to Table 9.3) based on the control and communication between EV and EVSE (Spöttle, et al., 2018). Europe's modes for EVSE are different from levels as the features for installation, communication, and protection are used to define modes.[3] There is no European counterpart for US Level 1 as 120-V voltage level is not prevalent in Europe, and *Mode 1* is the simplest possible charging by plugging the EV directly into a household power outlet. As this method is charging without any in-cable protection, it is discouraged in most geographies. In *Mode 2* generally, the EV is plugged in through a portable cable with an inbuilt protection and control device. In this case, there is a

TABLE 9.2
EVSE Levels

EVSE Levels	Maximum Power (kW)
Level 1	2.5
Level 2	24
Level 3	50
Level 4	150
Level 5	350

---

[1] As defined by the Infrastructure Working Council formed by Electric Power Research Institute (EPRI) and subsequently codified in the National Electric Code (NEC) under article 625
[2] The modes are defined in the international industry norm DIN/ IEC 61851
[3] In US the Levels were defined first based on output power and the associated installation and safety requirements were specified later. But in Europe, the output power, installation, communication and protection are used in defining the Modes of charging.

TABLE 9.3
Charging Modes

Modes	Description
Mode 1	AC charging in households at normal mains outlets without no protection devices in the charging cable
Mode 2	AC charging from normal mains with semi-active connection between EV and EVSE. Charger cable with integrated safety devices in an in-cable control box comprising residual current device, control pilot, and proximity detection.
Mode 3	AC charging at charging stations with active connection between EV and EVSE. No in-cable control as the safety equipment is a permanent part of the charging station required in the cable.
Mode 4	DC charging at charging stations with an active connection between EV and EVSE. Charging system can adopt the charging currents and voltages to suit battery requirements.

proximity detector, and control signalling between the in-cable control box and the EV. However, under this mode, the charger does not receive energy feedback from the vehicle. *Mode 3*, on the other hand, is an advanced charging mode for an EV as there is energy feedback. A Mode 3 charger can control the rate of the charging and hence enables smart charging functionality. The charging is done from a fixed outlet or a tethered cable that is capable of continuous control and communication between the EV and the charger. All DC charging, where the charger is located outside the EV, is classified under *Mode 4*. It is seen that with the progressive increase in charger power output from Mode 1 to Mode 3, the associated communication and protection protocols get more complex (Vesa, 2019; CCS, 2019).

d. *Types.* Chargers are classified into 'types' based on the types of the connectors used. In the United States and Japan, Type 1 connectors that conform to SAE J1772 standards are used for AC charging. In Europe and other geographies, Type 2 connectors that conform to IEC 62196 standard are commonly used for AC charging. In addition, there are four types of connectors associated with the most popular DC charging protocols: combined charging system (CCS), CHAdeMO, GB/T, and Tesla.[4] In the case of CCS and Tesla connectors, the same type of connectors can be used for AC and DC charging. For all the other protocols, AC charging and DC charging have two different sets of connectors. All types of connectors have power pins, earthing pin, and control pins (CHAdeMO, 2015; CCS, 2019).

9.4.3 CHARGING TECHNOLOGIES

There are primarily three technologies for charging of EVs. Charging of EV batteries can be performed through a wired connection, *i.e.*, by conduction, or wirelessly. In the case of conductive charging, the vehicle is supplied with either AC or DC. Battery swapping is the third charging technology where a fully charged battery replaces a

---

[4] These protocols are described in detail in DC charging.

Electric Taxi Fleets

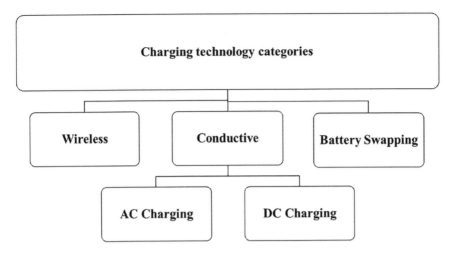

FIGURE 9.7 Categories of charging technologies.

depleted battery. These charging technologies are presented in Figure 9.7, and more details on the categories are captured in the subsequent sections.

9.4.3.1 AC Charging

An EV is charged by conduction by plugging it into EVSE. If the EV has an on-board charger, it is charged by plugging in into an AC power socket. The charging technology associated with this type of charging is called AC charging, as AC power is supplied to the vehicles. AC charging is commonly categorized into two levels: Level 1 and Level 2 (refer to Table 9.4). For a taxi fleet, better classification types are home and public charging.

9.4.3.1.1 Home Charging

Level 1 charging at 120 V and Level 2 charging at 230 V are commonly used for charging EVs at residences. The wall-mounted domestic socket provides a small amount of power to charge the EV (refer to Table 9.4) and results in prolonged charging times. However, this charging technology is relevant for a taxi fleet, as it enables

TABLE 9.4
Techno-Economic Specifications of AC Chargers

Specification	Level 1		Level 2
Input voltage AC (V)	120	230	380–480
Output voltage AC (V)	120	230	380–480
Maximum output current (A)	20	32	63
Output power range (kW)	1.4–2.4	3.3–7.4	11–43
Typical charger cost ($)	225–790		1,100–4,500
Installation and grid connection cost ($)	0–500		5,000

the cheapest charging for EVs using the existing electricity connection at residences of drivers. Case studies on EV charging also show that home charging is encouraged by most fleet operators. The cost estimates of the charger and installation are also shown in Table 9.4 (Spöttle, et al., 2018; Hall & Lutsey, 2020).

9.4.3.1.2 Public Charging

AC public charging facilities operate at Level 2 charging as shown in Table 9.4. Typically, all types of AC charging above 11 kW require the three-phase AC connections in most geographies. Though it is possible to charge vehicles till 43 kW using this method, the limiting factor for this kind of charging is the capacity of the on-board charger. For example, the on-board charger power of EVs in Europe is 22 kW, and the most common AC EVSEs in Europe are 480-V, 22-kW, three-phase chargers. High-power AC chargers may not be built in future for electric cars as there are no new electric car models with on-board charging power more than 22 kW. In case of India, most EVs have on-board chargers of 7.4 kW or less. Hence, most of the AC chargers in India are powered from single-phase 230-V supply. In China also, most of the public charging piles offer Level 2 charging at 220 V, 32 A. This type of charging is generally Mode 2 or Mode 3, and charging is performed with a protection and pilot control function (Spöttle, et al., 2018). The cost estimates of the EVSE and installation are shown in Table 9.4 (Spöttle, et al., 2018; Hall & Lutsey, 2020; SGCC, 2013).

9.4.3.2 DC Charging

DC charging is performed by supplying DC power directly to the battery as chargers are not on board in the EVs. The power output level of DC chargers is only restricted by the safe power limit the battery can accept. Thus, with the improvement in battery technology, the power output levels of DC chargers are on the rise. The existing DC charging facilities offer 50-kW power, whereas the upcoming facilities are 150 kW, 350 kW, or more. The costs of charging facilities also increase significantly with the increase in charging power. 150-kW and 350-kW charging facilities are typically 2 times and 5 times costlier than a 50-kW charging facility. The technical specifications and cost estimates for DC chargers currently available in the market suitable for charging are presented in Table 9.5 (Spöttle, et al., 2018; Hall & Lutsey, 2020).

As the charger is not colocated with the battery, the communication between the charger and the battery is of prime importance in DC charging. DC charging comes with standardized protocols governing communication and connectors. One of the

TABLE 9.5
Techno-Economic Specifications of DC Chargers

Specification	Level 3	Level 4	Level 5
Output voltage range DC(V)	200–500	200–500	200–1,000
Maximum output current (A)	100	300	500
Maximum output power (kW)	50	150	350
Typical charger cost ($)	8,500–11,000	17,000–22,000	45,000–67,000
Installation and grid connection cost ($)	5,000	11,000	>11,000

first DC charging protocols is the Charge-de-Move or CHAdeMO protocol, which supports the Nissan Leaf EV. Currently, this protocol offers charging to EVs from 6 to 400 kW power with the help of CAN-based communication between the charger and EV. The latest version of CHAdeMO protocol called 'ChaoJi' enables high-power DC charging up to 900 kW. The maximum charging current possible with this technology is increased to 600 A (CHAdeMO, 2020).

The most common DC charging protocol is perhaps the CCS protocol that originated in Europe. As the moniker denotes, the CCS connector is designed such that DC charging and AC charging are possible with the same plug, minimizing the cost. This protocol is used by popular EV manufacturers, including Volkswagen, General Motors, BMW, Daimler, Ford, Morris Garages, and Hyundai, that support CCS. Tesla also provides CCS charging ports for its EVs in Europe. The charging network established under the Electrify America programme provides CCS charging at 350 kW (CCS, 2019).

Another charging is used in the proprietary Supercharger network operated by Tesla. These Superchargers currently offer a maximum charging power of 120 kW and are expected to reach power levels as high as 350 kW in the near future (Hall & Lutsey, 2020; Hove & Sandalow, 2019).

China has established its own national DC charging standard protocol (GB/T) mandatory for all EVs. The maximum charging power offered by GB/T is currently 237.5 kW (950 V, 250 A). India adopted the GB/T standard to develop the Bharat DC charging protocol for its home-grown electric car models. The battery banks of these cars operate at low voltages that cannot be provided by the established charging protocols (Hove & Sandalow, 2019; CHAdeMO, 2020; Das, Sasidharan & Ray, 2020).

9.4.3.3 Wireless Charging

Wireless charging makes EV charging easier without a physical wire connection between the power supply and EV battery. Wireless transfer of power is possible by induction (short-range WPT) or magnetic resonance (mid-range WPT). There are three different modes of charging EVs wirelessly: stationary, quasi-dynamic, and dynamic. Wireless charging when the vehicle is parked or idle is called stationary charging. In the case of quasi-dynamic wireless charging, charging happens when the vehicle is moving slowly or in stop-and-go mode. It is also possible to transfer power dynamically when the vehicle is in motion. Among the three, dynamic charging is the most advanced and costliest option as it requires power transmitters embedded in the road to charge batteries as the vehicles move over it (Ahmad, Alam & Chabaan, 2017; Lukic & Pantic, 2013; Jang, Jeong & Lee, 2016).

To achieve wireless charging, power-transmitting and receiving equipment pieces are needed, apart from the chargers. Typically, the stationary transmitting equipment has an induction coil supplied from a high-frequency power supply, which gets magnetically coupled with the receiver. The EVs need to be equipped with receiver units with induction coils, and power conditioning equipment. One of the major concerns associated with wireless charging is the health impact from exposure to electromagnetic fields. There are a few standards for wireless charging that sets limits for the maximum power and distance at which energy can be transferred wirelessly (Qiu, Ching, & Liu, 2014; Lu et al., 2015; Musavi, Edington & Eberle, 2012).

The wireless charging standards that have been published or expected to be released soon are SAE J2954, ISO 19363, IEC 61980, and GB/T 38775. The Standard SAE J2954 establishes the specification of wireless charging between 3.7 and 11 kW (WPT 1-3) for unidirectional power transfer (SAE, 2019). This standard is designed for interoperability and supports both home and public charging installations. ISO 19363 standard specified the details of on-board vehicle equipment for unidirectional power transfer via magnetic field for passenger cars and light-duty vehicles (ISO, 2020). IEC 61980 standard is for magnetic field wireless power transfer to electric road vehicles when the EV is stationary. This standard is also applicable for wireless power transfer from on-site storage systems. China recently published a set of standards for EV wireless charging, based on the magnetic resonance technology (CEC, 2020). The standards for bidirectional power transfer or charging while EV is in motion are under development.

Wireless charging technologies have lower efficiencies in comparison with conductive charging technology. However, inductive charging is gaining popularity as a convenient charging option for taxis as it requires a minimal action from the driver. This technology is apt for opportunity charging at any predetermined location such as taxi bays. Norway is planning to install such wireless charging for electric taxis for the city of Oslo. WPT units of 75 kW power are expected to minimize waiting time and charging time for electric taxis. Charging will start when the driver parks the EV over the WPT units embedded in roads. Additionally, by eliminating cables and connectors, wireless charging improves the safety of the charging processes (Navigant, 2018; Statt, 2019; Reuters, 2019; Lukic & Pantic, 2013).

9.4.3.4 Battery Swapping

Battery swapping is often hyped as the future of EV charging, even though it is not a popular technology in practice for charging EVs. Battery swapping is considered as the best option to tackle the trilemma of long charging time, limited range, and high battery replacement cost. In swapping, depleted vehicle batteries are swapped with fully charged batteries in under 5 minutes. The battery swapping system has two components: the battery charging system and the battery swapping mechanism. A critical consideration in planning swapping operations is the battery inventory requirements. The number of spare batteries that the station requires depends on the demand rate for swapping at a station, which determines the power connection needed at the facility. Battery swapping is grid-friendly, with the ease in control of charging schedules and the ability to offer grid support. With respect to swapping technology, battery technological advancements that lower production costs will only have significant impacts if charging speed is fast enough (Rao et al., 2015; Sarker, Pandžić, & Ortega-Vazquez, 2014; Mak, Rong & Shen, 2013).

Only selected car models from Renault, NIO, and Tesla are designed with swappable batteries. Battery swapping emerged as a solution in a project for converting ICE vehicles to EVs. The retrofitted EV model Renault was used by Better Place in their battery swapping project. The company built a network of swapping facilities, but unfavourable economics for swapping operations led to project failure. The cost estimate for each swapping facility was $2 million, and EV owners had to incur a cost of $3,000 annually for subscribing to swapping services. A battery swapping-based taxi fleet pilot project in Tokyo in April 2010 also closed down with the downfall of the charging technology partner, Better Place. In the United States, Tesla tested battery swapping, but the solution was not commercially launched. Nevertheless, China is a keen believer that swapping technology is ideal for electric taxis. It is reported that around 200 battery swapping stations are operational to support electric taxis in 15 Chinese cities. Battery swapping is used for taxi fleet in the city of Hangzhou in China (Feldman, 2017; Chafkin, 2014; Kang et al., 2015; Bloomberg, 2020; Hove & Sandalow, 2019).

IEC 62840 and GB/T 29317 are the standards associated with battery swapping technology. A battery swapping station can promote charging at the low energy price periods of the day to in grid-to-battery mode (G2B), and enjoy the benefit of price arbitrage during the high-energy price periods by discharging in battery-to-grid mode (B2G). A battery swapping station is in effect an aggregator and with enough capacity can participate in capacity and ancillary markets (Wang et al., 2017).

9.4.4 Charging Technology Trends

There is hardly any consensus on the appropriate charging technology for fleet charging. Apart from charging technologies referred to in the section above, there are many upcoming options in the pipeline. Two of those options are selected and explained in brief in this section considering their suitability for commercial fleet applications.

9.4.4.1 Mobile EV Charging

Many charging service providers are exploring mobile charging services for parking lots and roadside assistance. These chargers have batteries that can provide AC or DC charging. These units may be beneficial for taxi fleet in ensuring access to charging and help avoid stranding of vehicles. For large fleet, it can help manage any spikes in demand. This technology can be a solution for the second life of lithium-ion batteries. However, the cost of supply will be higher with this charging technology as it entails additional losses in the storage. The standard governing moveable charging equipment is SAE J3068 (Hove & Sandalow, 2019; Freewire, 2020; Lee & Clark, 2018).

9.4.4.2 Solar EV Charging

Sustainable electrification of transportation goes hand in hand with renewable generation. Globally, there has been a rapid progress in the growth of solar photovoltaic (PV) power generation. With economies of scale, solar power generation has emerged as one of the cheapest sources of electricity in many markets. Fleet operators have benefited from coupling renewable energy to EV charging to reduce energy costs (Arora & Raman, 2019). The latest trend in the fleet electrification is the deployment of solar EV charging facilities, wherein power from the distributed solar plants is directly fed to EVs or stored in batteries for later use. Such solar EV charging solutions costing $60,000 are also being deployed in parking lots as part of the Electrify America scheme (Forbes, 2020; GT, 2020).

9.4.5 Charging Station Safety

One of the major concerns associated with charging facilities is its safety. Most EV markets have safety guidelines for EVSE installation and EV charging. In almost all of the countries, Mode 1 charging is not permitted as it is considered as a safety hazard. This is because there are no isolators or fault interrupters between the EV and EVSE in Mode 1 charging. A few nations, including the New Zealand, do not permit Mode 2 charging for public charging facilities. In other regions, safety restrictions for Mode 2 charging are applicable as a maximum current limit. All the standardized EV connectors have a proximity pilot wire, which ensures that the EV and EVSE are connected. When the proximity is not detected, the charging stops automatically. One of the most common safety precautions is with respect to protective earthing or an earth continuity conductor during charging. One of the key safety recommendations while charging is to maintain earthing. Most EVSEs and charging protocols have systems to monitor health of earthing connection continuously and disconnect the charging in the event of failure. Safety guidelines have provisions to prohibit the use of any connectors that is not recommended by the vehicle or EVSE manufacturer, which is considered as a safety violation. As expected, the safety norms for domestic charging include provision for a dedicated circuit with residual current protection for the EVSE. These residual current devices offer shock protection. For the public chargers, the safety norms are stricter, and it includes mandatory provisions for earthing monitoring system, i.e., residual current devices. Minimum ground clearance is specified for EVSEs that are charged under Mode 3 or Mode 4. After installation, procedures are developed to do the testing and inspection before the commissioning of the charging station. Periodic maintenance is another mandatory provision found in most of the safety guidelines. For public charging of EVSE or its installation in a damp location, all equipment pieces should have an adequate ingress protection. Any public chargers installed in locations such as parking sites should also have protection from mechanical damage (NEC, 2017; Worksafe, 2019; CEA, 2013; Glowacki, 2016).

9.5 CATEGORIES OF COMMERCIAL FOUR-WHEELER PASSENGER FLEET

Across the globe, four-wheeler passenger vehicles are employed for multiple private and commercial purposes. The commercial four-wheeler passenger (electric taxi) fleet can be categorized into two broad classes: ride-hailing fleet and corporate fleet. The ride-hailing fleet are the taxi fleet operated by taxi aggregators or car rental platforms for passenger transport from one location in an urban area to another. Corporate fleets are ride-sharing taxis that caters to the mobility needs of employees of an organization. The operations of these two types of taxi fleet differ based on the purpose of use/ mobility, origin and destination points of the trips, trip attraction/ generation models, service/ business catchment area, *etc.*

9.5.1 Ride-Hailing Fleet

Ride hailing is a service wherein commercial cars can be called upon for transporting passengers from one place to another in a city, which is the predominant form of intermediate between passenger and public transport in cities across the globe. This is due to the route flexibility offered by service, in addition to door-to-door mobility and other convenience factors. The service is normally backed by a mobile application equipped with matchmaking algorithms based on passengers' demand. However, simple roadside hailing by passengers is also common in the urban context. They operate in two ways primarily:

a. *Single node operation.* In this case, the starting point (node) for a taxi is fixed, which is also the point to which it will return after making one or multiple trips. The taxis may start their trip at the origin node or travel to a different customer-requested starting point (Rodrigue, Comtois & Slack, 2013).
b. *Double node-buffer zone operation.* Here, a taxi's movement generally centres around two nodes, *i.e.*, the starting point (taxi's initial location) and the ultimate destination point (often, it is the resting place as preset by the taxi driver). Matchmaking is done between the taxi and the rider(s) based on the preference of the driver and the requirement of the riders. For instance, if a taxi driver is travelling from point A to point B and a customer seeks a ride from point C to point D on the mobile application, then a matchmaking algorithm shall evaluate whether the travel route from C to D (along a reasonable measure of road lengths) falls within a predefined buffer zone of the route from A to B. In case the routes are in tandem with the requirements, the driver travels from point A to B, picks the passenger and goes to C, drops the passenger, and continues to D (Riejos, 2019; Schiller & Kenworthy, 2017).

9.5.2 Corporate Fleet

Such fleet ferries the employees of an organization between their workplaces (offices) and residences or to and from other prefixed drop points. The trips are mostly preplanned and the drivers are generally aware of the travel demand that may arise during a day. Such an operational model closely conforms to the single node operation model of cab (taxi)-on-demand services, as office fleet operates out of a single node and travels to other points on a road network based on passenger demand (Iles, 2005; Yaghoubi, 2017).

9.6 PLAUSIBLE LOCATIONS FOR CHARGING ELECTRIC TAXI FLEET

Public charging infrastructure is already identified as the key enabler for promoting the adoption of EVs. It has been estimated that there are almost 600,000 charging points in the world at the end of 2018. Both EV uptake and public charging infrastructure have experienced a growth rate of 60% in the 2013–2018 time period. It is not surprising that most of the public charging infrastructure is spread across 25 cities in the major EV markets in Europe, the United States, and China. The typical locations for installation of charging infrastructure other than homes and work places are public locations such as highway exits, fuelling stations, parking lots, and curbside (Hall & Lutsey, 2020).

9.6.1 CHARGING FACILITIES FOR TAXI FLEET

Facilitating zero-emission taxi fleet is understood as critical for sustainable urban transport. The recent trends in the charging infrastructure deployment showcase dedicated facilities to cater to the demand of electric taxi fleet. Taxi fleet often makes trips from certain popular hubs and requires charging facilities to reduce. It is envisaged that the identified commercial electric passenger fleet would depend on a host of charging options. Some of these charging facilities would also cater to the opportunity for charging requirement of private fleet. The charging facilities primarily differ on the locational aspect of the charging stations. Upon careful examination of the different possibilities of charging, two types of public charging facilities along with captive charging are important to support the mobility of commercial passenger vehicles.

9.6.1.1 Public Charging Hubs for En route Charging

Such facilities may be called the electrical counterparts of modern-day fuel refilling stations. As the name suggests, these stations are located adjacent to roads and are intended to serve 'en route' charging demand; i.e., they cater to charging demands of those EVs, which require charge mid-way in order to complete a trip between two locations in a city. In addition, such stations may be deployed for intercity travel. Erstwhile fuel refuelling stations equipped with chargers for EV can also fall in this category. Many cities, including New York and Amsterdam, are planning installation charging hubs for taxi cars. Stockholm is mapping priority areas for public charging facilities to suit the needs of commercial and taxi fleet its Charging Master Plan. In Vienna, the electricity distribution utility setup is done for charging infrastructure for the electric taxi fleet (Hall, Cui & Lutsey, 2018; Hall & Lutsey, 2020).

9.6.1.2 Charging Facility at Public Parking Spaces

Charging facilities are needed locations where vehicles are parked for an extended period of time to support fleet electrification. Parking access is already an established policy stance to promote EV adoption. It is logical to establish charging facilities at public parking spaces; clubbing parking access can be clubbed with charging facilities. In these locations as parking space is already available, space is only required for the placement of charging infrastructure. At public places like shopping malls, theatres, hospitals, parks, playgrounds, community centres, and transit nodes, charging facilities for EV charging can be set up. Charging piles can also be installed on curb areas. London is planning an installation of charging facilities for taxis in public hubs, commercial hubs, and semi-public depots. San Francisco and Oslo have mandated the installation of chargers in parking spaces (Hall, Cui & Lutsey, 2018; Hall & Lutsey, 2020).

9.6.1.3 Captive Charging Facilities

These are dedicated charging facilities that are set up for captive use by commercial-purpose fleet. Many charging service providers like Innogy, New Motion, Fortum, and ChargePoint have adopted business models supporting charging of fleet. For corporate fleet, the captive facilities are often set up at workplaces as they are the optimal locations for charging of fleet (Navigant, 2018).

9.6.2 Critical Factors for Siting Charging Facilities

The physical siting fleet charging facilities may pose a major challenge to the charging service providers and fleet operators. Finding an ideal site to establish a charging facility could be a challenging puzzle to solve. The effectiveness and the business viability of running a charging facility depend on its utilization, which, in turn, is contingent on the alignment of the preferences of fleet operators and charging service providers in terms of site selection. These preferences are believed to be tied with a set of factors or criteria for site selection for a public charging facility, and these are expected to vary between different types of charging facilities. Just taking note of these siting criteria may not help select a site for setting up a charging facility. One should be conscious of the fact that seldom one can find a site that would fully satisfy each criterion. Selection of a site based on the set criteria would involve certain trade-offs. Some criteria may take precedence over others during decision-making. In this regard, understanding the relative importance of the siting criteria would be very useful. The possible siting criteria are categorized under physical planning and grid connections, as listed in Table 9.6.

9.7 TECHNIQUES FOR LOCATING CHARGING FACILITIES

In order to maximize the return on investments in charging infrastructure, it is essential to select optimal locations for the installation of chargers. Taxis and other commercial fleet can provide initial high-usage and relatively predictable anchors to ensure a viable business case. Proper siting of charging facilities can also maximize the usage of chargers and minimize stress on the electricity grid simultaneously. The common strategy adopted to identify locations by a decision-making exercise considers multiple factors, including availability of land, taxi driving and parking patterns, and electricity grid capacity.

Multicriteria decision analysis, often called multicriteria decision-making (MCDM), deals with the process of making decisions in the presence of multiple objectives. MCDM is a versatile technique that is employed in EV charging problems (Liu, et al., 2018). Selection of suitable locations for charging stations is viewed as a MCDM problem. MCDM techniques, including analytical hierarchy process (AHP), technique for order preference by similarity to ideal solution (TOPSIS), are used for selecting locations for charging stations (Karaşan, Kaya, & Erdoğan, 2018; Wu, et al., 2016; Stojčić, et al., 2019).

9.7.1 Analytic Hierarchy Process (AHP)

AHP is a tool widely employed by the researchers in the MCDM problem. It helps quantify the relative priorities of a given set of criteria on a ratio scale, based on the judgement of decision-maker (Saaty, 1980, 2008). AHP relies on the judgement of experts to derive the priority of criteria influencing the decision. AHP has been widely used to

TABLE 9.6
Important Criteria for Selection of Sites for Setting up Different Types of Charging Facilities (Sasidharan, Ray & Das, 2019)

Types of Charging Facility	Criteria for Physical Planning	Criteria for Grid Connection
Charging hub	Road hierarchy[1] Land availability and cost Proximity to intersections or traffic signals Traffic volume on the road[2]	Proximity to distribution transformer, feeder or electric substations[5] Capacity of existing connection at the site
Charging facility at public parking space	Parking turnover ratio at the site Hierarchy of space at the site[3] Proximity to transit node[4]	Loading in the existing distribution network [5]) Power quality at the site [5])
Captive charging facility for ride-hailing fleet	Land availability and cost Proximity to transit node	
Captive charging facility at office premise	Number of employees availing the taxi services	

[1] Roads follow a hierarchy as they are composed of classes like arterial roads, subarterial roads, collector roads, and local roads, wherein each class of roads has a separate function, design, and usage rules. Road hierarchy determines the purpose, capacity, and functions of different types of roads in a city's/ town's road network. Different types of roads could impact the charging demand. It is understood that higher hierarchies of roads cater to higher volumes of traffic.

[2] The volume of traffic that flows on a particular road could impact the charging demand and thereby the usage of a charging facility. Traffic volumes on these roads should be compared for two sites. This is because an axiomatic linkage exists between traffic volume and potential charging demand. Post this comparison, it is suggested to gauge the proximity of the shortlisted sites to near-by traffic intersections, and the visibility of the sites from the adjoining roads. Closeness of a site to a traffic intersection and high visibility favour higher potential usage.

[3] A city has spaces where people agglomerate. The size, type, purpose, density, length of stay, etc. of such an agglomeration are governed by the hierarchy of space, where space could range from being a commercial centre to an educational hotspot, from being a recreational facility to a healthcare hub. It is common knowledge that higher the placement of space in that hierarchy, higher the volume of citizen and traffic it invites, which makes it an important parameter to consider from a business perspective when setting up a charging facility. For example, it would be more lucrative to invest on a large-scale charging facility in the central business district of an urban centre, instead of a convenience shopping centre which houses a milk booth, a stationery shop, and a grocery shop.

[4] Transit nodes are the locations in the transport network of a city where one mode of transport meets another, such as railway stations, where rail-based transport meets road-based transport, or metro stations, where rail-based transport meets the road- and walk-based transport. Parking turnover ratio determines the capacity utilization of a parking space. High traffic attraction of a parking area is desirable as it may potentially translate to high charging demand at the site.

[5] In case of electrical connection, three important aspects to examine would be the ease of getting a connection, network capacity, and the reliability of electricity supply. In this regard, the approach considers proximity of a site to a distribution equipment and the loading in the distribution network as the relevant indicators. The closer the site is to a distribution equipment, the lower would be the cost and time for obtaining a new electricity connection. The loading in the distribution network at the site and power quality are indicators for available network capacity and supply reliability at the site.

solve the real-world planning problems (Løken, 2007; Zyoud & Fuchs-Hanuch, 2017). This tool helps to determine the optimal alternatives when applied to the problems in planning. This is a convenient tool as it provides a methodology to calibrate the numeric scale for intangibles. It also includes a consistency index testing mechanism, which imposes discipline on the group decision-making (Sasidharan, Chandra & Das, 2019).

AHP helps to determine the important criteria by constructing a hierarchic structure and conducting pairwise comparisons. A complex decision-making problem is broken down into hierarchic levels descending from the overall goal, listing out the criteria and subcriteria in the successive levels. At each level, a pairwise comparison of components is to be made. An analytical process uses the results from pairwise comparison and determines the priority of components at each level by assigning weights through matrix operation. The individual scores from the experts would be aggregated and then used to identify the criterion weights (Sasidharan, Chandra & Das, 2019).

The criterion weights are calculated using the following formula:

$$W_r = \frac{1}{k}\left\{W_r^1 + W_r^2 + \cdots + W_r^k\right\}, \tag{9.1}$$

where

w_r^k is the weight of rth criterion for k decision-makers ($r = 1, 2,\ldots,n$; $k = 1,2,\ldots,k$).

The classic AHP algorithm necessitates experts to make pairwise comparisons and results in a complex questionnaire in the matrix format. Making pairwise comparison at times results in spurious scores, which could reduce the overall consistency index of AHP (Sasidharan, Chandra & Das, 2019).

9.7.2 Technique for Order of Preference by Similarity to Ideal Solution (TOPSIS)

TOPSIS is an MCDM matrix with intuitive and clear logic that represents the rationale of human choice. It is a technique with the ability to measure the relative performance for each alternative in a simple mathematical form. TOPSIS is a method that accounts for both the best and worst alternatives in scalar values. The process for the TOPSIS algorithm starts with the preparation of the decision matrix representing the relative importance of each criterion. Next, the matrix is normalized, and the values are multiplied by the criteria weights derived from AHP. Subsequently, the best (positive-ideal) and worst (negative-ideal) solutions are calculated. Later, the distance of each alternative to these solutions is calculated with a distance measure. Finally, the alternatives are ranked based on their relative closeness to the ideal solution (Lai, 1994; Guo & Zhao, 2015).

The TOPSIS technique is helpful for decision-makers to take informed decision based on analysis, comparisons, and ranking of the alternatives. The classical TOPSIS method solves problems only when there is no ambiguity in the data. However, most real-world problems, including planning, are complex and subject to face challenges in data availability. As a result, researchers have developed variants of TOPSIS method, which can deal with the lack of information and uncertainty. If the present study faces data limitations or uncertainty, a suitable variant of TOPSIS method would be applied for ranking.

9.7.3 Rationale for Using AHP and TOPSIS

AHP and TOPSIS, two techniques proposed to be employed, are complementary in nature. While TOPSIS is a purely numerical approach, AHP is a human-centric assessment technique. AHP helps to overcome the limitations of TOPSIS with provisions for weight elicitation, and consistency in checking of judgements. On the other hand, TOPSIS alleviates the weaknesses of AHP such as the requirement of pairwise comparisons and restrains imposed by the human capacity.

9.8 CONFIGURING A CHARGING FACILITY FOR AN ELECTRIC TAXI FLEET

9.8.1 Selection of Charging Technology

An important decision would be choosing the appropriate charging technology. The effectiveness and feasibility of deployment and the use of a charging technology hinge on a range of factors, both technical and economic. The selection of best-fit charging technology can also be treated as a decision-making problem, and suitable MCDM tools can be applied for optimizing diverse techno-economic objectives. A decision matrix is a simple, but sturdy MCDM tool that can be used for the demonstration of the ranking of alternatives in charging technology selection problem. An example of an application of the decision matrix to identify the 'best-fit' charging technology for any public charging facility is illustrated below.

9.8.1.1 Identification of Charging Technologies for Evaluation

The first step in a decision-making problem identification of plausible charging technologies is suitable for installation in public charging. This decision could depend on the maturity of charging technology and the availability of suitable EV models. As an example, battery swapping and inductive charging are not mature technologies for electric car charging in most EV markets. Level 1 charging is not applicable in regions where 120-V AC distribution is not available. There are hardly any cars suitable for Level 6 charging power.

9.8.1.2 Selection of Parameters for Decision-Making

The decision matrix consists of a set of techno-economic parameters which receive a rank and a weight. Some of the suggested decision-making parameters for the selection of charging technology are charging time, easiness of power connection, cost of charger, cost of installation and grid connection, and cost of electricity. While selecting the decision-making parameters, one should check the interdependencies of contesting parameters.

9.8.1.3 Deciding Relative Weights of Parameters

Each parameter is assigned a weight based on the assessed degree of importance using the interval scale in Figure 9.8. The relative weights of the parameters are a subjective call of the decision-maker, and could vary from location to location. For example, the importance of charging time will be different for charging at different locations. This parameter could be most important for a charging station operating

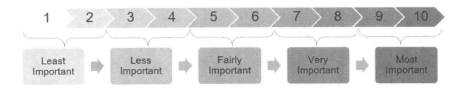

FIGURE 9.8 Scale for assessing the importance of a parameter.

like a regular refuelling station, whereas it could be fairly important for a charging facility located at a public parking space. It is possible to use other MCDM techniques such as AHP and TOPSIS to develop the weights for the ranking.

9.8.1.4 Ranking of Parameters

Each charging technology is ranked against individual parameters whereby the technology, which is closest to the ideal value for a parameter, is ranked highest. The ideal values of the parameters could be the best in the industry given the constraints. It is also important to ensure while ranking that all of the parameters are selected, such an ideal condition is represented either by maxima or by minima for all the parameters. For example, low charging time, low cost of charger, connection, and electricity cost are ideal conditions for charging technology selection. The ideal value of charging time is the minimum charging time seen, i.e., half an hour (refer to Figure 9.5). The relative ranks of charging times for the most common charging technologies used in public charging facilities are presented in Table 9.7 for easy reference for the decision-maker.

Similar rankings can be developed for other parameters for different charging technologies. It is essential to remember that the technical and financial aspects would vary in different geographical regions. Costs are dependent on other socio-economic factors such as maturity of the EV market and policy support for charging infrastructure. The easiness of grid connection is dependent on the design of the distribution network and the loading patterns. Indicative rankings are developed for the parameters using the details available in Tables 9.4 and 9.5.

9.8.1.5 Preparing Decision Matrix

Once the ranking is done and weights are decided, both are applied to a decision matrix. Justification for the weights used in the example is presented in Table 9.8. The individual weights are multiplied by the ranks for every parameter for all charging technologies. The total score of any ranking technology is obtained by normalizing and obtaining the weighted rank for all parameters. The charging technology

TABLE 9.7
Sample Ranks for Charging Time

Parameter	Level 2	Level 3	Level 4	Level 5
Charging power considered (kW)	7.4	22	50	150
Charging time	1	2	3	4

TABLE 9.8
Sample Weights Used in MCDM Example

S. No.	Parameter	Weight	Justification of Weight
1.	Charging time	10	This parameter is given the highest weight for fleet operation as time taken for charging is loss of revenue
2.	Ease of getting connection	8	This parameter is only secondary in importance to charging time as fleet charging energy and power demand for fleet charging are high
3.	Cost of charger	5	This parameter is given the lowest weightage as cost of charger is minimal in comparison with the cost of EV
4	Cost of installation and electricity connection	6	Cost of installation is also given a lower weightage as the cost of installation is minimal in comparison with the cost of EV

which notches up the highest normalized weighted rank would qualify as the most preferred option. The least normalized weighted rank would determine the least preferred option. According to the weights and ranks used in the example presented in Table 9.9, Level 3 charging technology is the most preferred option.

9.9 RECOMMENDATIONS FOR FLEET CHARGING

9.9.1 PUBLIC CHARGERS ARE REQUIRED TO SUPPORT FLEET

One of the key recommendations from the examination of electric taxi fleet operations is that public charging facilities are essential for fleet charging. MCDM tools can assist in the selection of suitable locations and best-fit charging technology for fleets as described in the section above. As an example, for charging hubs, the best possible chargers identified based on the selection criteria are Level 3 chargers. Once the suitable charging technology is identified, the next step is the estimation of the number of chargers at each location. Typically, the conservative estimates are required for DC fast charging, as they are relatively expensive than AC charging options. The decision-making factor is the number of vehicles that can be supported by one EV charger, and this factor is inherently dependent on the vehicle models. The number of vehicles that can be charged from Level 3 charger is around 10, 16, and 22 based on a utilization of 4, 6, and 8 hours in a day, respectively (Slowik, Wappelhorst & Lutsey, 2019). The most common car-to-fast charger ratio maintained by fleet operators in India is 5:1 (Das, Sasidharan & Ray, 2020).

9.9.2 ROLE OF AC CHARGING FOR FLEET

When the City of Los Angeles initiated a ride-sharing programme in 2015 for 80 EVs, the supporting charging facilities were all AC Level 2 (Kettles, 2016). With the

TABLE 9.9
Example of Decision-Making Matrix for Charging Technology Selection

Decision-Making Criteria	Weights	Level 2 Rank	Level 2 Rank × Weight	Level 3 Rank	Level 3 Rank × Weight	Level 4 Rank	Level 4 Rank × Weight	Level 5 Rank	Level 5 Rank × Weight
Charging time	10	1	10	2	20	3	30	4	40
Ease of getting connection	8	2	16	2	16	2	16	1	8
Cost of charger	5	4	20	3	15	2	10	1	5
Cost of installation and electricity connection	6	3	18	2	12	2	12	1	6
Sum	29		64		63		68		59
Normalized weighted ranks			2.20		2.17		2.34		2.03

advancement in charging technology and improvement in battery chemistries, fast charging infrastructure becomes more of a consideration when EV taxis is deployed. Yet, Level 2 AC charging is a viable option for overnight charging of commercial fleet. Research has also shown that charging electric ride-hailing vehicles at homes can lower the total cost of operation by 25% compared to using public rapid DC charging (Pavlenko, Slowik, & Lutsey, 2019).

9.10 GRID INTERACTION AND INTEGRATION OF FLEET

Fleet charging requires more energy and power in comparison with individual vehicle charging. Most of the chargers are connected to medium- and low-voltage distribution networks. These networks were not designed considering these kinds of loads. The grid impact from fleets will be more because of two factors: the number of vehicles and preference for high-powered charging. Though higher-powered charge points can reduce charge times, they cause a strain on electricity networks. The impact will be compounded by an increase in the number of vehicles charging at the same time and eventually lead to overloading and power quality issue parts of the network (Hu et al., 2016; Wu et al., 2011).

On the other hand, fleet charging also presents more opportunities for managing the charging in comparison with that of individual cars. The uncertainties in the charging behaviour is the biggest challenge to grid operation with EVs. There is a possibility to reduce the uncertainty in the charging of fleet vehicles, as it is easier to predict their charging needs based on operational patterns. Fleet charging infrastructure will have predictable energy requirements, typically high power, but at specific times of the day. Electric power utilities are critical partners in the roll-out of charging infrastructure as they have the best understanding of locations where there is available grid capacity for installing new stations and where upgrades would be needed (Sundström & Binding, 2010; Khodayar, Wu & Li, 2013).

The charging technology market has matured with controllable chargers and even bidirectional chargers. Fleet electrification thus opens a new market segment for grid services, which can help fleet operators and electricity distribution companies reduce their costs by managing energy in efficient ways. EVs have the capability of acting as demand response resources for the grid using grid-to-vehicle (V1G) functionality. EV fleets are potentially easier to bring in fleet charging management strategies. Fleet charging facilities can be configured to refill EV batteries with grid power when prices dip or when renewable energy generation is in excess. Fleet operators can benefit from charging at low energy rates (Nyugen et al., 2014; Heymann et al., 2017).

EVs, by virtue of being energy storage units, can potentially provide bidirectional flow of power using vehicle-to-grid (V2G) functionality. It is expected that the aggregation of EV fleets will help to promote V2G. Aggregated EV fleet can operate as virtual power plants and provide real and reactive power support to the local grid. Apart from that, by the versatile nature of being a distributed energy resource, they can assist in voltage and frequency support (DeForest, MacDonald & Black, 2018; Yi, Smart & Shirk, 2018).

NOMENCLATURE

AC:	Alternating current
AHP:	Analytical hierarchy process
B2G:	Battery-to-grid
CCS:	Combined charging system
CHAdeMO:	Charge de Move
DC:	Direct current
DoD:	Depth of discharge
EV:	Electric vehicle
EVSE:	Electric vehicle supply equipment
G2B:	Grid-to-battery
ICE:	Internal combustion engine
IEC:	International Electrotechnical Commission
LFP:	Lithium iron phosphate
LMO:	Lithium manganese oxide
MCDM:	Multicriteria decision-making
NCA:	Lithium nickel cobalt aluminium oxide
NMC:	Lithium nickel manganese cobalt oxide
PV:	Photovoltaic
SAE:	Society of Automotive Engineers
SoC:	State of charge
TCO:	Total cost of ownership
TOPSIS:	Technique for order preference by similarity to ideal solution
V1G:	Grid-to-vehicle
V2G:	Vehicle-to-grid
WPT:	Wireless power transfer

WORKS CITED

Ahmad, A., Alam, M. S., & Chabaan, R. (2017). A comprehensive review of wireless charging technologies for electric vehicles. *IEEE Transactions on Transportation Electrification*, 4(1), pp. 38–63.

Arora, N., & Raman, A. (2019). *Beyond Nagpur: The Promise of Electric Mobility*. New Delhi: Ola Mobility Institute.

Bloomberg. (2020). *China embraces battery-swapping system for electric vehicles*. Bloomberg News.

CCS. (2019). *Combined Charging System 1.0 Specification*. CHARIN. https://www.charinev.org/ccs-at-a-glance/ccs-specification/

CEA. (2013). *Draft Safety Provisions for Electric Vehicle (EVs) Charging Stations*. Central Electricity Authority. https://cea.nic.in/old/reports/others/enc/legal/cea_safety_regulations.pdf.

CEC. (2020). *GB / T 38775, Wireless Charging System for Electric Vehicles*. Chinese Electricity Council. https://www.chinesestandard.net/PDF.aspx/GBT38775.1-2020.

CHAdeMO. (2015). *CHAdeMO is published as IEEE standard*. Retrieved from: https://www.chademo.com/chademo-is-published-as-ieee-standard/.

CHAdeMO. (2020). *CHAdeMO 3.0 released: The first publication of ChaoJi, the new plug harmonised with China's GB/T*. Retrieved from: https://www.chademo.com/chademo-3-0-released/.

Chafkin, M. (2014). *A broken place: The spectacular failure of the startup that was going to change the world.* Retrieved from Fast Company: https://www.fastcompany.com/3028159/a-broken-place-better-place.

Clairand, J. M., Guerra-Terán, P., Serrano-Guerrero, X., González-Rodríguez, M., & Escrivá-Escrivá, G. (2019). Electric vehicles for public transportation in power systems: A review of methodologies. *Energies*, 12(16), p. 3114.

Coffin, D., & Horowitz, a. J. (2018, December). The supply chain for electric vehicle batteries. *The Journal of International Commerce and Economics.* Retrieved from https://www.usitc.gov/publications/332/journals/the_supply_chain_for_electric_vehicle_batteries.pdf.

Das, S., Sasidharan, C., & Ray, A. (2020). *Charging India's Four-Wheeler Transport.* New Delhi: Alliance for an Energy Efficient Economy.

DeForest, N., MacDonald, J. S., & Black, D. R. (2018). Day ahead optimization of an electric vehicle fleet providing ancillary services in the Los Angeles air force base vehicle-to-grid demonstration. *Applied Energy*, 210, pp. 987–1001.

Feldman, Y. (2017, August 31). *What Happened to Better Place's Electric Dreams?* Retrieved from The Jerusalem Post: https://www.jpost.com/Metro/A-story-that-should-be-told-503905.

Freewire. (2020). *Mobi EV charger.* Retrieved from: https://freewiretech.com/products/mobi-ev/.

Forbes. (2020). *Solar-powered electric vehicle charging stations are just around the corner.* Retrieved from: https://www.forbes.com/sites/kensilverstein/2020/02/10/solar-powered-electric-vehicle-charging-stations-are-just-around-the-corner/#5ff42f30320f.

GT. (2020). *Solar EV chargers simplify installation and adoption.* Retrieved from https://www.govtech.com/fs/infrastructure/Solar-EV-Chargers-Simplify-Installation-and-Adoption-.html.

Glowacki, P. (2016). *Developing Electrical Safety Standards to Introduce Electric Vehicles into Canada.* CSA Group, Canada.

Guo, S., & Zhao, H. (2015). Optimal site selection of electric vehicle charging station by using fuzzy TOPSIS based on sustainability perspective. *Applied Energy*, 158, pp. 390–402.

Hagman, J., & Langbroek, J. H. (2019). Conditions for electric vehicle taxi: A case study in the Greater Stockholm region. *International Journal of Sustainable Transportation*, 13(6), pp. 450–459.

Hall, D., & Lutsey, a. N. (2020). *Electric Vehicle Charging Guide for Cities.* Washington DC: International Council for Clean Transport.

Hall, D., Cui, H., & Lutsey, N. (2018). *Electric Vehicle Capitals: Accelerating the Global Transition to Electric Drive.* Washington DC: International Counçil for Clean Transport.

Heymann, F., Miranda, V., Neyestani, N., & Soares, F. J. (2017). December. Mapping the impact of daytime and overnight electric vehicle charging on distribution grids. In *2017 IEEE Vehicle Power and Propulsion Conference (VPPC)* (pp. 1–6). IEEE.

Hirschfeld, A. (2019) *Taxis go electric: Tesla joins cab fleet in New York city.* Retrieved from Observer https://observer.com/2019/11/tesla-taxi-electric-vehicles-cab-fleet/.

Hove, A., & Sandalow, D. (2019). *Electric Vehicle Charging in China and the United States.* Columbia, School of International and Public Affairs, Center on Global Energy Policy. Available online: https://energypolicy.columbia.edu/sites/default/files/file-uploads/EV_ChargingChina-CGEP_Report_Final.pdf.

Hu, J., Morais, H., Sousa, T., & Lind, M. (2016). Electric vehicle fleet management in smart grids: A review of services, optimization and control aspects. *Renewable and Sustainable Energy Reviews*, 56, pp. 1207–1226.

IEA. (2018). *Global EV Outlook 2018.* International Energy Agency, France.

Iles, R. (2005). Regulation of public transport services. *Public Transport in Developing Countries*, pp. 403–443. doi: 10.1108/9780080456812-019.

ISO. (2020). *ISO 19363:2020 Electrically Propelled Road Vehicles- Magnetic Field Wireless Power Transfer — Safety and Interoperability Requirement.* International Standards Organisation.

Jäger, B., Wittmann, M., & Lienkamp, M. (2017). Agent-based modeling and simulation of electric taxi fleets. *In 6. Conference on Future Automotive Technology.*

Jang, Y. J., Jeong, S., & Lee, M. S. (2016). Initial energy logistics cost analysis for stationary, quasi-dynamic, and dynamic wireless charging public transportation systems. *Energies*, 9(7), p. 483.

Jenn, A. (2019). *Electrifying ride-sharing: Transitioning to a cleaner future.* Retrieved from the University of California, Davis, https://escholarship.org/uc/item/12s554kd.

Jia, Y., Zhao, Y., Guo, Z., Xin, Y., & Chen, H. (2017). October. Optimizing electric taxi charging system: A data-driven approach from transport energy supply chain perspective. In *2017 IEEE Electrical Power and Energy Conference (EPEC)* (pp. 1–6). IEEE.

Kang, Q., Wang, J., Zhou, M., & Ammari, A. C. (2015). Centralized charging strategy and scheduling algorithm for electric vehicles under a battery swapping scenario. *IEEE Transactions on Intelligent Transportation Systems*, 17(3), pp. 659–669.

Karaşan, A., Kaya, İ., & Erdoğan, M. (2018). Location selection of electric vehicles charging stations by using a fuzzy MCDM method: A case study in Turkey. *Neural Computing and Applications*, 32(9), pp. 1–22.

Kettles, D. (2016). Electric vehicle fleet implications and analysis (No. FSEC-CR-2031-16).

Khodayar, M. E., Wu, L., & Li, Z. (2013). Electric vehicle mobility in transmission-constrained hourly power generation scheduling. *IEEE Transactions on Smart Grid*, 4(2), pp. 779–788.

Lai, Y. J. (1994). Topsis for MODM. *European Journal of Operational Research*, 76, pp. 486–500.

Lee, H., & Clark, A. (2018). *Charging the Future: Challenges and Opportunities for Electric Vehicle Adoption*, Harvard Kennedy School, Cambridge.

Lewis-Jones, A., & Roberts, J. (2017). *Electric private hire vehicles in London on the road, here and now, energy savings trust.* Retrieved from: https://energysavingtrust.org.uk/sites/default/files/reports/Uber%20EV%20Trial%20-%20Electric%20Private%20Hire%20Vehicles%20in%20London_1.pdf.

Liu, H.-C., Yang, M., Zhou, M., & Tian, G. (2018). An integrated multi-criteria decision making approach to location planning of electric vehicle charging stations. *IEEE Transactions on Intelligent Transportation Systems*, 20(1), pp. 362–373.

Løken, E. (2007). Use of multicriteria decision analysis methods for energy planning problems. *Renewable and Sustainable Energy Reviews*, 11(7), pp. 1584–1595.

Lu, X., Wang, P., Niyato, D., Kim, D. I., & Han, Z. (2015). Wireless charging technologies: Fundamentals, standards, and network applications. *IEEE Communications Surveys & Tutorials*, 18(2), pp. 1413-1452.

Lukic, S., & Pantic, Z. (2013). Cutting the cord: Static and dynamic inductive wireless charging of electric vehicles. *IEEE Electrification Magazine*, 1(1), pp. 57-64.

Mak, H. Y., Rong, Y., & Shen, Z. J. M. (2013). Infrastructure planning for electric vehicles with battery swapping. *Management Science*, 59(7), pp. 1557–1575.

Merkisz-Guranowska, A. and Maciejewski, M. (2015). *The Implementation of the Electric Taxi Fleet in the City of Poznan, Poland.* València: WIT Transactions on the Built Environment, pp. 243–254.

MIT, 2008, A Guide to Understanding Battery Specifications, MIT electric vehicle team http://web.mit.edu/evt/summary_battery_specifications.pdf.

Morrow, K., Karner, D., & Frankfort, J. (2008). *Plug-in hybrid electric vehicle charging infrastructure review.* US Department of Energy-Vehicle Technologies Program.

Musavi, F., Edington, M. and Eberle, W., (2012). Wireless power transfer: A survey of EV battery charging technologies. In *2012 IEEE Energy Conversion Congress and Exposition (ECCE)* (pp. 1804–1810). IEEE.

Navigant. (2018). *EV Charging Equipment Market Overview*. Boulder: Navigant Consulting, Inc.

NEC. (2017). *Article 625 Electric Vehicle Charging and Supply Equipment Systems*. National Electricity Code.

Nicholas, M., Slowik, P., & Lutsey, N. (2020). Charging infrastructure requirements to support electric ride-hailing in US cities, Washington DC: Publisher International Council for Clean Transport.

Pavlenko, N., Slowik, P., & Lutsey, N. (2019). *When Does Electrifying Shared Mobility Make Economic Sense?* Washington DC: The International Council on Clean Transportation.

Qiu, C., KT, C., Ching, T. W. & Liu, C. (2014). Overview of wireless charging technologies for electric vehicles. *Journal of Asian Electric Vehicles*, 12(1), pp. 1679–1685.

Rao, R., Zhang, X., Xie, J., & Ju, L. (2015). Optimizing electric vehicle users' charging behavior in battery swapping mode. *Applied Energy*, 155, pp. 547–559.

Reuters. (2019). *Oslo becomes the first city in the world to get wireless charging systems for EVs*. Retrieved from Auto NDTV: https://auto.ndtv.com/news/oslo-becomes-the-first-city-in-the-world-to-get-wireless-charging-systems-for-evs-2012267.

Riejos, F. A. (2019). *Urban Transport XXIV*. WIT Press, Spain.

Rodrigue, J.-P., Comtois, C., & Slack, B. (2013). *The Geography of Transport Systems*. Routledge, New York.

Rokhadiya, S., Wappelhorst, S., Pavlenko, N., & Bandivadekar, a. A. (2019). *Near-Term Incentives for Electrifying Electrifying Ride Sharing Fleets in India*. International Council of Clean Transportation, India.

Sasidharan, C., Chandra, D., & Das, S. (2019). Application of analytic hierarchy process for the siting of electric vehicle charging infrastructure in India. In *2019 IEEE Transportation Electrification Conference (ITEC-India)* (pp. 1–6). IEEE.

Sasidharan, C., Ray, A., & Das, S. (2019, October). Selection of charging technology for electric bus fleets in intra-city public transport in India. In *2019 Global Conference for Advancement in Technology (GCAT)* (pp. 1–8). IEEE.

Saaty, T. L. (1980). *The Analytic Hierarchy Process*. New York: McGraw-Hill.

Saaty, T. L. (2008). ` with the analytic hierarchy process. *International Journal of Services Sciences*, 1(1), 83–98.

SAE. (2019). *Wireless Power Transfer for Light-Duty Plug-in/Electric Vehicles and Alignment Methodology J2954*, Society for Automotive Engineers.

Sarker, M.R., Pandžić, H., & Ortega-Vazquez, M.A., (2014). Optimal operation and services scheduling for an electric vehicle battery swapping station. *IEEE Transactions on Power Systems*, 30(2), pp. 901–910.

Schiller, P. L., & Kenworthy, J. (2017). *An Introduction to Sustainable Transportation: Policy, Planning and Implementation*. Routledge, UK.

Slowik, P., Wappelhorst, S., & Lutsey, N. (2019). *How Can Taxes and Fees on Ride Hailing Fleets Steer them to Electrify?*, Washington DC.

Spöttle, M., Jörling, K., Schimmel, M., Staats, M., Grizzel, L., Jerram, L., & Gartner, J. (2018). *Research for TRAN Committee – Charging infrastructure for electric road vehicles*. European Parliament, Policy Department for Structural and Cohesion Policies. Brussels: European Parliament. Retrieved from http://bit.ly/2JBVvHq.

SGCC. (2013). *EV Infrastructure and Standardization in China, State Grid Corporation of China*, Washington DC.

Statt, N. (2019). *Norway will install the world's first wireless electric car charging stations for Oslo taxis*. Retrieved from The Verge: https://www.theverge.com/2019/3/21/18276541/norway-oslo-wireless-charging-electric-taxis-car-zero-emissions-induction.

Stojčić, M., Zavadskas, E. K., Pamučar, D., Stević, Ž., & Mardani, A. (2019). Application of MCDM methods in sustainability engineering: A literature review 2008–2018. *Symmetry,* 11(3), p. 350.

Sundström, O., & Binding, C. (2010) October. Planning electric-drive vehicle charging under constrained grid conditions. In *2010 International Conference on Power System Technology* (pp. 1–6). IEEE.

Teixeira, A. C. R, & Sodré, J. R. (2017). *A discussion on the application of electric vehicles to taxi cab fleets.* Retrieved from The European Energy Centre: https://www.euenergycentre.org/news/a-discussion-on-the-application-of-electric-vehicles-to-taxi-cab-fleets/.

Vesa, J. (2019). *SESKO.* Retrieved from https://www.sesko.fi/files/671/EV-charging_standards_may2016_Compatibility_Mode_.pdf.

Wang, H., & Cheu, R. L. (2013). Operations of a taxi fleet for advance reservations using electric vehicles and charging stations. *Transportation Research Record,* 2352(1), pp. 1–10.

Wang, H., Cheu, R. L., & Balal, E. (2016). Operations of electric taxis to serve advance reservations by trip chaining: Sensitivity analysis on network size, customer demand and number of charging stations. *International Journal of Transportation Science and Technology,* 5(2), pp. 47–59.

Wang, Y., Ding, W., Huang, L., Wei, Z., Liu, H., & Stankovic, J. A. (2017). Toward urban electric taxi systems in smart cities: The battery swapping challenge. *IEEE Transactions on Vehicular Technology,* 67(3), pp. 1946–1960.

WBCSD. (2019). *India Business Guide to EV Adoption.* World Business Council for Sustainable Development, India.

Worksafe. (2019). *Guidelines for Safe Electric Vehicle Charging.* New Zealand: Government of New Zealand.

Wu, Y., Yang, M., Zhang, H., Chen, K., & Wang, Y. (2016). Optimal site selection of electric vehicle charging stations based on a cloud model and the PROMETHEE method. *Energies,* 9(3), p. 157.

Wu, Q., Nielsen, A. H., Østergaard, J., Cha, S. T., & Ding, Y. (2011) December. Impact study of electric vehicle (EV) integration on medium voltage (MV) grids. In *2011 2nd IEEE PES International Conference and Exhibition on Innovative Smart Grid Technologies* (pp. 1–7). IEEE.

Yaghoubi, H. (2017). *Urban Transport Systems.* Croatia: BoD - Books on Demand.

Yi, Z., Smart, J., & Shirk, M. (2018). Energy impact evaluation for eco-routing and charging of autonomous electric vehicle fleet: Ambient temperature consideration. *Transportation Research Part C: Emerging Technologies,* 89, pp. 344–363.

Zyoud, S. H., & Fuchs-Hanuch, D. (2017). A bibliometric-based survey on AHP and TOPSIS techniques. *Expert Systems with Applications,* 78, pp. 158–181.

10 Machine Learning-Based Day-Ahead Market Energy Usage Bidding for Smart Microgrids

Mohd Saqib
Indian Institute of Technology (Indian School of Mines)

Sanjeev Anand Sahu
Indian Institute of Technology (Indian School of Mines)

Mohd Sakib
Aligarh Muslim University

Esaam A. Al-Ammar
King Saud University

CONTENTS

10.1 Introduction	250
10.2 Different Aspects of EVs	251
10.3 Description of Power Market Stakeholder Interaction Model	254
10.3.1 Activity Diagram	255
10.3.2 Data Flow Diagram	255
10.3.3 ER Diagram	256
10.4 AI Strategies	258
10.4.1 Artificial Neural Networks	258
10.4.2 Autoregressive Moving Average	258
10.4.3 Support Vector Machine	259
10.5 Overall Demonstration	260
10.6 Case Studies	262
10.6.1 Forecasting of Energy Price	262
10.6.2 Aggregate Demand-Supply System	262
10.6.3 xEV Market Analysis and Forecast	263
10.7 Result	263
10.8 Conclusion	263
Nomenclature	264
References	264

10.1 INTRODUCTION

Increment in the pollution has bound the government to take some strong decisions to implement zero-emission due to which many more xEVs come on the road. The high energy demand will be required due to the rapid growth of xEV and maybe become the reason of system failures or give rise to mismanagement. So it is the reason, in this era, we need an automated charging infrastructure that can operate by various entities related to the power grid. Furthermore, we know that the most paranoid part of the smart grid is the bidding process of the charging/electricity because like other resources of the energy (e.g., coal, petroleum, etc.), it cannot be stored. That is why we need to automate the bidding process for the system. As we are living in an era of artificial intelligence (AI), it can bring revolutionary changes in the energy market. To achieve these benefits, proposed work came into existence. In the proposed system, we introduce an AI-based cyberinfrastructure that automates and optimizes the bidding process in the day-ahead market.

The industries are now ready to establish a new system and move on from our antiquated, human-oriented, traditional gasoline engine vehicles because of lots of problems created by such vehicles [1]. To reduce emission, many gasoline engines are transferring to electric vehicles [2]. This [3] study shows different government programs to implement zero-emission and needed to achieve CO_2 targets soon. xEVs can be performed as a dependable, sustainable, and finest powertrain alternative for transportation soon, but its adoption in the present infrastructure can pose difficulty for management unless it is smartly integrated. The increasing number of electric vehicles caused new challenges like tracking the nearest charging station, bidding price, queuing delay, etc. The massive demand for electric vehicles in the present transport sector requires a conventional system and coordination mechanisms to operate. As we are aware, BigData is time-demanding and complex to handle, which needs a powerful system to maintain good accuracy. The future demand for EV is rapidly growing, but it is hard to estimate the upcoming load, so it is impractical to develop an optimal scheduling scheme for EVs [4]. Also in Ref. [5], the authors explain difficulties with the new computer-oriented power plants and discussed proper solutions for it as well. However, in Ref. [6], researchers discuss to formulate the real-time charging and pricing system of xEVs based on the Internet of Things (IoT) and also conclude with a global optimal scheduling method. A synchronized approximate dynamic programming model for coordinated charging control is introduced in Ref. [7], and similarly, a scheduling paradigm has been represented in Ref. [8] to minimize electricity depletion. Also, Information and Communication Technology (ICT) opens new doors for improvement in the scenario of power grid management. In Refs [7] and [9], the authors have presented a smart infrastructure model to handle xEV market in India using ICT. Refs [7] and [10] present the need from xEV users to xEV charging control mechanism based on hierarchical aggregators, which can provide coordination between demand and distribution of power through real-time charging infrastructure. In Ref. [8], a genetic algorithm-based charging controller has been demonstrated to maintain the system threshold voltage within permissible limits, and to perform this tedious job, various inputs like SOC, minimum voltage, and price of energy are loaded. A mathematical model has been employed by the researcher in this Ref. [11] study to manage energy demand near a highway entry-exit port. This study also overcomes the further issues of peak energy demand for xEVs on the highway. A sliding window optimization algorithm has been presented in this Ref. [12] to achieve low-cost charging in an automated grid. In Ref. [13], the dynamic programming paradigm is adopted to manage multidimensional energy flow.

Although ICT brings revolutionary changes in the energy market and tackled many engineering problems, we are living in the era of AI, machine learning (ML), and deep learning (DL), which may take the energy market to the next level. This [14] article explains all the existing problems with ICT-based systems, for example, the complexity of solutions, demand of quality programming and computer experts, threats of cyber tools, etc. and also speculates on the regulation of various xEV-related forecasting.

And this [15] research presented how AI can transform the energy market and also introduce the platform architectural logic that circumscribes technical and financial architecture of AI-based smart grid platforms for the upcoming energy market. Same as in Ref. [16], a framework to forecast electricity prices is proposed using DL. Also, these [17–20] studies proposed an artificial neural network (ANN)-based algorithm to forecast energy price and load as well.

The manuscript is organized as follows. First of all, the manuscript starts with a detailed section of "**Different aspects of EVs**" followed by the "**Description of Power Market Stakeholder Interaction Model**" which explains the software-working model for the stakeholder entities. The section "**AI Strategies**" describes the AI techniques that can be formed by automated bidding system followed by "**Overall Demonstration**", which explains the overall system and different connectivity. In the next section, three case studies have been given to justify results namely "**Case Studies**". And results are described in the last section "**Result**" followed by the conclusion in "**Conclusion**".

10.2 DIFFERENT ASPECTS OF EVs

As we have understood from the literature survey, vehicle to grid (V2G) and grid to grid (G2G) infrastructure has been executed and analyzed in an inadequate number of pilots, and various problems need to be solved before wide adoption. While newly developed charging infrastructures also have some issues, solutions are under process. In Feb 2017, around 580 thousand were sold in the U.S.A. due to a lack of proper regulation [20]. To overcome all these issues, various other technologies need to be discussed apart from AI and ML as follows [21]:

- *Faster charging stations.* Slow charging stations are the major inconvenience, and it is one of the reasons most users do not feel free to take out EVs for long-distance roots. Various industries have already head-started on deploying global supercharging stations, for example, Tesla.
- *Battery quality improvement.* Using lithium is not a better way to make batteries because it is hard to find. As in this [22], the authors discussed a system to enhance battery life as well as battery performance.
- *Autonomous driving.* Autonomous driving is the need of today to make roads safer. Tesla and many others are leading in this technology.

- *Software and mobile apps.* There are so many apps developed to manage and monitor real-time data related to charging stations and EVs. For example, in Ref. [23], the author discussed cloud-based monitoring software to access real-time information.

EVs have several **environmental benefits** compared to gasoline engine cars. Carbon dioxide (CO_2) is produced by burning gasoline, which causes greenhouse gas and leads to climate change issues [3]. Using EVs instead of gasoline vehicles reduces the amount of CO_2 about 4,096 pounds per year [24]. It is also estimated by Stanford that the amounts of carbon emissions will reduce to roughly $4,506 over 10 years. Furthermore, EVs also show significant benefits from the methods used for electricity generation. The problem to obtain gasoline is carbon-intensive production in pumping, mining, and refining tasks. So the conclusion is EVs also will be beneficial even for those countries that produce electricity by burning fossil fuels.

So many studies represented the **economic benefits** of EVs [25]. Surely, the arrival of EV infrastructure will result in job losses in the conventional gasoline and mechanical industries, but many other new doors will open for research and development fields, automation industries, AI companies, and battery manufacturers. Also, EVs cost less to maintenance and operational tasks as summarized in Table 10.1 [25].

Apart from the economic and environmental aspects of EVs, they also have various **safety aspects** as discussed in Ref. [26]. The first thing that should be kept in mind is the safety of the electrical system. Different ranges of voltage used in EVs (small vans vary from 48 to 120 V, large vans from 90 to 250 V, buses from 250 to 500 V, and AC vehicles more than 250 V) are considered dangerous for human safety. Therefore, there is a need to prevent electric shocks in direct and indirect touches. Second, the functioning of EV systems is also very sensitive, e.g., range anxiety, real-time information regarding CS locations, nearby areas, queuing delay, etc. Apart from all these regulations and monitoring safety, cyber-security is also an obstacle. Automated EVs need special cyber securities, otherwise it may cause very serious accidents on the roads. Third, battery safety is the most critical part of EVs. It may cause several electrical, chemical, and mechanical complications and an explosion [26]. There is a need for strong locking connections when multiple batteries are implemented to prevent short circuits. There should be surveillance measures

TABLE 10.1
Comparisons of Traditional Vehicle and EV

Factors	Gasoline Vehicle (Price in $)	EV (Price in $)
Brakes	400.00	200.00
Oil	600.00	0
Spark plugs and wires	200.00	0
Transmission fluid	60.00	0
Tires	700.00	700.00
Muffler	180.00	0
Total	2,140.00	900.00

to avoid explosion, which especially occurs at the end of charging. At the time of charging, EVs are connected with the main distributed network, so there should be precautions to avoid electric shock.

After all these discussions, let's come on major government programs empowering EV infrastructures in various markets and highlight some emerging **best practices** that increase the stock of EVs, subsidies, public-private partnerships, etc. as discussed in this [27] detailed report of the International Council on Clean Transportation, U.S.A.

- *China.* Energy stockholders in China, including government sectors, have been fast-moving toward developing EV infrastructures in the country. In the upcoming years, it will enhance its market position. In the time frame of 2020, China wants to set new goals to sell EVs that are in the embryonic stage to achieve the target of 300,000 EVs [27]. Also, it has announced EVs having 20% increment in production by 2025. Furthermore, the numbers of charging stations are also rapidly growing in China especially in the 88 designated pilot cities led by Shanghai, Beijing, and Shenzhen. There is a plan to build around 100,000 smart charging stations and more than 400,000 total public stations by 2020. The goal of the program is to provide 8 EVs per charging stations and farther than 1 km any point within the center area of the city.
- *Japan.* Since the introduction of modern EVs in Japan in 2011, at a very early stage, energy stockholders and government sensed that the key requirement for EV sales is charging infrastructure, so they supported [27]. In 2013, the government sector worked on advance charging architecture projects to developed CSs around the cities, and also the Bank of Japan cooperated with Nissan, Honda, and Toyota to develop Nippon Charge Service (NCS). Now, 7,500 CSs are connected to this project with regular funding.
- *United States.* The U.S. EV sales are rapidly growing due to government support, zero-emission vehicle regulatory policy, public-private relationship, etc. In California, a zero-emission policy is expected to increase electric vehicles in the market from more than sixty thousand to several million in decay. To complete the requirement of guiding the charging deployment process, California developed a model that projects needs of CSs by 2025 [28].
- *France.* In France, the government has initiated a plan to move all vehicle sales to electric by 2040. Also, the target is to operate 12,000 charging stations in the city. To achieve this goal, many funds are releasing, and the recipients must ensure to develop at least 20 char points and also make slots for free parking. The federal government's strong role is evident in empowering the charging infrastructure in France.
- *Germany.* Germany has a goal to reach 1 million by 2020 and 5 million by 2030. At the start of 2009, In Hamburg and Saxony, the country started to support more than 200 projects in eight "model regions" with €130 million, boosting charging infrastructure. Furthermore, the country planned to promote EVs, including €300 million earmarked for public charging infrastructure through 2020.

Europe. xEV charging infrastructure has been regularized by private as well as government bodies. To empower the sales of EVs and charging infrastructure, many countries within the European Union have created funding schemes. Some of them are using charging infrastructure for several years like Norway and Netherland and some are recently launched like Germany. Under the initiative for increasing the number of CSs in the city, a program, TEN-T/CEF-T, was launched [27]. And under this program, the European Commission is ensuring that charging stations are in public reach without any suffering of range anxiety.

So, not only in China, Japan, and the U.S.A that the EVs became the major transport, but also in Europe, more than 70% of transport responsibility depends on them [29,30]. Besides, it has the following issues:

- *Range anxiety.* Although technically it sounds good, but still users of EVs have the problem of range anxiety due to technical limitations of the battery.
- *Lack of understanding.* People still have less awareness and understanding of e-mobility.
- *Charging system.* Although government is working so far to build a reachable charging infrastructure, there is still a lack of reachability, and users are facing monitoring problems.

The government is still struggling to build up sufficient infrastructure and implement policies favoring e-mobility systems in urban areas [30]. In 2018, a meeting was conducted between energy experts in Stockholm to discuss on different initiatives and solution to operating and adoption of e-mobility, and they have concluded the following seven points [30]:

- Increment in charging points/stations
- User-friendly systems
- Smart solutions based on ICT and AI
- Enhancement in the number of the training-testing scenario of the projects related to smart charging infrastructure
- Capacity enhancement of power grid
- More focus and regularization in urban areas
- Spread awareness of e-mobility among urban areas

10.3 DESCRIPTION OF POWER MARKET STAKEHOLDER INTERACTION MODEL

This section used different diagrams of Unified Modeling Language like activity diagram used to represent different entities of the system and its activities, entity-relation (ER) diagram constructed to demonstrate the relationship among the entities that later used to create database schema of the software, and data flow diagram that explains the flow of data during different tasks of the system.

10.3.1 Activity Diagram

Activity diagrams also known as use case diagrams illustrate to show an association between different entities and their actions (use cases). In other words, it illustrates to which functionality (s) the entities/actors in the system are involved. Each use case diagrams manifest some relevant, beneficial, and observable results to the software entities or other stakeholders of the software. It shows the individual work of each entity and manifests very useful information to the system developer. For example, in our case, Figure 10.1 represents the task of xEV users like registration, login, bidding for the energy, information updates, changing bid, etc. and in the same way also shows the task of charging stations and power grids like registration, login, view all bidding, etc. This diagram also helps the programmer to identify coupling among methods of entities.

10.3.2 Data Flow Diagram

A very essential need of a developer to construct a stable design of the software is to understand the flow of information and distribution of the data among entities. Here, data flow diagram (DFD) comes into the picture. DFD is a diagrammatical representation of database modeling and relationship with the methods of entities (see Figure 10.2). In the proposed system, mainly three tables of different energy stakeholders are developed – xEVs_Users_DB (to keep track of all the information of the xEV User), CSP_DB (storing charging service provider details), and EP_DB (hold

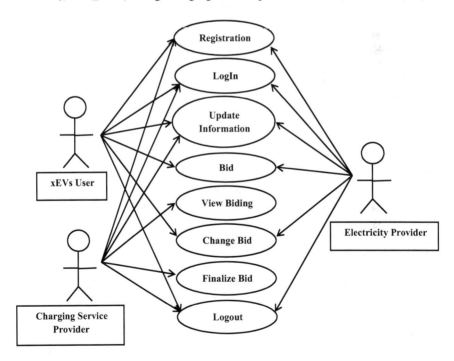

FIGURE 10.1 Activity diagram.

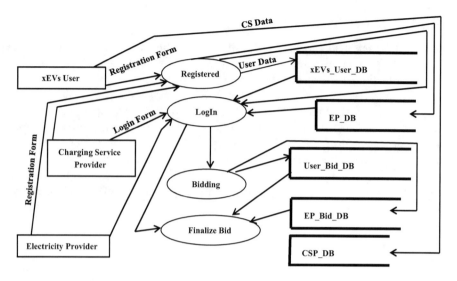

FIGURE 10.2 DFD.

information regarding different power plants associated with the system). Another two tables are User_Bid_DB and EP_Bid_DB. These two tables store all the information regarding the bidding by users and power plants and provide insight for the charging station admin to finalize the bidding.

10.3.3 ER Diagram

Developers use ER diagram to store real-world objects into the database using its parameters and attributes; e.g., in this proposed system, xEV users, charging service provider, and power plants are the entities of the system. In the system, an entity can be defined and distinguished by its attributes (Figure 10.3). All the entities and their attributes are the following:

- *xEV users.* xEV user is the consumer of energy who bids for it by using an online process, but it has to be authenticated on the system before accessing. xEV user needs to be registered by filling the online form and providing the details, e.g., xEV vehicle identification number (VIN), name, contact number, email, address, type of xEV battery, etc. After providing all the necessary details, xEV users will receive USER_ID and self-generated password to login to the system.
- *Charging service providers.* Charging service provider is the main entity of the system. It is also registered in the same way as xEV users. It has different attributes, e.g., name, contact detail (address, contact number, fax, email, etc.), price-quantity of energy, capacity, the places where xEV user can relax during charging. It can log in on the system by system/self-generated CSP_ID and password.

Day-Ahead Market Energy Usage Bidding 257

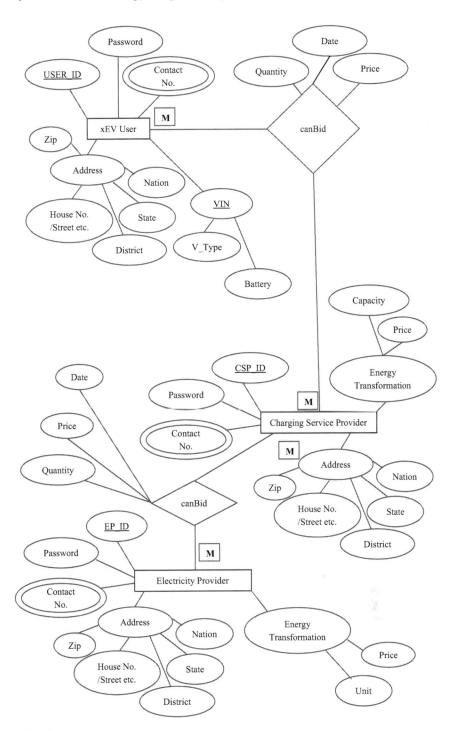

FIGURE 10.3 ER diagram.

- *Electricity provider.* It is the main entity of the system, which is responsible to originate the energy and having attribute similar to the charging service provider.

10.4 AI STRATEGIES

Energy bidding is a very sensitive process because of the nature of electricity that cannot be stored like other sources of energy, coal, petroleum, etc. There is a need for optimal bidding to be chosen for both (Energy Provider and xEV Users), and this process is nothing but a time series forecasting, which can achieve any AI strategy. Some of them are the following:

10.4.1 Artificial Neural Networks

Since energy demand in day-ahead market depends on previous demand, an ANN model is implemented for prediction of the future energy price. In this algorithm model train by using previous year data, and after that we can predict the upcoming rate for bidding [31]. Previous years' bidding data need to be split into training and testing datasets. Training data can be used for model construction, and testing data are for validating the results. Many metadata like weights, learning rates, the numbers of hidden layers need to be initialized after splitting the datasets. Minimum error and epochs are initialized. Many activation functions can be used for regression time series forecasting. As Figure 10.4 demonstrates, vector $X = [x_1, x_2, x_3, \ldots x_n]$ defines an input layer which can be previous energy prices. And y will be the predicted price of the upcoming day. In the last step, we need to get the accuracy of the model, which can be achieved using either the mean square error method (MSEM) or the mean absolute error method (MAEM).

10.4.2 Autoregressive Moving Average

In this [32] paper, the researcher has used to predict a load of power system using autoregressive moving average (ARMA). In the same way, we can also work on forecasting bidding price quantity for the electricity. The predictions take place based on a data sequence over time. Once model parameters are obtained from training set, bidding price can be predicted for futuristic demand. For the time series, observations are recorded based on time in the following format:

$$\ldots, \ldots, \ldots, Z_{t+l+1}, Z_{t+l}, Z_{t+l-1}, \ldots, \ldots, \ldots \quad (10.1)$$

The present term can be represented in the form of previous terms as follows:

$$Z_{t+l+1} = \varnothing_1 Z_{t+l} + \varnothing_2 Z_{t+l-1} + \cdots + \varnothing_p Z_{t+l-p} - \theta_1 a_{t+l} \cdots - \theta_q a_{t+l-q} + a_{t+l+1} \quad (10.2)$$

where $\varnothing_1, \varnothing_2 \cdots \varnothing_p$, and $\theta_1, \ldots \theta_q$ are the constants. The model task is to find the values of these constants to reduce prediction error up to a certain limit.

Day-Ahead Market Energy Usage Bidding

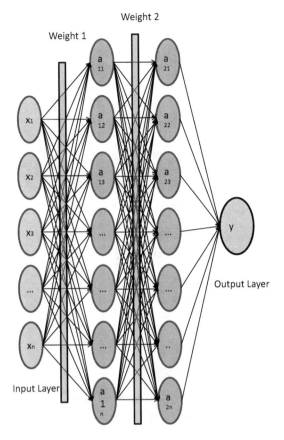

FIGURE 10.4 Demonstration of ANN.

10.4.3 Support Vector Machine

This ML algorithm is used for classification and regression problems, but its model can be extended feasibly to the time series prediction [33,34]. The following function is used to solve regression problem:

$$f(x_i) = w.\varnothing(x_i) + b \tag{10.3}$$

Where x_i is the previous bidding price as input data, b is the bias, and w is the weight vector. And $\varnothing(x_i)$ is a Kernel function (KF), which has many types, but the most used KF is radial basis function kernel:

$$\varnothing(x_i, \bar{x}_i, \sigma) = \exp\left(-\frac{\overline{x_i - \bar{x}_i}^2}{2\sigma^2}\right) \tag{10.4}$$

where $\bar{x}_i$ denotes the mean, and σ represents the standard deviation of the input values. The regression line is best defined to that line which minimizes the following cost function (CF):

$$\text{CF} = \frac{1}{2}\|w\|^2 + c\sum L^{\varepsilon}(x_i, y_i, f) \tag{10.5}$$

Using constraints,

$$y_i - wx_i - b \leq \varepsilon + \xi_i \tag{10.6}$$

$$wx_i + b - y_i \leq \varepsilon + \xi_i^* \tag{10.7}$$

$$\xi_i, \xi_i^* \geq 0 \tag{10.8}$$

where c is the positive constant, ε is the deviation, and ξ_i, ξ_i^* the slack variables. In Ref. [33], the author used support vector machine (SVM) to predict the short-term load on the grid.

10.5 OVERALL DEMONSTRATION

The proposed system is developed using many programming languages and environments. The entire server-related processing tasks and database interaction code are written in PHP. On the client-side, interactive web pages are designed using HTML, CSS, and Ajax. An android app is also designed to reduce the load on the server, making it more user friendly, which is written in JAVA. The main task of the system is to predict both, electricity provider and xEV users. Python of version 3.x is used to implement AI predictions. The proposed expert system for the bidding process of day-ahead market for energy using ML strategies illustrates the important ICT-based paradigm focused to deploy a conventional system and coordination mechanisms to operate a virtual energy market. It demonstrates the Bid and auction mechanism that undertook between xEV users and charging stations stack holders working as energy providers as well as power system stack holders with charging stations working as energy consumers. The real-time information regarding predicted energy tariff, nearest charging station, ensuring least queuing delay, etc. is easy and can be monitored using the proposed system. A prototype of the model is represented in Figure 10.5. The whole scenario is divided into four layers as represented in Figure 10.6. All the layers have some significant tasks.

- *Layer 1* is the first layer of the system, which contains all the end-users of the system and has access to Layer 2 via client-side web pages or android app.
- *Layer 2* has server-side codes and APIs which interact with databases on the cloud.

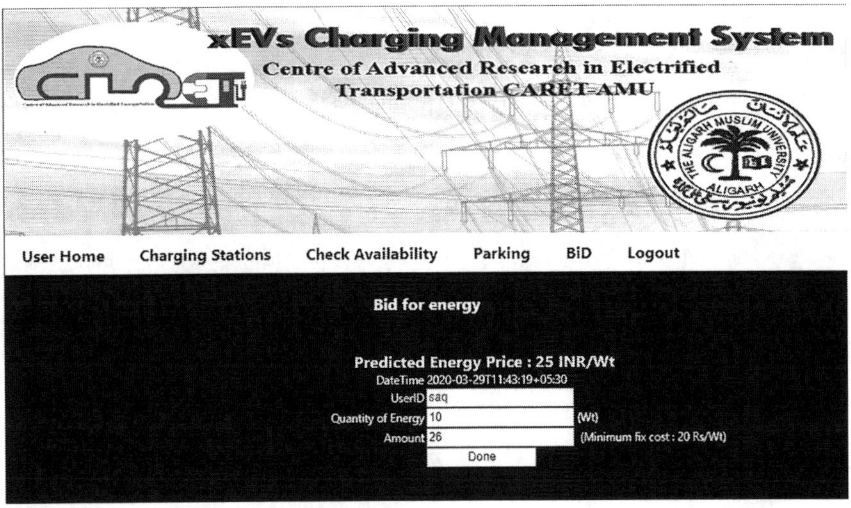

Day-Ahead Market Energy Usage Bidding   261

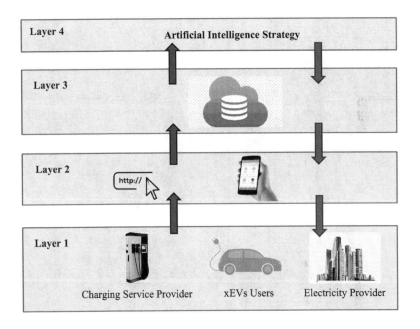

FIGURE 10.5 Prototype of the model.

FIGURE 10.6 Overall demonstration.

- *Layer 3* is the layer where a centralized database comes into existence, which is available on the cloud. This layer provides all the data needed to process AI activity, which happens on Layer 4.
- *Layer 4* has AI algorithms and libraries that work with available data and predict the information.

10.6 CASE STUDIES

To justify the results and importance of the system, we have taken three case studies – Forecasting of energy price, Aggregate Demand-Supply System, and xEV Market Analysis and Forecast.

10.6.1 Forecasting of Energy Price

According to Ref. [15], AI is rapidly growing in the fields related to smart grid and energy markets. In this study, two methods are introduced to predict price and price spike, ARMA, and ANNs. In this case study, the model applied real data from the Finnish Nord Pool Spot day-ahead energy market and compared both original price and predicted price.

10.6.2 Aggregate Demand-Supply System

Figure 10.7 presents a chart showing one-day (29th March 2020) market scenario (Buy & Sell). The total no. of buy bids on 29th March 2020 is 221 MW, and the total no. of sell bids is 107 MW. All the informative data such as the no. of bidder, seller, cleared volume, market-clearing price, and locations to withdraw and inject the bided power and the details of the customer behavior based on previous traded transactions will be stored in the cloud system to execute the proper day-ahead as well as term-ahead energy trading mechanism [35].

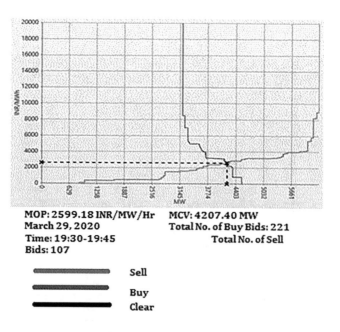

FIGURE 10.7 Aggregate demand-supply system.

10.6.3 xEV Market Analysis and Forecast

According to Ref. [36], xEV market was evaluated at $162.34 billion in 2019 and predicted to be $802.81 billion by 2027. In the world, Asia-Pacific was on the top of revenue contribution of $84.84 billion in 2019, which is predicted to reach $357.81 billion by 2027 with a compound annual growth rate (CAGR) of 20.1%. Even though EV infrastructure is in a nascent phase in India, EV market is expected to grow with a CAGR of 43.13% and installation of charging Infrastructure to grow 42.38% during the time from 2019 to 2030. The various aspects such as increasing demand for fuel, high performance, and zero-emission government rule supplement the demand for xEVs. We need to be ready to adopt this rapid growth of EVs and have to develop an intelligent management system as proposed in this study.

Now from analysis of these case studies, we conclude that the applications of AI may help a lot in the energy market to solve various engineering problems, and we also observed that bidding for the energy is so frequent, which may cause a system failure. That is why in this era of ICT and AI, we need to bring revolutionary changes in the energy market via using the proposed expert system.

10.7 RESULT

The proposed work demonstrated an AI-based smart grid interface for the bidding process of day-ahead market. The framework describes different parameters and circumstances to predict the future price for energy tariff and detailed ML methods suitable for the task, e.g., ANN, ARMA, and SVM. Also, the system successfully describes the various objectives of different stakeholder-specific infrastructures that handle massive demand for electric vehicles into the present transport sector through implementing conventional and coordination mechanisms to operate. The model provides an extremely good analytical and computational power to operate BigData because of many powerful tools and libraries introduced to compile ML algorithms, e.g., Python, Sklearn, and cloud-based 3-tier architect database showing that after the prediction, we can implement an optimal scheduling scheme for xEV users. Moreover, the software also does dedicated jobs for xEV users, charging providers, and other entities to evolve to the charging-bidding process. The proposed system keeps track of previous data of biddings and predicts for the current situation. The work also ensures the cyber security to maintain personal records of energy stockholders.

10.8 CONCLUSION

The proposed work presented an AI-based automated bidding expert system for the day-ahead market of energy. The application provides stakeholder-specific interface and module for optimal bidding for electricity. At the xEV user side, it provides an efficient and user-friendly GUI (Graphical User Interface) for bidding, and for power providers, it provides a suggestion for optimal service quotes to enhance the chance for their selection. All these suggestions are based on AI strategies using and analyzing previous years' bidding data. Moreover, it can reduce a load of processing on charging service providers by implementing some AI algorithms and will start working more and more accurately, as well as data grow on the cloud.

NOMENCLATURE

b: Bias
C: Positive constant
$o(tk)$:   Output training attribute at time t_k
W: Weight
$x(tk)$:   Input training attribute at time t_k
X_i: i^{th} input vector
$\overline{x_i}$:   Mean
Z_{t+l}: l observations recorded based on time t
$\varnothing_p$: p^{th} Constant of time series
$\varnothing(x_i)$:   Kernel value for X_i input vector
σ: Standard deviation
ξ_i, ξ_i^*:   Slack variables
L^ε: Regularization term
ε: Deviation

REFERENCES

1. A. Holms and R. Argueta, "A Technical Research Report: The Electric Vehicle," in *Business*, 2010, pp. 1–12.
2. C. Johansson et al., "Science of the total environment impacts on air pollution and health by changing commuting from car to bicycle," *Sci. Total Environ.*, vol. 584–585, pp. 55–63, 2017.
3. G. Hill, O. Heidrich, F. Creutzig, and P. Blythe, "The role of electric vehicles in near-term mitigation pathways and achieving the UK ' s carbon budget," *Appl. Ener.*, vol. 251, no. July 2018, p. 113111, 2019.
4. I. Abaker et al., "The rise of 'big data' on cloud computing : Review and open research issues," *Inf. Syst.*, vol. 47, pp. 98–115, 2015.
5. Y. Zhang, "Demand-Responsive Virtual Plant Optimization Scheduling Method Based on Competitive Bidding Equilibrium," *Energy Procedia*, vol. 152, pp. 1158–1163, 2018.
6. R. Aburukba, "Role of internet of things in the smart grid technology," *J. Comput. Commun.*, no. May, pp. 229–233, 2015.
7. M. Saqib, M. M. Hussain, M. S. Alam, M. M. S. Beg, and A. Sawant, "Public Opinion on Viability of xEVs in India," in *ISGW 2017: Compendium of Technical Papers*, 2018, pp. 139–150.
8. M. S. Alam, "Key Barriers to the Profitable Commercialization of Plug-in Hybrid and Electric Vehicles," *Adv. Automob. Eng.*, vol. 2, no. 2, pp. 1–2, 2013.
9. A. K. Digalwar and G. Giridhar, "Interpretive Structural Modeling Approach for Development of Electric Vehicle Market in India," *Procedia CIRP*, vol. 26, no. December, pp. 40–45, 2015.
10. M. Alonso, H. Amaris, J. G. Germain, and J. M. Galan, "Optimal Charging Scheduling of Electric Vehicles in Smart Grids by Heuristic Algorithms," *Energies*, vol. 7, no. 4, pp. 1–27, 2014.
11. S. Bae and A. Kwasinski, "Spatial and Temporal Model of Electric Vehicle Charging Demand," *IEEE Trans. Smart Grid*, vol. 3, no. 1, pp. 394–403, 2012.

12. I. S. Bayram, G. Michailidis, I. Papapanagiotou, and M. Devetsikiotis, "Decentralized Control of Electric Vehicles in a Network of Fast Charging Stations," *Globecom 2013-Symposium Selected Areas in Communications (GC13 SAC)*, Atlanta, GA, pp. 2785–2790, 2013.
13. H. Carter, P. Traynor, and K. R. B. Butler, "CryptoLock (and Drop It): Stopping Ransomware Attacks on User Data," *2016 IEEE 36th International Conference on Distributed Computing Systems (ICDCS)*, Nara, Japan, 2016, pp. 303–312.
14. R. Weron, "Electricity Price Forecasting: A Review of the State-of-the-Art with a Look into the Future," *Int. J. Forecast.*, vol. 30, no. 4, pp. 1030–1081, 2014.
15. Y. Xu, P. Ahokangas, and J. Louis, "Electricity Market Empowered by Artificial Intelligence: A Platform Approach," *Energies*, vol. 12, no. 21, p. 4128, 2019.
16. J. Lago, F. De Ridder, and B. De Schutter, "Forecasting Spot Electricity Prices: Deep Learning Approaches and Empirical Comparison of Traditional Algorithms," *Appl. Energy*, vol. 221, no. January, pp. 386–405, 2018.
17. T. A. Nakabi and P. Toivanen, "An ANN-Based Model for Learning Individual Customer Behavior in Response to Electricity Prices," *Sustain. Energy Grids Netw.*, vol. 18, p. 100212, 2019.
18. T. Y. Kim and S. B. Cho, "Predicting Residential Energy Consumption Using CNN-LSTM Neural Networks," *Energy*, vol. 182, no. 1, pp. 72–81, 2019.
19. M. A. R. Biswas, M. D. Robinson, and N. Fumo, "Prediction of Residential Building Energy Consumption: A Neural Network Approach," *Energy*, vol. 117, no. 1, pp. 84–92, 2016.
20. J. Massana, C. Pous, L. Burgas, J. Melendez, and J. Colomer, "Short-Term Load Forecasting in a Non-Residential Building Contrasting Models and Attributes," *Energy Build.*, vol. 92, no. 1, pp. 322–330, 2015.
21. R. Udas, "How can technology boost EV infrastructure?" *Express Computer*. [Online]. Available: https://www.expresscomputer.in/features/can-technology-boost-electric-vehicles-infrastructure-in-india/49380/. [Accessed: 05-May-2020].
22. A. A. Abdullah Al-karakchi, G. Lacey, and G. Putrus, "A Method of Electric Vehicle Charging to Improve Battery Life," *2015 50th Int. Univ. Power Eng. Conf.*, vol. 1, no. 1, pp. 1–3, 2015.
23. M. Saqib, M. M. Hussain, M. S. Alam, M. M. Sufyan Beg, and Amol Sawant. "Smart Electric Vehicle Charging through Cloud Monitoring and Management," *Technol. Econ. Smart Grids Sustain. Energy*, vol. 2, p. 18, 2017. https://doi.org/10.1007/s40866-017-0035-4.
24. "Environmental aspects of the electric car," *Wikipedia*. [Online]. Available: https://en.wikipedia.org/wiki/Environmental_aspects_of_the_electric_car. [Accessed: 05-May-2020].
25. I. Malmgren, "EVS29 symposium quantifying the societal benefits of electric vehicles," *World Electr. Veh. J.*, vol. 8, no. June, pp. 996–1007, 2016.
26. T. A. Stevan Kjosevski, Aleksandar Kostikj, and A. Kochov, "Risks and Safety Issues Related to Use of Electric and Hybrid Vehicles," *Mach. Technol. Mater.*, vol. II, pp. 169–172, 2017.
27. E. W. Wood, C. L. Rames, A. Bedir, N. Crisostomo, and J. Allen, "California Plug-In Electric Vehicle Infrastructure Projections: 2017–2025 - Future Infrastructure Needs for Reaching the State's Zero Emission-Vehicle Deployment Goals," Natl. Renew. Energy Lab. (NREL), Golden, CO, pp. 1–58, 2018.
28. G. R. C. Mouli, V. Prasanth, and P. Bauer, "Future of Electric Vehicle Charging," *19th International Symposium on Power Electronics, Ee 2017*, vol. 2017 -December, pp. 1–7, 2017.
29. "Recommendations on Electric Vehicle Charging Infrastructure" [Online], Available: https://www.interregeurope.eu/policylearning/news/5039/ev-energy-s-recommendations-on-electric-vehicle-charging-infrastructure/. [Accessed: 08-May-2020].

30. I. Strategic, M. Conference, O. F. Demirel, and S. Zaim, "Forecasting Electricity Consumption with Neural Networks and Support Vector Regression," *8th International Strategic Management Conference*, vol. 58, pp. 1576–1585, 2012.
31. J. Chen, W. Wang, and C. Huang, "Analysis of an Adaptive Time-Series Autoregressive Moving-Average (ARMA) Model for Short-Term Load Forecasting," *Electr. Power Syst. Res.*, vol. 34, no. 3, pp. 187–196, 1995.
32. Y. Fu, Z. Li, H. Zhang, and P. Xu, "Using Support Vector Machine to Predict Next Day Electricity Load of Public Buildings with Sub-metering Devices," *Procedia Eng.*, vol. 121, pp. 1016–1022, 2015.
33. A. S. Ahmad, M. Y. Hassan, M. P. Abdullah, H. A. Rahman, F. Hussin, H. Abdullah, and R. Saidur, "A Review on Applications of ANN and SVM for Building Electrical Energy Consumption Forecasting," *Renew. Sustain. Energy Rev.*, vol. 33, pp. 102–109, 2014.
34. "Aggregate Demand Supply Curves," Available: https://www.iexindia.com/marketdata/demandsupply.aspx. [Accessed: 08-May-2020].
35. A. Singh, "Electric Vehicle Market" [Online], *Allied Market Research*, Available: https://www.alliedmarketresearch.com/electric-vehicle-market. [Accessed: 05-May-2020].

11 Smart Microgrid-Integrated EV Wireless Charging Station

Aqueel Ahmad and Yasser Rafat
Aligarh Muslim University

Samir M. Shariff
Taibah University

Rakan Chabaan
Hyundai Kia America Technical Center Inc

CONTENTS

11.1 Introduction 267
11.2 Solar PV Module Configuration with a Wireless Charging System 268
11.3 Solar to EV Battery Feasibility Analysis 269
11.4 Wireless Charging System for EVs 269
11.5 Finite Element Analysis Modeling and Simulation of the WPT Coils
 for Magnetic Analysis 272
11.6 Results and Discussion 273
11.7 Conclusion 277
Acknowledgment 278
References 278

11.1 INTRODUCTION

The growing public concern on environmental problems and rising fuel costs has led to growing emission-free, eco-friendly means of transportation [1]. Hence, EVs and PHEVs are emerging as an alternative to ICEVs [2]. Various international and national government such as (China, Japan, France, Germany, USA, UK, Netherland, Norway, etc.) have been passed resolutions, owing to these environmental concerns such as air quality, global warming, etc. and allocated significant funds to promote PEV and EV implementation and deployment [3]. Enduring scenario planning specifies the global vehicle fleet capturing by 2050 especially attracted by renewable energy sources, to overcome the worst case of global climate change scenario [4].

The EV deployment has been directed toward challenges such as limited fuel storage (i.e., small battery size or lower battery capacity) and range anxiety, which restricts EV deployment due to a limited number of commercial EV charging stations in the long driving range of EVs. The conductive charging faces challenges such as charger connectivity, manual connectivity, and limited life of contact charger. However, wireless charging technology can address most of the challenges confronted by the EV charging infrastructure. The EV wireless charging includes the convenience of charging during rest as well as motion, safe due to noncontact operation. Hence, EV charging using renewable energy will lead to pollution-free, economic, and efficient transportation [5].

The solar energy received from the sun around the earth in 1 hour can provide sufficient energy to fulfill a full year of utilization [6]. Solar energy is the cleanest and most sustainable form of energy received by the earth. It will take estimated another 40 years for the solar photovoltaic (SPV)-based energy generation to reach its maximum potential [7]. The measurement of sunlight intensity on any plane surface can be analyzed as energy per unit time per unit area, i.e., irradiance or power per unit area. Approximately the planet earth receives about four million exajoules (1 EJ = 1,018 J) of annual solar energy, out of which practically harvestable amount of energy has been claimed as 5.104 EJ [8].

In this manuscript, a solar-based wireless charging system has been proposed, where the power is transferred using the principle of magnetic induction, which is based on Faraday's law and Lenz's law. Where the magnetic field can be generated by a time-variant current-carrying conductor, the generated magnetic field is linked with a secondary coil conductor to generate a time-varying magnetic field [9]. The receiver is connected to the load to close the circuit; hence, the power is transferred wirelessly.

In this manuscript, SPV has been utilized to receive solar energy as a DC power supply. This power is wirelessly transferred to charge the EV.

11.2 SOLAR PV MODULE CONFIGURATION WITH A WIRELESS CHARGING SYSTEM

The block diagram of the solar PV system, grid, and EV wireless charging system is shown in Figure 11.1. During daytime, the solar energy can be utilized in charging EVs, and extrasolar power can be utilized to store the energy for off-day time utilization. However, if during daytime, the SPV generation is lower than the demand of the EV wireless charging system, then it is recommended to use the grid supply for charging the EV. The wireless charging makes charging very convenient, and the automation with this system may lead to a human-free system.

The major components of the solar-based wireless charging system are as follows:

1. *Solar PV system.* Depending on the demand, the capacity of the solar panel can be decided; however, this manuscript presents the 3.3 kW solar-based wireless charging system.
2. *Solar inverter and Maximum Power Point Tracking (MPPT) controller.* The solar power generated from PV varies with the intensity of the solar radiation. Hence, an inverter embedded with the MPPT controller is required.

Integrated EV Wireless Charging Station

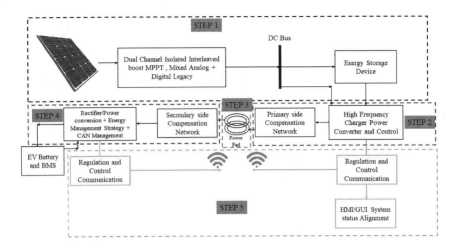

FIGURE 11.1 Block diagram of the solar, energy storage, and EV system.

3. *Energy storage system.* Solar power cannot be utilized directly to charge the EV; hence, an energy storage system consisting of batteries of very high capacity is required. Firstly, the solar energy will be utilized to store the energy in the battery and will be utilized to charge the EV. Simultaneously, the storage system can be utilized to store the extra energy generated during daytime and can be utilized during nighttime.

4. *Wireless charging system.* The wireless charging system will receive the DC power from the battery, the supply will be converted to a high-frequency AC, passing through transmitter and receiver coil, and the power is again rectified to charge the EV.

11.3 SOLAR TO EV BATTERY FEASIBILITY ANALYSIS

The solar energy utilized for charging EV is very efficient, sustainable, and environmentally friendly. Many of the researchers have performed the solar-based analysis to charge EVs such as in Refs [1–6]. However, there is none that discussed the solar-based wireless charging system. Most of them have already determined the efficient solar energy utilization, calculation of SPV requirement based on demand to charge EVs [10], optimization of solar panel, and energy demand utilization.

In this manuscript, the wireless charging system has been discussed in detail, which receives the input from the SPV system. Figure 11.2 presents the overall concept of the proposed system.

11.4 WIRELESS CHARGING SYSTEM FOR EVs

The fundamental principle of application of wireless power transfer (WPT) is similar to an air-cored transformer where the transmitter and receiver of the WPT system are considered as primary and secondary coil transformers [12]. Figure 11.3 shows

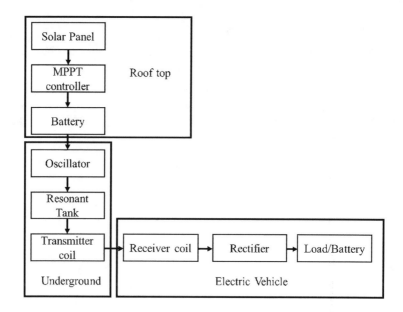

FIGURE 11.2 Block diagram of the overall wireless charging system.

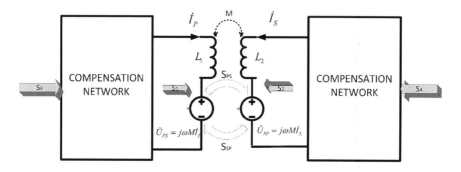

FIGURE 11.3 Simplified block diagram of two-coil WPT system.

a separate compensation topology of WPT. A simple WPT analysis has been performed neglecting coil resistances and magnetic losses. Hence, the exchanged complex power from L_P and L_s can be evaluated as follows.

For the two-coil system, the primary to the secondary exchange of apparent power is given by (11.1)

$$S_{PS} = -\dot{U}_{PS}\,\dot{I}_S^* = -j\omega M \dot{I}_P \dot{I}_S^* = \omega M I_P I_S \sin\varphi_{PS} - j\omega \cos\varphi_{PS} \quad (11.1)$$

Further, for the two-coil system, the secondary to the primary exchange of apparent power is given by (11.2)

Integrated EV Wireless Charging Station

$$S_{SP} = \dot{U}_{SP}\dot{I}_P^* = -j\omega M \dot{I}_S \dot{I}_P^* = -\omega M I_P I_S \sin\varphi_{PS} - j\omega M I_P I_S \cos\varphi_{PS} \quad (11.2)$$

Furthermore, the active power transfer from the primary side to the secondary side can be expressed by (11.3)

$$P_{PS} = \omega M I_P I_S \sin\varphi_{PS} \quad (11.3)$$

Hence, for the overall WPT system, the total reactive power (Q) developed between the primary and secondary coils is (11.4)

$$Q = \omega\left(L_P I_P^2 + L_S I_S^2 + 2M I_P I_S \cos\varphi_{PS}\right) \quad (11.4)$$

For the maximum efficiency of the WPT system, the ratio between the active power (P_{PS}) and reactive power (Q) should be maximized as shown in Equation 11.5:

$$\frac{|P_{PS}|}{Q} = \left|\frac{\omega M I_P I_S \sin\varphi_{PS}}{\omega L_P I_P^2 + \omega L_S I_S^2 + 2\omega M I_P I_S \cos\varphi_{PS}}\right| \quad (11.5)$$

To achieve the maximum value of $f(\varphi_{PS})$, Equation 11.6 can be solved as follows:

$$\frac{\partial}{\partial \varphi_{PS}} f(\varphi_{PS}) = 0 \quad \frac{\partial^2}{\partial^2 \varphi_{PS}} f(\varphi_{PS}) < 0 \quad (11.6)$$

Since we have neglected magnetic losses, it is considered as a traditional transformer, and hence, the value of coupling coefficient (k) is close to 1 if the current induced by $\dot{I}_p$ to the secondary side is $\dot{I}_s$, and the value of k is near 1; hence, the $\cos\varphi_{PS}$ is also close to 1. However, there is approximately 180° phased difference between $\dot{I}_p$ and $\dot{I}_s$. Further, in the case of the WPT system, the coupling coefficient is near 0. Hence, the maximum value of $f(\varphi_{PS})$ is at $\sin\varphi_{PS} = 1$. Thus, WPT is maximized, and the phase difference between the $\dot{I}_p$ and $\dot{I}_s$ is 90° instead of 180°. Hence, the difference between the loosely coupled system and a tightly coupled system is observed. Furthermore, the designing of a compensation network depends on the coupling degree. For the series-series coupling, there are two ways to design a resonant capacitor. Firstly, for tight coupling such as WPT, where $k > 0.5$, the $f(\varphi_{PS})$ has to be increased to achieve high efficiency. Further, for the self-coil inductance, the resonance can be achieved if $\varphi_{PS} = \pi/2$ and $f(\varphi_{PS})$ is low. Due to an increase in magnetic loss, the above condition is not recommended. The traditional transformer can work with resonance compensation with leakage inductance if there is an increase in $f(\varphi_{PS})$; however, the overall WPT system would not perform at the resonant condition. For wireless power system, i.e., at loosely couple transformer system where $k < 0.5$ for the resonance condition, the tuning of the capacitor with self-inductance is required for the maximum WPT. For this system, electrical energy is stored between the primary and secondary windings in the form of a magnetic field. The copper loss is proportional to the square of the conducting current. To achieve maximum efficiency, the current induced in secondary coil current $\dot{I}_s$ must lag to the current induced to the primary current $\dot{I}_p$ by 90°. $\dot{I}_s$ and U_{PS} should be in phase since U_{PS} and $\dot{I}_p$ lag by 90° on the receiving coil. The pure resistive characteristics are found on the secondary side.

For maximum power transfer, the induced current in the secondary coil $\dot{I}_s$ must lag to primary induced current $\dot{I}_p$ by 90°. U_{PS} and $\dot{I}_s$ should be in phase since U_{PS} and $\dot{I}_p$ lag by 90° on the receiving coil. On the secondary side, the pure resistive characteristic is seen. Meanwhile, at the primary side, the apparent power S_3 must be reduced. When $\cos\varphi_{PS} = 0$, the complex power $\left(\dot{S}_1\right)\pm$ is

$$\dot{S}_1 = j\omega L_P I_P^2 + \omega M L_P I_S \tag{11.7}$$

Maximum efficiency is $\eta_{max} = \dfrac{k^2 Q_P Q_S}{\left(1+\sqrt{1+k^2 Q_P Q_S}\right)^2}$, which is achieved at

$$\alpha_{\eta\,max} = \sqrt{1+k^2 Q_P Q_S} \tag{11.8}$$

The maximum efficiency for various types of compensation topologies has been derived by many authors. For the efficiency evaluation of static wireless charging, if the primary and secondary coils have a quality factor of 300 and coupling coefficient between coils around 0.1–0.25, the theoretical calculation states that WPT efficiency will be approximately 96.7%.

11.5 FINITE ELEMENT ANALYSIS MODELING AND SIMULATION OF THE WPT COILS FOR MAGNETIC ANALYSIS

The analysis performed in Section 11.4 presents the factors affecting the WPT system such as the coupling coefficient, transmitter side quality factor, and receiver side quality factor. The coupling coefficient is related to the designing of the primary and secondary coils, and quality factors are related to the power electronics circuit design. In this section, we are focusing on the coupling coefficient. Hence, to analyze the coupling coefficient, finite element analysis (FEA) has been performed for the transmitter and receiver coil. The modeling of the FEA model has been designed in the Ansys Maxwell® as shown in Figure 11.4.

The circular coils for the transmitter and receiver have been taken for the analysis. For the performance analysis, the airgap and horizontal misalignment have also been varied to present the robustness of the system. The vertical air gap has been varied from 100 to 150 mm, whereas the horizontal misalignment from 0 to 50 mm in a single direction. The overall parameters for the FEA model have been listed in Table 11.1. The frequency of analysis is 85 kHz. The maximum number of passes is 10, with a percentage error of 1%. The convergence parameters are refinement per pass 30%, minimum converged passes 1, and the minimum number of passes 2. Nonlinear solver with residual on 0.01 has been taken. The solution type is magnetostatic, which shows the finite element model for the circular coil WPT system.

Integrated EV Wireless Charging Station

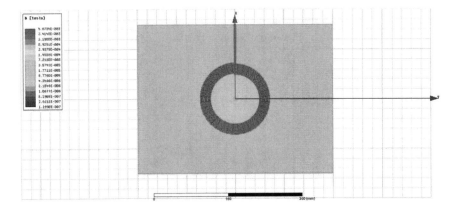

FIGURE 11.4 Circular coil model for a wireless charging system.

TABLE 11.1
Simulation Parameters for the Analysis

Parameters	Symbol	Values
Number of turns in the transmitter coil	N_T	10
Number of turns in the receiver coil	N_R	10
The outer radius of the coil	r_o	75 mm
Internal radius of the coil	r_i	50 mm
Conductor diameter	D	2 mm
The air gap distance between the coils	d	100 mm
Current in the receiver coil	I_R	8 A
Current in the transmitter coil	I_T	8 A

11.6 RESULTS AND DISCUSSION

The FEA results present a very small coupling coefficient of about 0.05. However, this can be improved by further improvement in the power pad design, impedance matching, and performing the WPT at resonance condition. The coupling coefficient for horizontal misalignment and vertical height variation has been presented in Figure 11.5. The graph shows the reduction in the coupling coefficient while increasing the air gap between the transmitter and receiver, vertically as well as horizontally.

A very important factor that determines the WPT efficiency is the coupling coefficient. Figure 11.5 shows the coupling coefficient of the presented WPT model. The results show the variation of the coupling coefficient when there is an increase in the air gap and when the horizontal alignment between the power pads has been varied. Figures 11.6 and 11.7 present the magnetic field plot around the transmitter coil, which verifies the reduction in the magnetic field with an increase in the distance from the coil.

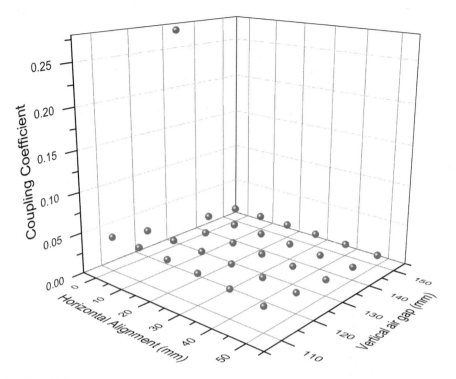

FIGURE 11.5 A 3D graphical presentation of the results from FEA analysis.

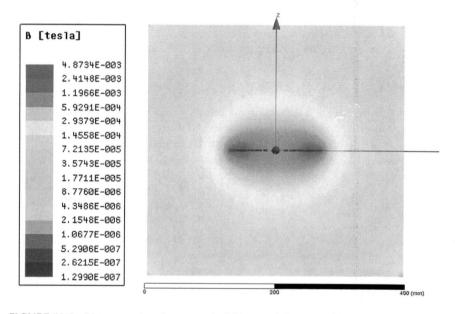

FIGURE 11.6 Plot presenting the magnetic field around the transmitter coil.

Integrated EV Wireless Charging Station

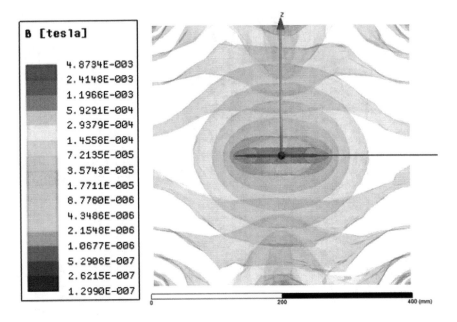

FIGURE 11.7 Stepwise magnetic flux pattern around the transmitter coil.

The results from the above analysis have been presented in Table 11.2 where the coupling coefficient is 0.53645. The value of the coupling coefficient is very small as compared to the transformer application, due to the air as medium and larger air gap and since no core has been used.

The results present the feasibility of the WPT system for the implementation of the solar-based wireless charging system. Figure 11.8 presents the overall wireless charging system magnetic field plot in the air as a medium. The power transfer efficiency of the available WPT system can be more than 85%, which is less than the conductive wireless charging.

TABLE 11.2
Result for the FEA Analysis at Zero Misalignments and 100 mm Height

Parameters	Symbol	Values
Mutual inductance	M	1.024816 µH
Self-inductance of the receiver coil	L_R	19.130200 µH
Self-inductance of the transmitter coil	L_T	19.077310 µH
Coupling coefficient	k	0.053645
Magnetic flux of transmitter	φ_T	0.000145 Wb
Magnetic flux of receiver	φ_R	0.000144 Wb

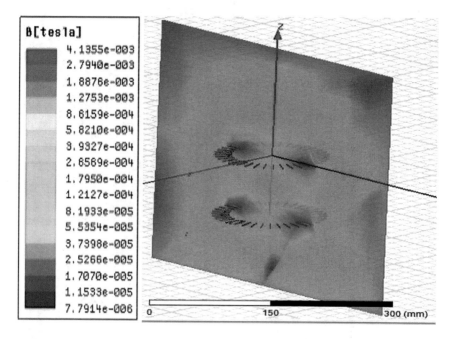

FIGURE 11.8 Overall plot of the WPT system.

The Ansys Maxwell model has been exported to Ansys Simplorer and through magnetostatic analysis. Using series-series compensation topology, the input and output power has been analyzed and recorded. Figure 11.9 shows the Simplorer model, and Figure 11.10 shows the output waveform.

The FEA model has been fed to the Ansys Simplorer to analyze circuit analysis of the wireless charging system. The Simplorer circuit parameters are listed in Table 11.3. Figure 11.10 presents the input and output power waveforms.

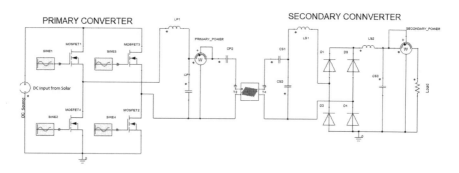

FIGURE 11.9 Simplorer model for transient simulation of a solar-powered wireless charging system with LCC topology.

Integrated EV Wireless Charging Station

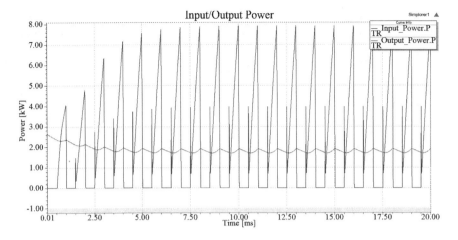

FIGURE 11.10 Input and output power across the wireless charging system.

TABLE 11.3
Electrical and Circuit Parameter for the Solar-powered Wireless Charging System

S. No	Parameter	Values
1.	V_S	220 V
2.	R_L	4.61 Ω
3.	L_P	62 µH
4.	L_S	43 µH
5.	L_f	100 µH
6.	C_{rp}	54.6 nF
7.	C_P	9.02 µF
8.	C_S	42.67 nF
9.	C_{rs}	84.2 µH
10.	C_f	500 µF
11.	R_L	18 Ω

11.7 CONCLUSION

The manuscript presents a sustainable, environmentally friendly, and economic idea of charging EVs using solar energy. The circuit analysis for the wireless charging system has been performed to present the factors that affect the efficiency of the WPT system where the coupling coefficient is found to be a major factor that influences the efficiency of the WPT system. Further, the coupling coefficient for the WPT system has been analyzed using FEA modeling and simulation. The results present the variation in the coupling coefficient by varying the alignment between the transmitter and receiver coil. However, the perfect alignment may lead to more than 85% efficiency of the WPT. Hence, the EV charging using solar energy is a very efficient and sustainable solution for the clean environment. However, for the countries having solar energy available throughout the year, the presented idea is a very efficient and viable solution for the deployment of EV technology.

ACKNOWLEDGMENT

This work was supported by the Center of Advanced Research in Electrified Transportation, Aligarh Muslim University, through the Science and Engineering Research Board (SERB)-approved IMPRINT-2 Project (IMP/2018/001267).

REFERENCES

1. A. Ahmad, M. S. Alam, and R. Chabaan, "A comprehensive review of wireless charging technologies for electric vehicles," *IEEE Trans. Transp. Electrif.*, vol. 4, no. 1, pp. 38–63, Mar. 2018.
2. Z. Duan, B. Gutierrez and L. Wang, "Forecasting Plug-In Electric Vehicle Sales and the Diurnal Recharging Load Curve," in *IEEE Trans. on Smart Grid*, vol. 5, no. 1, pp. 527–535, Jan. 2014.
3. D. Hall and N. Lutsey, "Emerging best pratices for electric vehicle charging infrastructure," *ICCT White Pap.*, October, 2017.
4. S. Khan, A. Ahmad, F. Ahmad, M. Shafaati Shemami, M. Saad Alam, and S. Khateeb, "A comprehensive review on solar powered electric vehicle charging system," *Smart Sci.*, vol. 6, no. 1, pp. 54–79, Jan. 2018, doi: 10.1080/23080477.2017.1419054.
5. "Electric Vehicles | International Council on Clean Transportation." [Online]. Available: https://theicct.org/electric-vehicles. [Accessed: 14-Jan-2020].
6. "Solar Power Information and Facts." [Online]. Available: https://www.nationalgeographic.com/environment/global-warming/solar-power/. [Accessed: 14-Jan-2020].
7. A. Ahmad and M. S. Alam, "Magnetic analysis of copper coil power pad with ferrite core for wireless charging application," *Trans. Electr. Electron. Mater.*, vol. 20, no. 2, pp. 165–173, Apr. 2019, doi: 10.1007/s42341-018-00091-6.
8. L. Fara and D. Craciunescu, "Output analysis of stand-alone PV systems: Modeling, simulation and control," in *Energy Procedia*, 112, pp. 595–605, 2017.
9. E. Kabir, P. Kumar, S. Kumar, A. A. Adelodun, and K. H. Kim, "Solar energy: Potential and future prospects," *Renew. Sustain. Energy Rev.*, 82, pp. 894–900, 2018.
10. G. Badea et al., "Design and simulation of Romanian solar energy charging station for electric vehicles," *Energies*, vol. 12, no. 1, p. 74, 2019.
11. A. Ahmad, M. S. Alam, and A. A. S. Mohamed, "Design and interoperability analysis of quadruple pad structure for electric vehicle wireless charging application," *IEEE Trans. Transp. Electrif.*, vol. 5, no. 4, pp. 934–945, Dec. 2019, doi: 10.1109/TTE.2019.2929443.
12. A. Dhianeshwar, P. Kaur and S. Nagarajan, "EV: Communication Infrastructure Management System," *2016 First International Conference on Sustainable Green Buildings and Communities (SGBC)*, Chennai, 2016, pp. 1–6, doi: 10.1109/SGBC.2016.7936090.
13. W. Khan, A. Ahmad, F. Ahmad, and M. Saad Alam, "A comprehensive review of fast charging infrastructure for electric vehicles," *Smart Sci.*, vol. 6, no. 3, pp. 1–15, Mar. 2018, doi: 10.1080/23080477.2018.1437323.

12 Shielding Techniques of IPT System for Electric Vehicles' Stationary Charging

Ahmed A. S. Mohamed
National Renewable Energy Laboratory (NREL)

Ahmed A. Shaier
Zagazig University

CONTENTS

12.1 Introduction 279
12.2 Components of Transmitter and Receiver Pad 281
    12.2.1 Conductive Wires 281
    12.2.2 Flux Concentrator 281
    12.2.3 EMF Shielding 283
        12.2.3.1 Passive Shielding 283
        12.2.3.2 Active Shielding 285
        12.2.3.3 Reactive Shielding 287
12.3 Conclusion 288
Acknowledgment 289
References 289

12.1 INTRODUCTION

The transportation sector is one of the primary consumers of fossil fuels in the world, which makes it the biggest contributor to the greenhouse gases (GHGs) [1]. Clean transportation technologies are crucial to reduce the dependency of fossil fuels and emission of GHGs. Electric vehicles (EVs) are one of the primary players in this space, due to the associated advantages related to performance, emission, and safety. Charging infrastructure of EVs is one of the main challenges that slows down the EV penetration market. Among the charging technologies, inductive charging shows promising features for EV because of being automatic, convenient, reliable in harsh environment, durable against vandalism, and flexible (can be implemented on the road, public parking, and private parking) [2,3]. Inductive charging methods present a new revolution in the EV industry. Unlike conventional plug-in tethered to a charger, no power connection is needed and, instead, an EV can charge its battery remotely, either during long-term parking (stationary), driving (dynamic), or transient stops (quasi-dynamic). The conventional configuration of an EV inductive charger is depicted in Figure 12.1. The system includes two galvanically isolated sides: primary (grid) and secondary (vehicle). The former includes a transmitter coil, which is connected to the power source through a grid-rectifier, high-frequency (HF) inverter, and compensation network. The latter consists of a receiver coil, which is coupled to the EV battery though a compensation network and rectifier. The transmitter is embedded in the road and is responsible for generating HF electromagnetic fields (EMFs) that are coupled to the receiver when the vehicle exists above the system. The linked EMFs induce power in the secondary circuit, which is rectified and stored into the battery. The system operates at high frequency (79–90 kHz) [4], which helps to reduce the system components' size (transmitter, receiver, power converters, etc.) and enhance the power transfer [5,6]. For high-power and high-efficiency operation, resonance capacitors are connected to both the transmitter and receiver coils. These capacitors help to compensate for the large leakage inductances due to the large airgap and provide the required reactive power for magnetizing this airgap. These capacitors can be connected in series, or in parallel, and can be a combination of LC circuits [7–9].

In indictive power transfer (IPT) systems, the power transfers from the transmitter to a receiver by transmitting a significant amount of magnetic field through a large airgap. Part of these fields is coupled with the receiver coil to represent the useful power, and the other part leaks in the air around the system. If these leakage fields exceed the safe limits, they have the potential to present safety concern to the living objects in the proximity of the system [10]. These fields may result in high induced currents inside human organs, which lead to exposure of the body tissues to heat stress and create serious risks to human health [3,11]. In addition, these leakage EMFs may have an adverse impact on the portable medical devices, such as pacemakers, by interrupting their operation [12,13]. Several international entities have defined safe limits for the leakage EMFs at different operating frequencies. The J2954 committee recommended considering the 2010 International Commission on Non-Ionizing Radiation Protection (ICNIRP 2010) limits for IPT systems [14]. The ICNIRP 2010 recommends limits of external magnetic field density (B) of 27 µT for

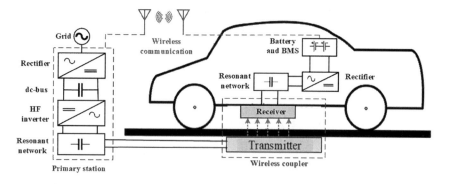

FIGURE 12.1 Inductive charger system components. (Generated by authors).

humans and 15 µT for pacemakers [15,16]. In order to meet these limits, the transmitter and receiver need to be carefully designed, optimized, and shielded.

12.2 COMPONENTS OF TRANSMITTER AND RECEIVER PAD

The inductive pad is the most sensitive part of IPT system, as it is responsible for the energy transfer from the source to the vehicle. An IPT system includes two pads (transmitter and receiver), and each pad incorporates three main parts: conductive wires, flux concentrator, and EMF shield.

12.2.1 Conductive Wires

The conductive wires carry HF currents that are responsible for generating magnetic fields. The high operating frequency of IPT system results in high eddy current losses in the windings due to skin and proximate effects, which leads to high coil ac resistance. High coil resistance means less quality factor and efficiency [17]. Therefore, there is an urgent need for using special types and structures of wires in IPT systems to minimize the conductive losses and enhance the system efficiency. Several classes of wires have been tested, demonstrated, and reported in the literature for IPT systems that show low ac resistance, such as litz wire [18,19], magneto-plate wire [20,21], magneto-coated wire [22,23], tubular conductor [24,25], REBCO wire [26], and Cu-clad-Al wire [18,27], as indicated in Figure 12.2. Analyzing the different wires shows a tradeoff between performance and cost, as summarized in Table 12.1 [28].

12.2.2 Flux Concentrator

As HF currents flow in the wires, very fast pulsating magnetic fields are generated in form of closed loops around the source coil. The strength of these loops diminishes

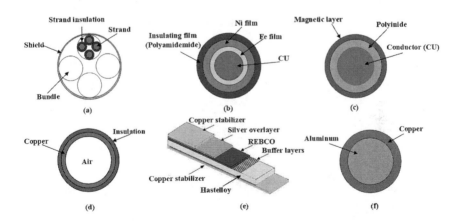

FIGURE 12.2 Types of conductive wires; (a) litz wire, (b), magneto-plate wire (c) magneto-coated wire, (d) tubular conductor, (e) REBCO wire, and (f) Cu-clad-Al wire. (Generated by authors).

TABLE 12.1
Characteristics of Different Types of Wire

Characteristic	Litz Wire	LMPW	LMCW	Tubular Copper	REBCO	CCA
R_{skin}	Medium	Very low	Very low	Low	Low	Medium
R_{prox}	Medium	Very low	Very low	Low	Low	Medium
R_{hys}	-	Low	Low	-	-	-
R_{ac}	Medium	Very low	Very low	Low	Medium	Slightly high
Cost	High	Very high	Slightly high	Medium	High	Low
Complexity	Medium	High	Medium	Low	High	Low
Density (m³/kW)	Medium	High	High	Slightly high	Medium	small
Flexibility	High	Low	Low	Medium	Slightly high	High
References	[18,19]	[20,21]	[22,23]	[24,29]	[26,30]	[18,27]

with distance, which limits the charging distance (< 1 m). Flux concentrators are typically used in IPT systems to direct the flux lines from the transmitter toward the receiver [31]. These concentrators help to enhance the coupling performance and system efficiency and reduce the leakage EMFs around the system. Flux concentrators are typically made of a magnetic material that is highly conductive for magnetic fields. Ferrite cores are the most used as flux concentrator due to their high magnetic permeability and low electrical conductivity [32]. In Refs [33,34], a ferromagnetic nanoparticle material was proposed for the IPT system, which offers higher power transfer capability, improves the shielding performance, and reduces the system weight [35]. Utilization of flexible magnetic cores for IPT system was discussed in Refs [36,37], which leads to less core losses and better-quality factor and system performance. Ferrite, magnetic nanoparticle, and flexible core are more convenient for vehicle pad; however, they are not ideal for transmitter pad as they are incompatible with the road. Any cracks in the road will damage the transmitter pad. Therefore, a magnetizable concrete was developed and proposed for a transmitter design in Refs [38,39]. It is very cheap since the magnetic particles can be made of recycled materials [40]. A comparison among the different magnetic materials for flux concentrator is presented in Table 12.2.

TABLE 12.2
Characteristics of Magnetic Materials

Material	Cost	Flexibility	Weight	References
Ferrite	High	Hard and fragile	High	[33,41]
Magnetizable concrete	Very low	High	Very high	[39,40]
Flexible magnetic	Low	Very high	Low	[36,37]
Nanoparticles	Medium	Medium	Low	[33,34]

12.2.3 EMF Shielding

IPT-based EV charger involves high power (up to few hundreds of kilowatts) to be transferred through a relatively large distance based on magnetic fields. Strong EMFs are typically generated around the system during charging. These fields may surpass the safe limits reported in the international standards and guidelines [10,42]. EMF shielding is typically used in IPT systems to minimize these leakage EMFs, thus improving the coupling performance leading to better efficiency and quality factor [43]. Different types of shield have been reported in the literature, such as passive (magnetic, conductive, or both), active, and reactive, as indicated in Figure 12.3.

12.2.3.1 Passive Shielding

Passive shielding is achieved by adding a passive component (conductor or/and magnetic) that helps to block and/or direct the EMFs and reduce the leakage part. Magnetic passive shielding is manufactured from nonconductive materials with high magnetic permeability to direct magnetic flux lines, enhance self and mutual inductance, improve the system performance, and reduce the leakage flux [31,42]. Ferrite flux concentrators discussed in Section 12.2.2 provide this shielding functionality [43]. These cores are installed in the pad as a plate, multiple bars, or multiple tiles, as in Refs [44–46]. The design of an inductive pad with the use of ferrite only as a magnetic shield is illustrated in Figure 12.4. Although using magnetic materials for shielding shows good performance, it increases the system weight and cost. Therefore, another direction was taken to minimize the magnetic materials in the system for flux concentration and add a lighter and less expensive conductor for shielding. Therefore, the trend of IPT system design is to combine a magnetic material with another type of shield, whether conductive passive, active, or reactive [47,48].

The conductive passive shield combined with magnetic shield is depicted in Figure 12.5. It is made of copper or aluminum [49,50] and installed under the magnetic core in the case of transmitter or over the core in the case of receiver. It also can be attached directly to the coil in case of air-core coil, as indicated in Figure 12.6. It is a conductive plate that acts as another coil, in which eddy currents are generated due to the HF EMFs [51]. These eddy currents produce magnetic fields that oppose the original fields and reduce them. This helps to minimize the leakage flux around the system but adversely affects the system performance by reducing the coupling factor and efficiency [31,47,52,53]. Several studies are presented in the literature to

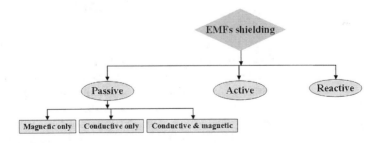

FIGURE 12.3 Types of EMF shielding in the IPT system. (Generated by authors).

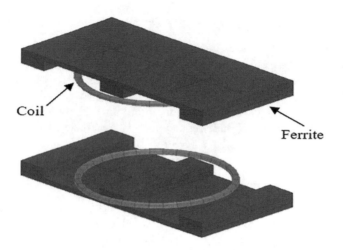

FIGURE 12.4 Inductive pad with the use of ferrite only as a magnetic shield. (Generated by authors).

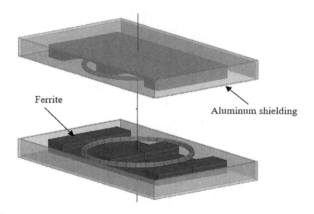

FIGURE 12.5 Inductive pad with the use conductive passive shield. (Generated by authors).

optimize the shield design for reducing its negative impact on the system efficiency, considering different materials, such as copper [54], aluminum [55,56], different dimensions [57,58], and different positions [49,59,60]. In Refs [31,48], the impact of magnetic and conductive passive shield on the system performance (self-inductance, coupling factor, efficiency, and EMF emissions) was investigated. It was concluded that the use of ferrite core increases the power transfer, transmission efficiency, and coupling coefficient between the two coils. In addition, the consideration of the vehicle chassis as an additional passive conductive shielding and investigating its impact on the IPT system performance were presented in Refs [41,49,61].

In Ref. [42], a 100 kW IPT system with matched DD coils has been laboratory-tested at 85 kHz frequency with an aluminum plate only as shielding, which shows a 25% increase in EMF emissions. The study proposed an extra magnetic loop of

Shielding Techniques of IPT System

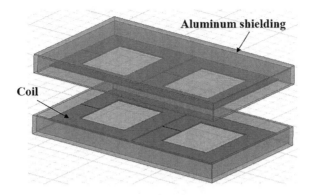

FIGURE 12.6 Inductive pad with the use of conductive shield. (Generated by authors).

ferrite bars to be added around the coil for minimizing the leakage flux, which was able to reduce the emission by 60%. In Ref. [10], a comparison was made between two types of passive shields for circular pad: the traditional aluminum plate (Figure 12.7a) and a thin aluminum plate surrounded by a copper shield-ring (Figure 12.7b). It was concluded that the use of a copper ring with aluminum plate as a shield reduces the losses by 21% compared to the traditional aluminum.

12.2.3.2 Active Shielding

For high-power IPT systems (>100 kW), it is a challenge to manage the leakage EMFs around the vehicle and keep them within the safe limits using the conventional passive shielding [42]. Therefore, active shielding, as a more effective shielding technique, was investigated for the IPT system [62]. In this case, extra turns are added to each coil and wounded with reverse polarities, as depicted in Figure 12.8. When current passes through the original coils, the same current will pass in the shielding turns that generate intentional EMFs that have the same frequency and value of the original fields but oppose them [47,52] acting as cancelling fields to minimize the leakage EMFs. This type shows an effective shielding performance

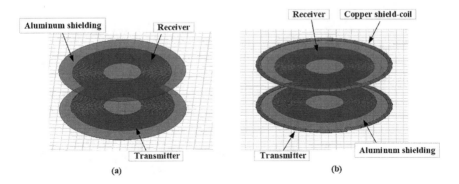

FIGURE 12.7 (a) traditional aluminum plate and (b) copper ring-based shield. (Generated by authors).

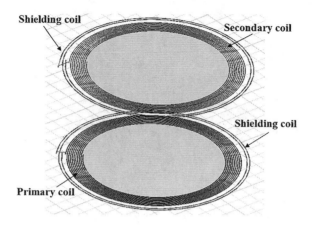

FIGURE 12.8 Principle of active shield coil system. (Generated by authors).

compared to the passive one; however, the negative impact on the original field and the system performance is worse. In addition, adding extra turns increases the system cost, weight, and coil losses. Both passive conductor and active shield are used with magnetic shield [47].

In Ref. [63], an active coil, placed on the primary side only, was used for shielding of the planar coils, which resulted in a significant reduction in the leaked EMF around the system. Two active coil structures were explored and compared: traditional active coil (Figure 12.9a) and adopted active coil (Figure 12.9b). The system was tested at 85 kHz frequency and 7.7 kW output power. The impact of the two shields on the self-inductance and mutual inductance and coupling coefficient was studied. The adopted active coil shows a significant reduction in the leakage EMF around the system.

In Ref. [64], planar circular coils were tested by applying active shielding coil on the primary and secondary side. Two structures were presented based on the connection of the active coils: the inductive-based structure, in which the shield coils are electrically isolated, and the shield currents are induced using magnetic induction,

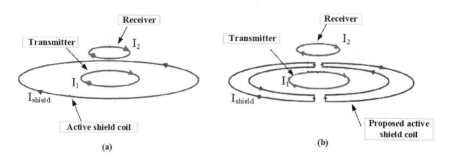

FIGURE 12.9 Active coil system, (a) traditional active shielding, and (b) proposed active shielding system. (Generated by authors).

Shielding Techniques of IPT System

as depicted in Figure 12.10a. The other structure involves shield coils connected in series with the main coils at opposite polarities, as described in Figure 12.10b. The impact of different number of turns at the shield coil on the transmission efficiency, coupling coefficient, and EMF distribution was analyzed. The study showed that with a small number of turns in the inductive structure, the system was able to provide a better shielding effect than the case of using a large number of turns with the series-connected structure.

12.2.3.3 Reactive Shielding

For reducing the negative impact of the active shield on the system performance, the resonant reactive shield was proposed in Refs [47,52]. It depends on the use of a passive compensation loop coil with a resonant capacitor, as indicated in Figure 12.11. This type doesn't need to be powered to produce an opposite intentional field. However, an extra shield coil is added near the original one and wound in a way such that when the original magnetic field passes through the shielding coil, an induced voltage is generated, which results in a HF current in the shield coil that generates the opposite magnetic field to cancel the original one. This structure shows a better system efficiency than the active shield case [52].

In Ref. [47], reactive shield was considered in an inductively charged electric bus to reduce the EMF around the system which consists of two circular pads. The impact of the shield on the self-inductance and mutual inductance and magnetic field was investigated. It was found that the reactive shielding was able to reduce the magnetic field by 64%, with minimal impact of the system parameters. Furthermore, the study concluded that reactive shielding is more effective and efficient than the passive conductive shielding of the same size. In Ref. [60], a different configuration for the shield capacitor was presented and compared with the conventional one. Four phase shifter capacitors are used in the shield and called double reactive shield, as depicted in Figure 12.12. The use of phase shifter capacitors leads to the generation of a magnetic field that opposes the original field and is more effective and efficient in reducing leakage magnetic field around the system. In Ref. [65], various types of shielding (passive, active, and reactive) were analyzed and compared for circular coils.

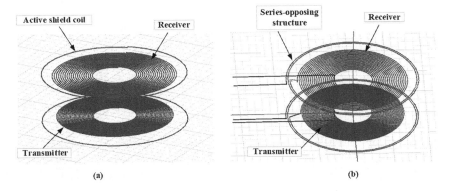

FIGURE 12.10 Shielding coils: (a) inductive structure and (b) series-opposing structure. (Generated by authors)

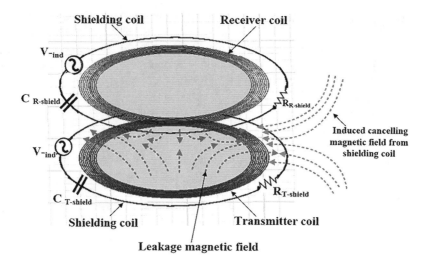

FIGURE 12.11 Principle of reactive shield system. (Generated by authors).

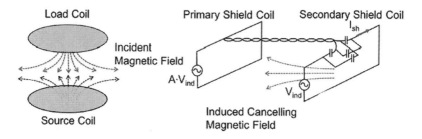

FIGURE 12.12 Double reactive shield. (Generated by authors).

Their impact on the system losses, EMF, and self-inductance and mutual inductance was investigated. Different operating cases were explored, including no shielding case, the use of the vehicle's body as shield, using passive shield, applying a combination of active and passive shield, and finally when passive, active, and reactive shields are applied together.

12.3 CONCLUSION

This chapter presents an inclusive study and review for the current state-of-the-art of the shielding techniques in inductive power transfer systems for EV charging. Different types of shielding (passive, active, and reactive) are presented and compared in detail. As a conclusion, passive shields are appropriate for low- and medium-power IPT systems (<100 kW), and they should be able to bring the EMF levels to be below the safe limits. In addition, they are cheaper, simpler in implementation and design, and more robust. For high-power IPT systems, active shielding and reactive shielding are more promising for the system to comply with the standard limits of EMFs. This chapter provides comprehensive guidelines for researchers, students, and engineers who are interested in designing an appropriate shield for inductive charging technology.

ACKNOWLEDGMENT

The National Renewable Energy Laboratory (NREL) is the current address for the first author only. NREL and the U.S. Department of Energy (DOE) did not contribute to this work.

REFERENCES

1. US Environmental Protection Agency, "Sources of Greenhouse Gas Emissions," *US EPA*, Dec. 29, 2015. https://www.epa.gov/ghgemissions/sources-greenhouse-gas-emissions (accessed Oct. 28, 2019).
2. M. Budhia, G. A. Covic, and J. T. Boys, "Design and optimization of circular magnetic structures for lumped inductive power transfer systems," *IEEE Trans. Power Electron.*, vol. 26, no. 11, pp. 3096–3108, 2011.
3. A. A. S. Mohamed, A. Meintz, P. Schrafel, and A. Calabro, "In-Vehicle Assessment of Human Exposure to EMFs from 25-kW WPT System Based on Near-Field Analysis," *2018 IEEE Vehicle Power and Propulsion Conference (VPPC)*, Chicago, IL, 2018, pp. 1–6, doi: 10.1109/VPPC.2018.8605011.
4. X. Lu, P. Wang, D. Niyato, D. I. Kim, and Z. Han, "Wireless charging technologies: Fundamentals, standards, and network applications," *IEEE Commun. Surv. Tutor.*, vol. 18, no. 2, pp. 1413–1452, 2015.
5. A. A. S. Mohamed, D. Allen, T. Youssef, and O. Mohammed, "Optimal Design of High Frequency H-Bridge Inverter for Wireless Power Transfer Systems in EV Applications," *2016 IEEE 16th International Conference on Environment and Electrical Engineering (EEEIC)*, Florence, Italy, 2016, pp. 1–6, doi: 10.1109/EEEIC.2016.7555646.
6. Shuo Wang and D. Dorrell, "Review of Wireless Charging Coupler for Electric Vehicles," *IECON 2013 - 39th Annual Conference of the IEEE Industrial Electronics Society*, Vienna, Austria, 2013, pp. 7274–7279, doi: 10.1109/IECON.2013.6700342.
7. A. A. S. Mohamed, A. Berzoy, F. G. N. de Almeida, and O. Mohammed, "Modeling and assessment analysis of various compensation topologies in bidirectional IWPT system for EV applications," *IEEE Trans. Ind. Appl.*, vol. 53, no. 5, pp. 4973–4984, Sep. 2017, doi: 10.1109/TIA.2017.2700281.
8. W. Zhang and C. C. Mi, "Compensation topologies of high-power wireless power transfer systems," *IEEE Trans. Veh. Technol.*, vol. 65, no. 6, pp. 4768–4778, Jun. 2016, doi: 10.1109/TVT.2015.2454292.
9. J. L. Villa, J. Sallan, J. F. Sanz Osorio, and A. Llombart, "High-misalignment tolerant compensation topology for ICPT systems," *IEEE Trans. Ind. Electron.*, vol. 59, no. 2, pp. 945–951, Feb. 2012, doi: 10.1109/TIE.2011.2161055.
10. A. A. S. Mohamed, A. Meintz, P. Schrafel, and A. Calabro, "Testing and assessment of EMFs and touch currents from 25-kW IPT system for medium-duty EVs," *IEEE Trans. Veh. Technol.*, vol. 68, no. 8, pp. 7477–7487, Aug. 2019, doi: 10.1109/TVT.2019.2920827.
11. F. Wen and X. Huang, "Human exposure to electromagnetic fields from parallel wireless power transfer systems," *Int. J. Environ. Res. Public. Health*, vol. 14, no. 2, p. 157, 2017.

12. C. Xiao, K. Wei, D. Cheng, and Y. Liu, "Wireless charging system considering eddy current in cardiac pacemaker shell: theoretical modeling, experiments, and safety simulations," *IEEE Trans. Ind. Electron.*, vol. 64, no. 5, pp. 3978–3988, May 2017, doi: 10.1109/TIE.2016.2645142.
13. T. Campi, S. Cruciani, F. Palandrani, V. De Santis, A. Hirata, and M. Feliziani, "Wireless power transfer charging system for AIMDs and pacemakers," *IEEE Trans. Microw. Theory Tech.*, vol. 64, no. 2, pp. 633–642, Feb. 2016, doi: 10.1109/TMTT.2015.2511011.
14. International Commission on Non-Ionizing Radiation Protection, "Guidelines for limiting exposure to time-varying electric and magnetic fields (1 Hz to 100 kHz)," *Health Phys.*, vol. 99, no. 6, pp. 818–836, Dec. 2010, doi: 10.1097/HP.0b013e3181f06c86.
15. International Commission on Non-Ionizing Radiation Protection. "Guidelines for limiting exposure to time-varying electric, magnetic, and electromagnetic fields (up to 300 GHz)," *Health Phys.*, vol. 74, no. 4, pp. 494–522, 1998.
16. S. Cruciani, T. Campi, F. Maradei, and M. Feliziani, "Wireless Charging in Electric Vehicles: EMI/EMC Risk Mitigation in Pacemakers by Active Coils," *2019 IEEE PELS Workshop on Emerging Technologies: Wireless Power Transfer (WoW)*, London, Jun. 2019, pp. 173–176, doi: 10.1109/WoW45936.2019.9030634.
17. H. Takanashi, Y. Sato, Y. Kaneko, S. Abe, and T. Yasuda, "A Large Air Gap 3 kW Wireless Power Transfer System for Electric Vehicles," *2012 IEEE Energy Conversion Congress and Exposition (ECCE)*, Raleigh, NC, Sep. 2012, pp. 269–274, doi: 10.1109/ECCE.2012.6342813.
18. D. Barth, B. Klaus, and T. Leibfried, "Litz Wire Design for Wireless Power Transfer in Electric Vehicles," *2017 IEEE Wireless Power Transfer Conference (WPTC)*, Taipei, 2017, pp. 1–4, doi: 10.1109/WPT.2017.7953819.
19. H. Rossmanith, M. Doebroenti, M. Albach, and D. Exner, "Measurement and characterization of high frequency losses in nonideal litz wires," *IEEE Trans. Power Electron.*, vol. 26, no. 11, pp. 3386–3394, Nov. 2011, doi: 10.1109/TPEL.2011.2143729.
20. T. Mizuno, T. Ueda, S. Yachi, R. Ohtomo, and Y. Goto, "Dependence of efficiency on wire type and number of strands of litz wire for wireless power transfer of magnetic resonant coupling," *IEEJ J. Ind. Appl.*, vol. 3, no. 1, pp. 35–40, 2014, doi: 10.1541/ieejjia.3.35.
21. H. Shinagawa, T. Suzuki, M. Noda, Y. Shimura, S. Enoki, and T. Mizuno, "Theoretical analysis of AC resistance in coil using magnetoplated wire," *IEEE Trans. Magn.*, vol. 45, no. 9, pp. 3251–3259, Sep. 2009, doi: 10.1109/TMAG.2009.2021948.
22. Y. Konno, T. Yamamoto, Y. Chai, D. Tomoya, Y. Bu, and T. Mizuno, "Basic characterization of magnetocoated wire fabricated using spray method," *IEEE Trans. Magn.*, vol. 53, no. 11, pp. 1–7, Nov. 2017, doi: 10.1109/TMAG.2017.2719047.
23. T. Yamamoto, Y. Konno, K. Sugimura, T. Sato, Y. Bu, and T. Mizuno, "Loss reduction of LLC resonant converter using magnetocoated wire," *IEEJ J. Ind. Appl.*, vol. 8, no. 1, pp. 51–56, 2019, doi: 10.1541/ieejjia.8.51.
24. A. M. Jawad, R. Nordin, S. K. Gharghan, H. M. Jawad, M. Ismail, and M. J. Abu-AlShaeer, "Single-tube and multi-turn coil near-field wireless power transfer for low-power home appliances," *Energies*, vol. 11, no. 8, p. 1969, Aug. 2018, doi: 10.3390/en11081969.
25. Z. Pantic and S. Lukic, "Computationally-efficient, generalized expressions for the proximity-effect in multi-layer, multi-turn tubular coils for wireless power transfer systems," *IEEE Trans. Magn.*, vol. 49, no. 11, pp. 5404–5416, Nov. 2013, doi: 10.1109/TMAG.2013.2264486.
26. N. Sekiya and Y. Monjugawa, "A novel REBCO wire structure that improves coil quality factor in MHz range and its effect on wireless power transfer systems," *IEEE Trans. Appl. Supercond.*, vol. 27, no. 4, pp. 1–5, Jun. 2017, doi: 10.1109/TASC.2017.2660058.

27. C. R. Sullivan, "Aluminum windings and other strategies for high-frequency magnetics design in an era of high copper and energy costs," *IEEE Trans. Power Electron.*, vol. 23, no. 4, pp. 2044–2051, Jul. 2008, doi: 10.1109/TPEL.2008.925434.
28. A. A. S. Mohamed, A. A. Shaier, H. Metwally, and S. I. Selem, "A comprehensive overview of inductive pad in electric vehicles stationary charging," *Appl. Energy*, vol. 262, p. 114584, Mar. 2020, doi: 10.1016/j.apenergy.2020.114584.
29. D. Filipović and T. Dlabač, "A closed form solution for the proximity effect in a thin tubular conductor influenced by a parallel filament," *Serbian J. Electr. Eng.*, vol. 7, no. 1, pp. 13–20, 2010.
30. Y. Shi, A. R. Dennis, K. Huang, D. Zhou, J. H. Durrell, and D. A. Cardwell, "Advantages of multi-seeded (RE)–Ba–Cu–O superconductors for magnetic levitation applications," *Supercond. Sci. Technol.*, vol. 31, no. 9, p. 095008, 2018.
31. M. Feliziani and S. Cruciani, "Mitigation of the Magnetic Field Generated by a Wireless Power Transfer (WPT) System without Reducing the WPT Efficiency," *2013 International Symposium on Electromagnetic Compatibility*, Brugge, Belgium, 2013, pp. 610–615.
32. A. A. S. Mohamed and O. Mohammed, "Physics-based co-simulation platform with analytical and experimental verification for bidirectional IPT system in EV applications," *IEEE Trans. Veh. Technol.*, vol. 67, no. 1, pp. 275–284, Jan. 2018, doi: 10.1109/TVT.2017.2763422.
33. A. Delgado, J. A. Oliver, J. A. Cobos, J. Rodriguez, and A. Jiménez, "Optimized Design for Wireless Coil for Electric Vehicles Based on The Use of Magnetic Nano-articles," *2019 IEEE Applied Power Electronics Conference and Exposition (APEC)*, Anaheim, CA, 2019, pp. 1515–1520, doi: 10.1109/APEC.2019.8721998.
34. A. Delgado, G. Salinas, J. Rodríguez, J. A. Oliver, and J. A. Cobos, "Finite Element Modelling of Litz Wire Conductors and Compound Magnetic Materials based on Magnetic Nano-particles by means of Equivalent Homogeneous Materials for Wireless Power Transfer System," *2018 IEEE 19th Workshop on Control and Modeling for Power Electronics (COMPEL)*, Padua, Italy, Jun. 2018, pp. 1–5, doi: 10.1109/COMPEL.2018.8460012.
35. T.-J. Yoon, W. Lee, Y.-S. Oh, and J.-K. Lee, "Magnetic nanoparticles as a catalyst vehicle for simple and easy recycling," *New J. Chem.*, vol. 27, no. 2, pp. 227–229, 2003, doi: 10.1039/B209391J.
36. X. Sun, Y. Zheng, X. Peng, X. Li, and H. Zhang, "Parylene-based 3D high performance folded multilayer inductors for wireless power transmission in implanted applications," *Sens. Actuators Phys.*, vol. 208, pp. 141–151, Feb. 2014, doi: 10.1016/j.sna.2013.12.038.
37. X. Sun, Y. Zheng, Z. Li, X. Li, and H. Zhang, "Stacked Flexible Parylene-Based 3D Inductors with Ni80Fe20 Core for Wireless Power Transmission System," *2013 IEEE 26th International Conference on Micro Electro Mechanical Systems (MEMS)*, Taipei, Taiwan, 2013, pp. 849–852, doi: 10.1109/MEMSYS.2013.6474376.
38. M. Esguerra and R. Lucke, "Application and production of a magnetic product," US6696638B2, Feb. 24, 2004.
39. R. Tavakoli, A. Echols, U. Pratik, Z. Pantic, F. Pozo, A. Malakooti, and M. Maguire, "Magnetizable Concrete Composite Materials for Road-Embedded Wireless Power Transfer Pads," *2017 IEEE Energy Conversion Congress and Exposition (ECCE)*, Cincinnati, OH, Oct. 2017, pp. 4041–4048, doi: 10.1109/ECCE.2017.8096705.
40. C. Carretero, I. Lope, and J. Acero, "Magnetizable Concrete Flux Concentrators for Wireless Inductive Power Transfer Applications," *IEEE J. Emerg. Sel. Top. Power Electron.*, vol. 8, no. 3, pp. 2696–2706, 2019, doi: 10.1109/JESTPE.2019.2935226.
41. M. Ibrahim, "Wireless Inductive Charging for Electrical Vehicles: Electromagnetic Modelling and Interoperability Analysis," PhD Thesis, Paris 11, 2014.

42. A. A. Z. A. Khan, M. SaadAlam, Y. Rafat, A. A. Khan, A. A. Deshpande, and R. C. Chabaan, "Analytical Review of xEVs Standards," *IEEE Transactions on Transportation Electrification*, March 08, 2016, p. 29.
43. J. Zhou, Y. Gao, C. Zhou, J. Ma, X. Huang, and Y. Fang, "Optimal Power Transfer with Aluminum Shielding for Wireless Power Transfer Systems," *2017 20th International Conference on Electrical Machines and Systems (ICEMS)*, Sydney, Australia, 2017, pp. 1–4, doi: 10.1109/ICEMS.2017.8056143.
44. M. Mohammad, J. Pries, O. Onar, V. P. Galigekere, G. Su, S. Anwar, J. Wilkins, U. D. Kavimandan, and D. Patil, "Design of an EMF Suppressing Magnetic Shield for a 100-kW DD-Coil Wireless Charging System for Electric Vehicles," *2019 IEEE Applied Power Electronics Conference and Exposition (APEC)*, Anaheim, CA, 2019, pp. 1521–1527, doi: 10.1109/APEC.2019.8722084.
45. N. E. Kazantseva, J. Vilčáková, V. Křesálek, P. Sáha, I. Sapurina, and J. Stejskal, "Magnetic behaviour of composites containing polyaniline-coated manganese–zinc ferrite," *J. Magn. Magn. Mater.*, vol. 269, no. 1, pp. 30–37, Feb. 2004, doi: 10.1016/S0304-8853(03)00557-2.
46. T. Kim, S. Yoon, J. Yook, G. Yun, and W. Y. Lee, "Evaluation of Power Transfer Efficiency with Ferrite Sheets in WPT System," *2017 IEEE Wireless Power Transfer Conference (WPTC)*, Taipei, 2017, pp. 1–4, doi: 10.1109/WPT.2017.7953894.
47. K. Hata, T. Imura, and Y. Hori, "Dynamic Wireless Power Transfer System for Electric Vehicles to Simplify Ground Facilities - Power Control and Efficiency Maximization on the Secondary Side," *2016 IEEE Applied Power Electronics Conference and Exposition (APEC)*, Long Beach, CA, 2016, pp. 1731–1736, doi: 10.1109/APEC.2016.7468101.
48. Y. Matsuda, H. Sakamoto, H. Shibuya, and S. Murata, "A non-contact energy transferring system for an electric vehicle-charging system based on recycled products," *J. Appl. Phys.*, vol. 99, no. 8, p. 08R902, Apr. 2006, doi: 10.1063/1.2164408.
49. S. Kim, H.-H. Park, J. Kim, J. Kim, and S. Ahn, "Design and analysis of a resonant reactive shield for a wireless power electric vehicle," *IEEE Trans. Microw. Theory Tech.*, vol. 62, no. 4, pp. 1057–1066, Apr. 2014, doi: 10.1109/TMTT.2014.2305404.
50. H. Kim, J. Cho, S. Ahn, J. Kim, and J. Kim, "Suppression of Leakage Magnetic Field from a Wireless Power Transfer System Using Ferrimagnetic Material and Metallic Shielding," *2012 IEEE International Symposium on Electromagnetic Compatibility*, Pittsburg, Aug. 2012, pp. 640–645, doi: 10.1109/ISEMC.2012.6351659.
51. L. Tan, J. Li, C. Chen, C. Yan, J. Guo, and X. Huang, "Analysis and performance improvement of WPT systems in the environment of single non-ferromagnetic metal plates," *Energies*, vol. 9, no. 8, p. 576, Aug. 2016, doi: 10.3390/en9080576.
52. Y. Kitano, H. Omori, T. Morizane, N. Kimura, and M. Nakaoka, "A New Shielding Method for Magnetic Fields of a Wireless EV Charger with Regard to Human Exposure by Eddy Current and Magnetic Path," *2014 International Power Electronics and Application Conference and Exposition*, Shanghai, China, 2014, pp. 778–781, doi: 10.1109/PEAC.2014.7037956.
53. J. Kim, J. Kim, S. Kong, H. Kim, I. Suh, N. P. Suh, D. Cho, J. Kim, and S. Ahn, "Coil design and shielding methods for a magnetic resonant wireless power transfer system," *Proc. IEEE*, vol. 101, no. 6, pp. 1332–1342, Jun. 2013, doi: 10.1109/JPROC.2013.2247551.
54. D. B. Geselowtiz, Q. T. N. Hoang, and R. P. Gaumond, "The effects of metals on a transcutaneous energy transmission system," *IEEE Trans. Biomed. Eng.*, vol. 39, no. 9, pp. 928–934, Sep. 1992, doi: 10.1109/10.256426.
55. X. Zhang, Z. Yuan, Q. Yang, H. Meng, Y. Jin, Z. Wang, and S. Jiang, "High-Frequency Electromagnetic Force Characteristics on Electromagnetic Shielding Materials in Wireless Power Transmission System," *2017 IEEE PELS Workshop on Emerging Technologies: Wireless Power Transfer (WoW)*, London, May 2017, pp. 1–5, doi: 10.1109/WoW.2017.7959369.

56. F. Wen and X. Huang, "Optimal magnetic field shielding method by metallic sheets in wireless power transfer system," *Energies*, vol. 9, no. 9, p. 733, Sep. 2016, doi: 10.3390/en9090733.
57. S. Bandyopadhyay, V. Prasanth, P. Bauer, and J. A. Ferreira, "Multi-Objective Optimisation of a 1-kW Wireless IPT Systems for Charging of Electric Vehicles," *2016 IEEE Transportation Electrification Conference and Expo (ITEC)*, Dearborn, MI, 2016, pp. 1–7, doi: 10.1109/ITEC.2016.7520210.
58. G. Ke, Q. Chen, L. Xu, S.-C. Wong, and C. K. Tse, "A Model for Coupling under Coil Misalignment For DD Pads and Circular Pads of WPT System," *2016 IEEE Energy Conversion Congress and Exposition (ECCE)*, Milwaukee, WI, Sep. 2016, pp. 1–6, doi: 10.1109/ECCE.2016.7854706.
59. R. Bosshard, U. Iruretagoyena, and J. W. Kolar, "Comprehensive evaluation of rectangular and double-D coil geometry for 50 kW/85 kHz IPT system," *IEEE J. Emerg. Sel. Top. Power Electron.*, vol. 4, no. 4, pp. 1406–1415, Dec. 2016, doi: 10.1109/JESTPE.2016.2600162.
60. A. Tejeda, C. Carretero, J. T. Boys, and G. A. Covic, "Ferrite-less circular pad with controlled flux cancelation for EV wireless charging," *IEEE Trans. Power Electron.*, vol. 32, no. 11, pp. 8349–8359, Nov. 2017, doi: 10.1109/TPEL.2016.2642192.
61. H. Moon, S. Kim, H. H. Park, and S. Ahn, "Design of a resonant reactive shield with double coils and a phase shifter for wireless charging of electric vehicles," *IEEE Trans. Magn.*, vol. 51, no. 3, pp. 1–4, Mar. 2015, doi: 10.1109/TMAG.2014.2360701.
62. M. Mohammad, E. T. Wodajo, S. Choi, and M. E. Elbuluk, "Modeling and design of passive shield to limit EMF emission and to minimize shield loss in unipolar wireless charging system for EV," *IEEE Trans. Power Electron.*, vol. 34, no. 12, pp. 12235–12245, Dec. 2019, doi: 10.1109/TPEL.2019.2903788.
63. S. Y. Choi, B. W. Gu, S. W. Lee, W. Y. Lee, J. Huh, and C. T. Rim, "Generalized active EMF cancel methods for wireless electric vehicles," *IEEE Trans. Power Electron.*, vol. 29, no. 11, pp. 5770–5783, Nov. 2014, doi: 10.1109/TPEL.2013.2295094.
64. T. Campi, S. Cruciani, F. Maradei, and M. Feliziani, "Active Coil System for Magnetic Field Reduction in an Automotive Wireless Power Transfer System," *2019 IEEE International Symposium on Electromagnetic Compatibility, Signal & Power Integrity (EMC+SIPI)*, New Orleans, LA, July 2019, pp. 189–192, doi: 10.1109/ISEMC.2019.8825202.
65. Y. Wang, B. Song, and Z. Mao, "Analysis and experiment for wireless power transfer systems with two kinds shielding coils in EVs," *Energies*, vol. 13, no. 1, p. 277, 2020, doi: 10.3390/en13010277.
66. R. Vaka and R. K. Keshri, "Evaluation and selection of shielding methods for wireless charging of e-rickshaw," *Electr. Eng.*, vol. 102, no. 6, pp. 1005–1019, Jan. 2020, doi: 10.1007/s00202-020-00929-4.

13 Economic Placement of EV Charging Stations within Urban Areas

Ahmed Ibrahim AbdelAzim
Ethos Esco Consultancy

CONTENTS

13.1 Introduction 295
13.2 The Problem of Choosing Charging Stations' Locations 297
13.3 Methodologies for Placing Charging Stations 299
13.4 Economics of Charging Station Placement 302
13.5 Case Study: Applying an Agent-Based Network Graph Placement
        Method on Cairo, Egypt 305
References 311

13.1 INTRODUCTION

The environmental footprint of the transportation sector is huge; it consumes around 19% of the world total energy use, with light-duty vehicles (including passenger cars) being the most significant contributor [1]. In 2015, governments around the world endorsed 17 sustainable development goals (SDGs), including SDG 7, which is to secure affordable and clean energy. The promotion of leaner transport technologies, most importantly electric vehicles (EVs), contributes to achieving SDG 7 and is a cornerstone of the efforts to reduce transportation sector-related emissions [2]. The United Nations Environment Programme estimates that 20% of all road vehicles must be electric-powered by 2030 to stay within the global 2°C climate scenario [3]. EVs reduce fossil fuel consumption and GHG emissions, especially when using electricity generated from renewable resources.

The increasingly stringent regulations, the growing public awareness of environmental challenges, and the volatility of oil prices have encouraged car makers not only to reduce their conventional vehicles' emissions but also to venture into producing hybrid and fully battery-powered EVs [4]. The number of EVs reached 5.1 million globally in 2018, with China and Norway leading the market in sales and market share. Forecast scenarios, which account for the impact of announced policy ambitions and pledges, estimate that global EV sales will be between 23 and 43 million in 2030, while the total EV stock is expected to be between 130 and 250 million (not including two and three wheelers) [5].

Potential adopters of any new product (or service) usually inquire about two main aspects: practicality and pricing. People compare new products to the ones they are currently using based mainly on these two aspects. Any marketing efforts to convince the users to adopt the new product or service instead of what they are currently used to (which is called *conversion* in marketing terminology) must typically address practicality and price among other things.

Like any new technology, the potential adopters are often skeptic about practicality, as compared to the established technology. In the case of EVs, people simply look around and observe that there are more gasoline fueling stations than EV charging stations on the streets. As such, they might be worried that the EV's battery could run out of charge before reaching either his or her destination or a charging point. This perception is known as "range anxiety", and it is a psychological barrier to large-scale adoption of EVs. Solving the lack of conveniently placed EV charging infrastructure is therefore an important issue, which can significantly encourage people to adopt EV cars. If people see more and more EV charging stations, their perception will change, and range anxiety might ease off.

In terms of pricing, and with battery technology advancement being a significant driver, EV prices are dropping. The price of lithium-ion (Li-ion) battery storage dropped around 87% between 2010 and 2019 (see Figure 13.1). Consequently, passenger EV car initial prices, although still generally higher, became comparable with conventional fuel-powered cars.

In terms of operating costs, EVs are significantly cheaper; maintenance is mainly related to changing brake fluid, coolant, or tires, as opposed to the more costly internal combustion engine maintenance. The American Automobile Association

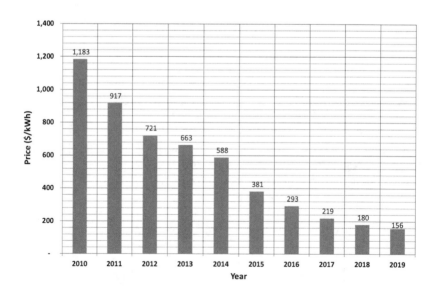

FIGURE 13.1 Li-ion battery price trend (2010–2019).

Economic Placement of EV Charging Stations

estimates the annual maintenance and repair costs of gasoline cars to be around 1,200 $ per annum, versus around 450 $ for EVs. Figure 13.2 below shows a life cycle cost (LCC) comparison of a gasoline and electric compact passenger car, based on a gasoline price of 2.6 $/gal, an electricity tariff of 13.2 $cents/kWh, and around 10,200 miles of driving per annum.

Another important aspect is the environmental aspect. Conventional vehicles, despite the efficiency improvement imposed by laws, still inevitably emit exhaust gases. For illustration, the gasoline-fueled Volkswagen Golf in the above LCC analysis emits around 2.9 tons of CO_2 equivalent per annum.

It can be predicted that EV growth will first happen in urban centers, where daily trip distances are not very long, and consequently, large EV battery ranges are not needed. Early EV adopters living in cities may therefore be encouraged to purchase and use EVs if the charging infrastructure is well established.

13.2 THE PROBLEM OF CHOOSING CHARGING STATIONS' LOCATIONS

Charging stations have to be placed at convenient locations that are frequent enough to cover all common routes of EV users. Yet, increasing the number of charging stations also increases their capital and operational expenditures. Therefore, while ideally a charging station could be placed at each and every block within a large city, the investment required would be practically and economically prohibitive. Moreover, policies and subsidies play an important role in supporting the deployment of charging infrastructure, as the public and governmental sector will most likely not be able to keep up with the rapid pace of EV expansion alone. The financial burden of the EV charging infrastructure development on national budgets may not be attractive to

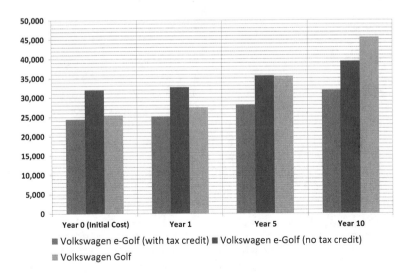

FIGURE 13.2 Total LCC comparison of electric and gasoline-fueled Volkswagen Golf.

policy and decision-makers. Indeed, the amounts required would better go to other sectors, which are already in dire need such as education or health. Private sector participation in EV infrastructure development can then be encouraged through public policies, subsidies, and incentives. For example, laws can be put in place to allow private sector participation through various concessionary or public–private partnership schemes, including [6] the following:

- *Build, Own, and Operate (BOO)*, where a private sector investor finances, owns, builds, and then operates a project for a specific concession period (e.g. 10–20 years) while selling the project's output to the government at predetermined contract rates (and sometimes to other customers at different rates). The investor recovers investment, operation, and maintenance (O&M) costs from the project's revenues.
- *Build, Operate, and Transfer (BOT)*, where the private sector investor enters into a concession with a governmental entity to finance build then operates a project for a specific period, recovering costs through the project revenue minus a predetermined percentage to the governmental entity. At the end of the concession period, the project assets' ownership and operational responsibility are wholly transferred from the concession entity (special purpose company) to the governmental entity.
- *Build, Own, Operate, and Transfer (BOOT)*, same as BOT, except that the private sector investor owns the project during the concession period and therefore pays no annual percentage to the concessionary governmental entity until the ownership transfer.
- *Build, Lease, and Transfer (BLT)*, where the private sector investor finances, owns, and builds the project and then leases it to the governmental entity for a specific period (in years) in exchange for a periodic lease amount allowing the investor to partly or fully recover the project investment. At the end of the leasing period, the project ownership is transferred to the governmental entity at a predetermined price, which may be zero.

Such schemes have proven to be generally more effective in developing public infrastructure by ensuring higher quality while reducing the necessary public sector investment. As private sector players are mainly profit-driven, they tend to seek optimum utilization of their capital investment for profit maximization. Consequently, they would target to minimize the required capital expenditure and maintain an acceptable level of service to users while still selling at a profitable rate. In EV charging station context, this simply means that private sector investors will seek to deploy the minimum number of charging stations required to effectively cover a given target service area.

Another hurdle is the profitability of EV charging stations themselves. With the number of EV users still growing, it might not be financially attractive to deploy and operate charging stations to serve such a *niche* market. This hurdle can be largely overcome by public sector incentives such as targeted subsidies or tax credits which, when factored into the project's financials, might turn the investment profitable to the private sector investors.

Economic Placement of EV Charging Stations

Technically, the EV charging stations' deployment problem can be broken down into three subproblems, namely:

1. *Placement.* Selecting the locations of the charging stations in order to minimize their number and optimize their accessibility for EV users.
2. *Power supply.* Choosing the source of electricity for the charging stations, be it from the grid or from on-site generation (renewable or conventional), and whether energy storage should be used.
3. *Queuing.* Deciding on how many charging points to be placed in each station and analyzing the cost-benefit tradeoff of adding or removing charging points and how this affects EV users' waiting times and usage patterns.

Clearly, the first subproblem (placement of stations) is the most influential; it affects in whole or part the other two. The location of a charging station may mandate the electricity source to be used and, depending on the number of possible users, may suggest increasing the number of charging points. Based on this, the first subproblem is tackled in this chapter: the placement problem.

13.3 METHODOLOGIES FOR PLACING CHARGING STATIONS

To illustrate the basic strategy for EV charging station placement, consider the following simple case: an EV user only drives on a single road between two points: from home to work in the morning and then vice versa in the evening. Where would a charging station be placed so as to serve her or him best? To answer this question, we break it down into few simpler questions:

1. How long is the road between these two destination points?
2. If fully charged, what is the driving range of the user's EV?
3. What is the EV state of charge at the beginning of the trip?

A trivial case would be that the road distance (L) is equal to or less than half of the range of the fully charged EV (R_{EV}). In this scenario, if the EV user starts the day with a fully charged EV battery, she or he would be able to drive to work in the morning then back home in the evening before depleting the EV battery. Consequently, the suggested solution for this scenario is to place an EV charger at the user's home. The EV user would plug the EV into the charger upon arriving in the evening until it fully charges the next day. We can express this scenario as follows:

$$2 * L \leq R_{EV} \tag{13.1}$$

Suppose the road distance is longer than half the EV range but still less than or equal to the full EV range. In that case, starting the day with a full charge, the EV user would then drive to work normally in the morning but run out of charge while driving back in the evening before reaching her or his home. For this scenario, a solution would be to place one charger at the user's home and another at work premises. The EV user would then plug her or his EV into the charger at work upon arriving to fully charge during the workday and then drive home and plug it again into his home charger to charge for the next day. This scenario can be expressed as follows:

$$R_{EV} < 2*L \leq 2*R_{EV} \tag{13.2}$$

Interestingly, if we know for a fact that the user can always maintain the EV state of charge at 50% or more, we could replace the two chargers in this scenario with a single public fast charger in the middle of the road. That way, the user can charge the EV halfway at least once throughout the morning and evening trips.

Now consider one last scenario where the road distance is longer than the whole EV range but less than or equal to its double:

$$2*R_{EV} < 2*L \leq 4*R_{EV} \tag{13.3}$$

In that case, even if the user maintains a starting state of charge of 100% by using a home charger, at least two additional chargers would still be needed: one halfway through the road and another at work premises. Adding even more charging stations along the road, e.g., two stations on the road (plus those at home and work premises) would reduce the required starting state of charge to 66.7%. This can be linked to the concept of range anxiety, which was discussed earlier; adding more charging stations reduces the starting state of charge to the level required to reach only the first charging station on the driving route. This would give more confidence to the EV user that she or he can recharge during his trip and therefore would alleviate range anxiety.

Although considerable research efforts have been directed toward the charging station placement problem, only few seem to be practically applicable to real-life situations, especially in metropolitan settings. Shi, Pan, Wang, and Cai [7] investigated the minimum EV battery size required to satisfy 100% of the travel distance demands of taxis and private vehicles. This is done by analyzing the historical travel data of both

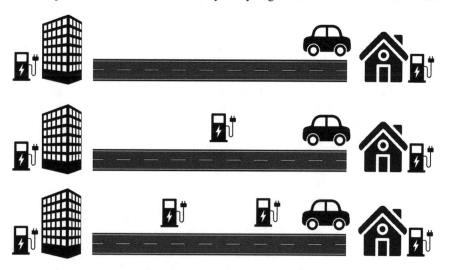

FIGURE 13.3 Illustration of the discussed simple case scenarios.

taxi fleets and private vehicles in Beijing, China. The dataset included trip start and end locations and times and travel distance, as well as total trip travel time. The paper concluded that the battery technology is unlikely to be a major bottleneck to EV adoption. The dataset analysis in the paper shows that current EV battery ranges (which the paper estimates to a maximum of 480 km at 0.217 kWh km^{-1} consumption) are sufficient for the majority of travel patterns of taxis, as well as private vehicles, which were found to require even lower battery ranges compared to taxis.

Predicting that EV growth will mainly happen in urban areas where battery range is not the biggest concern, Cui, Weng, and Tan [9] attempted a cost optimization approach to the EV charging station placement problem. The paper formulated the placement problem as a mixed-integer objective function with constraints added one by one to preserve convexity. The effect of these cost constraints, which included the cost of the charging station, distribution network expansion, protective device upgrade, and voltage regulation costs, is presented, and a global minimum of this objective function is investigated. Finally, the paper conducted a sensitivity analysis and demonstrated, through numerical results, the trade-off relationships between different factors affecting this optimization problem.

ElBanhawy and Nassar [10] discussed the influence of early adaptors' perceptions about EVs (including range anxiety) on the market penetration level and proposed a movable charging unit (MCU) as a solution to charging points' placement problem in metropolitan residential areas. The proposed MCU would be capable of both fast charging and battery swapping and would move through a hypothetical neighborhood using a shortest-path algorithm to "home-deliver" charging to EV users, especially those with no domestic charging facilities.

Xiong et al. [12] addressed the charging station placement problem using an approach that mixes game theory and algorithmic analysis. In their paper, the authors formulated the placement problem as a charging optimization game, where EV drivers' individual strategies are to minimize their own charging cost, while their collective behaviors affect the road network traffic conditions. The paper used a suggested algorithm (called OCEAN-C) to compute the optimal solution to the placement problem and validated the approach using real dataset from Singapore. Wagner, Götzinger, and Neumann [13] proposed a placement approach based on proximity to points of interest such as parks, restaurants, and banks. An "attractiveness" metric is defined for charging stations, based on the usage frequency of a charging station (as the more a charging station is frequented, the higher importance it has) as well as how close it is to points of interest. The paper performs a regression analysis on data collected over a period of several months for approximately 230 charging stations in Amsterdam, Netherlands. A location model is then derived, which provides infrastructure planners with optimal locations for new charging stations based on point of interest locations. Beside Amsterdam, the paper also presented a case study for Brussels, Belgium.

The paper by Franke et al. [14] thoroughly examined the perception of comfort in EV driving ranges among EV users. A 6-month field study was conducted where forty EV users were surveyed, with an emphasis on psychological stress-buffering personality traits and coping mechanisms with regard to range anxiety. The study introduces the term "safety buffer", which quantifies the increase in the range needs perceived by the EV user. More importantly, the paper highlights that the perceived EV range barriers, including suboptimal range utilization, may be overcome using psychological interventions including user training and EV interface design. The study also suggests that a reliable usable EV range may be more important than enhancing the maximum possible range. In the paper by Yuan et al. [13], the notion that EV drivers tend to have a higher range anxiety compared to conventional vehicle drivers is highlighted. The paper conducts a survey on EV drivers in China to assess and quantify their actual range anxiety level, as well as their resulting behavioral changes. The paper also made use of the concept of a "safety buffer" introduced by Franke et al. [12], which is defined as the minimum difference between the EV's remaining mileage and the trip range. Accessibility to charging stations, driver experience, and emotional confidence were found to be significant factors to the perceived range anxiety level. The paper analyzes the data for Chinese EV drivers based on three trip scenarios: short (10 km), medium (30 km), and long (60 km) and concludes that the average safety buffer for the three trip scenarios is approximately 47%, which is quite high. Finally, the authors suggest that EV human machine interface design should address range anxiety by incorporating driving style and charging strategy detection and recommendation algorithms.

13.4 ECONOMICS OF CHARGING STATION PLACEMENT

Establishment of public EV charging infrastructure is a costly investment, and the costs are broadly categorized into capital costs, comprising hardware and installation costs and operational costs comprising electricity and maintenance costs. Capital expenditure is usually referred to as CAPEX, while operational expenses are referred to as OPEX. To discuss the economics of public charging stations' placement, we shall use a LCC approach, which aims to quantify the expenses throughout the expected lifetime of operating the charging station. Table 13.1 below lists typical CAPEX and OPEX cost components related to public charging stations.

IEC 61851 defines four modes of charging, namely [14]

- *Mode 1.* Domestic slow AC charging, from a regular electric socket cord.
- *Mode 2.* Domestic AC charging with EV-specific protective device.
- *Mode 3.* Slow or fast AC charging on a dedicated circuit using specific EV multipin sockets and EV-specific protective features.
- *Mode 4.* Fast direct current (DC) charging using a special connector and special charger technology.

TABLE 13.1
Typical EV Charging Station CAPEX and OPEX Cost Components

CAPEX	OPEX
Charging hardware equipment	Electricity supply costs
Material (cabling, mounting, etc.)	Maintenance and repair
Installation labor	Taxes
Permits and licenses	

Similarly, the US Society of Automotive Engineers (now known as SAE International) defines three levels of charging: level 1 (slow household charging on 110–120 VAC), level 2 (upgraded household or commercial charging on 208–240 VAC), and fast DC charging, which is sometimes referred to as level 3, although not fully standardized (charging on 208–480 VDC). The IEC 62196 also standardizes the EV charging multipin connectors into three configurations, [15] with type 2 being the most widely used. We will mainly talk about public charging infrastructure serving users in metropolitan areas. Therefore, the charging stations that we will discuss are mainly level 2 in the US (or IEC 61851 Mode 3 or 4 in Europe).

In order to simplify discussing the economics of charging infrastructure deployment, few assumptions have to be made. First, it is assumed that only 1 charger is installed (i.e. single charging port), and therefore, economies of scale are ignored. Second, the charger's capacity (i.e. kW rating) is ignored and, for the sake of practicality, shall be assumed not to be a deciding factor in the placement decision. More conveniently, this could be thought of as if the charger's rating is large enough that the charging time required by an EV to fully charge is consequently reasonably short enough to be done during morning and evening trips in the previously discussed one-road simple scenario. Another important simplifying assumption is that power source is readily available at the charging station's location, and therefore, no utility upgrades or special power connection provisions are required. One final assumption is to ignore land ownership or leasing arrangements, that is to assume the charging station's installation location is already owned by the infrastructure developer, and therefore, no leasing, renting, or usufruct fees are paid. In practice, many factors affect the installation cost of EV charging station, including the following:

- Distance from the power connection point
- Number of charging ports to be installed
- Whether the installation is indoor or outdoor
- Labor cost at the installation site

The analysis presented here uses CAPEX figures estimated by Electric Power Research Institute (EPRI) for the US in 2013, but the purpose is to illustrate the general framework of analyzing the economics of EV charging station deployment [16]. The EPRI study is particularly useful because it provides the total average costs per charging station as well as a breakdown by labor, materials (including the charger), permits, and taxes. Figure 13.4 below shows the split of these cost categories in percentages of the total.

According to EPRI, the average total CAPEX per 1-port charging station in the US is approximately 4,484 $. We will therefore take the rounded-up figure of 4,500 $ as the investment required to install one charging station. The OPEX consists mainly of the cost of electricity consumed by the charging station (i.e., to charge EVs of customers) plus any amounts spent on maintenance and repairs. The electricity tariff is assumed to be 13.2 $cents/kWh, and we will estimate the maintenance and repair to consume 30% of the required CAPEX each year, i.e., around 1,350 $/year.

One of the first questions the investor would tackle is: How much should we charge customers for using the EV charging station? The answer to this question

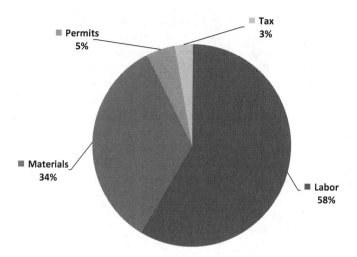

FIGURE 13.4 Installation cost split of EV charging station installation in the US.

requires a market study to estimate the number of customers that may use the charging station, and the more thorough this study is conducted, the more accurately the rate can be determined. Estimating the potential number of customers allows the selling rate to be calculated so as to cover the costs (or "break-even"), recover the CAPEX amount over a certain number of years, and achieve a certain profit margin to the investor(s).

This exercise also usually begins with calculations for a single unit, i.e., one charging station, and once the unit economics are settled upon, the numbers are simply applied to the required total number of charging stations to be deployed. Consequently, we shall consider a single charging station, and the calculations are in Table 13.2 below:

So the charging station revenue needs to equal at least 10,850 $/year to cover its estimated yearly expenses. The investor would also require recovering the CAPEX over, say, three years. This means that the revenue needs to increase by an additional 1,500 $/year (4,500 $/3 years) to become 12,350 $/year. This revenue

TABLE 13.2
Example Economic Calculations for a Single Port Charging Station

Estimated number of users (from market study):	300 users (monthly), with 40 kWh EV battery at an average charge state of 50%
CAPEX:	4,500 $
Yearly cost of electricity	= Yearly electricity demand (kWh) × 13.2 $cents/kWh
	= 300 users/month × 12 months × 40 kWh × 50% × 13.2 $cents/kWh ≈ 9,500 $/year
Maintenance and repair	= 30% × 4,500 $ = 1,350 $/year
Total yearly OPEX	= 9,500 + 1,350 = 10,850 $/year

figure would then be divided by the estimated yearly demand to yield a selling rate of 17.2 $cents/kWh. Figure 13.5 below shows the cash flow for the first 8 years of operation (note the break-even point at year 3). After the break-even point, the difference between revenue and costs is net profit (assuming zero tax).

At this point, it is useful to consider multiple scenarios for the number of customers, the required CAPEX payback period, or both. Table 13.3 below shows the variation of the selling rate with varying estimates for the number of customers and the CAPEX recovery period.

13.5 CASE STUDY: APPLYING AN AGENT-BASED NETWORK GRAPH PLACEMENT METHOD ON CAIRO, EGYPT

This section will describe an approach to solving the placement problem of EV charging stations in a large metropolitan area. The method that will be used combines graph theory (GT) with agent-based modeling (ABM) and will be applied to Greater Cairo area (in Egypt) as a case study.

GT is concerned with the study of the relations and properties of complex systems by abstracting them into graphs. A graph is a mathematical network structure comprising a set of nodes (or "vertices") interconnected by a set of links (also called "edges"). The links of a graph can be either directed, which signifies a one-way, asymmetrical relation between the interconnected nodes, or undirected, signifying a two-way, symmetrical relation. The graph links can also be assigned numerical weights (or costs) so as to represent the modeled system or network more meaningfully.

ABM is the computational approach of simulating entities (i.e., the agents) and studying their interactions with each other as well as with the environment.

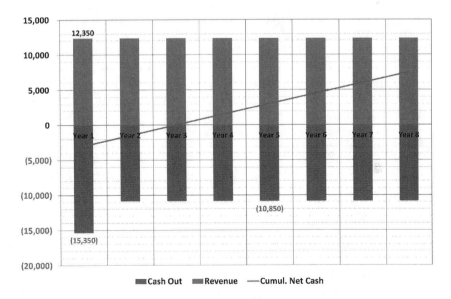

FIGURE 13.5 Example 8-year cash flow of EV charging station operation.

TABLE 13.3
Variation of Selling Rate (in $cents/kWh) with Users and Recovery Period

		\multicolumn{4}{c}{CAPEX Recovery Period (in years)}
		3	4	5	6
No. of monthly users	300	17.2	16.6	16.3	16.1
	450	15.8	15.5	15.3	15.1
	600	15.2	14.9	14.8	14.7
	750	14.8	14.6	14.5	14.4

An individual agent is an autonomous, programmatic entity defined using specified attributes, which influence its behavior. The agent attributes typically incorporate random elements to simulate real-life complexity.

Although each of the two methods was previously used separately to tackle the placement problem of EV charging stations, both approaches are combined here into a hybrid distance- and agent-based one. A research paper by Sun, Zhao, He, and Li [17] identifies the lack of charging stations as the main obstacle facing EV adoption and presents a methodology to deploy charging stations within a road network that is modeled as a directed graph. The paper proposed a statistical model to describe the spatiotemporal characteristics of the charging demand of EVs. Charging stations are then placed so as to maximize the number of EV trips passing at least one charging station, while not exceeding the electricity supply grid capacity.

Sweda and Klabjan [18] proposed an agent-based decision support system for deploying new charging infrastructure. The proposed model used road network data to identify EV ownership and driving patterns in Chicago metropolitan area. One main aim of the paper is to quantify the market demand for and transition to EV usage in order to attract investors to build charging stations. Agents are assigned attributes including income, preferred vehicle class, and range anxiety level, and their EV purchasing behavior is simulated based on their interaction. The results of the model compare a base case (20 existing charging stations) to two proposed charging station deployment scenarios, each with 70 additional charging stations. The paper also illustrates how projected EV adoption is affected by gasoline price adjustment, with higher gasoline prices accelerating EV adoption trend.

For the proposed method, the Python programming language was used to implement the models of both the graph and the agents. Python is a high-level, object-oriented programming language widely used in scientific computing. The road network of the Greater Cairo area in Egypt is modeled using Python programming language as an undirected network graph, with its link weights representing the approximate driving distances between different network nodes. As such, nodes simply represent crossroads in the road networks, and we can imagine users navigating between starting and ending nodes through this road network. The road network graph extends between downtown Cairo and eight of its surrounding edge cities (see Figure 13.6 below).

Economic Placement of EV Charging Stations 307

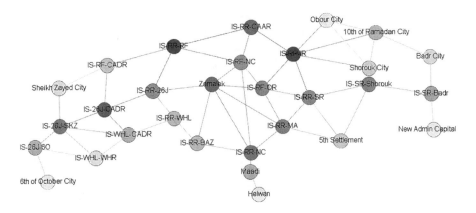

FIGURE 13.6 Network graph representation of Greater Cairo metropolitan area (including edge cities).

The graph includes 30 nodes and 53 links, and each link is assigned a cost attribute equal to its approximate real driving distance. Most graph nodes have more than one link to surrounding nodes, which results in multiple alternate routes between any two given nodes. Table 13.4 below lists the eight edge cities included in the graph model.

The drivers of EVs are modeled as agents with defining attributes including home node, destination node, EV charge state, and range anxiety level. The driver agents are implemented as a Python class objects, and the object's home and destination attributes are randomly selected from the graph network model. Similarly, charge state and anxiety level are initialized to a random percentage between 0 and 100%. Figure 13.7 below presents the attributes and actions of the driver agent.

TABLE 13.4
The 8 Edge Cities Surrounding Cairo

City	Approximate Distance to Downtown Cairo (km)
10th of Ramadan city	63
5th Settlement	27
6th of October city	46
Badr city	59
New Administrative Capital city	75
Obour city	40
Sheikh Zayed city	32
Shorouk city	47

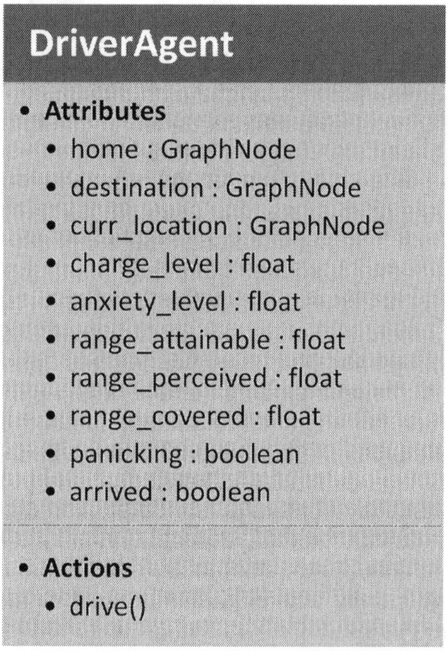

DriverAgent

- **Attributes**
  - home : GraphNode
  - destination : GraphNode
  - curr_location : GraphNode
  - charge_level : float
  - anxiety_level : float
  - range_attainable : float
  - range_perceived : float
  - range_covered : float
  - panicking : boolean
  - arrived : boolean

- **Actions**
  - drive()

FIGURE 13.7 Class representation of the driver agent.

The agent object (named DriverAgent) has the following attributes and methods:

home:	The graph node from where the driver agent starts the trip.
destination:	The graph node to which the driver agent makes the trip.
curr_location:	The current location of the driver agent within the graph network.
charge_level:	The level of charge of the EV, defined as a percentage between 0% and 100%.
anxiety_level:	The range anxiety level of the driver agent and which affects the driving range perceived as possible. It is defined as a percentage between 0% and 100%.
range_attainable:	The actual attainable driving range based on the charge level.
range_perceived:	The perceived driving range based on the actual attainable range and the range anxiety level of the agent
range_covered:	The actual range driven by the agent so far.
panicking:	A flag triggered when *range_covered* exceeds *range_perceived*.
arrived:	A flag triggered when the agent reaches *destination*.
drive():	This is the main action of the driver agent, which advances him through the path between *home* and *destination*.

The *range_attainable* is calculated as follows:

$$range\_attainable = charge\_level\,(\%) * ev\_battery\_capacity(kWh) * ev\_mileage(km/kWh) \quad (13.4)$$

Economic Placement of EV Charging Stations

And the *range_perceived* is calculated as follows:

$$range\_perceived = range\_attainable(km) * (1 - anxiety\_level(\%)) \quad (13.5)$$

The main concept of the proposed approach is to randomly instantiate a large number of EV driver agents, let them drive within the network graph through the shortest path while tracking their charge state and anxiety level, and then register the locations where they actually run out of charge before reaching their destinations. These locations are then surveyed to identify and rank them based on their frequency of "out-of-charge" incidences (OCIs). The rationale is that the more EVs run out of charge at or around a specific location, the more suitable it is for placing an EV charging station. There are few assumptions made in order to simplify the model, and they are as follows:

1. The EV drivers use compact to subcompact cars, which are less expensive and more suitable for urban environments. Consequently, the battery capacity of the EVs is assumed to be 24 kWh.
2. The mileage of the EVs is assumed at an average of 6.0 km/kWh, which is obtained from Environmental Protection Agency MPG data based on city cycle[1].
3. The only criterion for routing through the graph network is the driving distance, and therefore, traffic congestion and any associated time impact are ignored.
4. Driver agents can start their trip from any node on the graph network, including nodes that represent real-life road intersections.

The driver agents' trip routes are determined using a custom implementation of the Dijkstra shortest path algorithm. Dijkstra's algorithm is an established algorithm used to find the shortest path between two given graph nodes by performing a "best-first" search. For each driver agent, the shortest path of the current trip is computed using the algorithm, and then the agent is iteratively simulated to drive through it until the agent either reaches the trip destination or runs out of charge. If the agent runs out of charge en route to the destination, the location is saved to an array. The array of possible locations is then rearranged so as to remove duplicates and count their frequency of occurrences, and finally, the locations and their frequencies are exported to a comma-separated file (CSV) for analysis.

For the simulation, 5,000 driver agents were instantiated. The exact number of driver agents did seem to affect the statistical distribution of the OCIs over the graph links, but it was found that this effect diminishes around 2,000–2,500 agents. Consequently, the links with the most frequent OCIs tend to remain the same above this range, and the 5,000 figure was chosen as both large enough to minimize the statistical distortion and small enough to be simulated within a practically short code running time.

The output CSV file lists the graph links and the number of OCIs, and the list was reordered in a descending no. of OCI frequencies to obtain a "priority" list of links where an EV charging station was recommended. Table 13.5 below lists the top 10 links prioritized by the algorithm for EV charging station placement.

---

[1] https://www.fueleconomy.gov/feg/download.shtml

TABLE 13.5
Links Prioritized for EV Charging Station Deployment (Results of Simulation)

Link From	Link To	OCI Frequency (%)	Cumulative OCI Frequency (%)
Intersection – Ring Road – Suez Road	Intersection – Suez Road – Shorouk City Entrance	9.5%	9.5%
Intersection – Rode El Farag Axis – Orouba Road	Zamalek	6.6%	16.1%
Intersection – Suez Road – Shorouk City Entrance	Intersection – Suez Road – Badr City Entrance	6.1%	22.2%
Intersection – Ring Road – Mosheer Axis	Intersection – Ring Road – Nile Corniche (Maadi)	5.5%	27.7%
Intersection – 26th July Axis – Cairo Alex Desert Road	Intersection – Ring Road – 26th July Axis	5.3%	33.0%
Intersection – Ring Road – Ismailiya Road	Intersection – Rod El Farag Axis – Orouba Road	4.7%	37.7%
Intersection – Suez Road – Badr City Entrance	New Administrative Capital City	4.6%	42.2%
Intersection – Ring Road – Al-Bahr Al-Aazam	Intersection – Ring Road – Wahat Link	4.1%	46.3%
10th of Ramadan City	Intersection – Ring Road – Ismailiya Road	4.0%	50.4%
Helwan	Maadi	3.3%	53.7%

Those familiar with Cairo will quickly note that the first graph network link prioritized by the algorithm for charging station placement is in a suburban area: the link between the ring road/Suez road intersection and the entrance of Shorouk City. This result is not intuitive, as it was expected that the most frequented graph links would be more central to Greater Cairo area that is within downtown area. On the other hand, the last link in the list (the link between Maadi and Helwan) is in line with the intuition as the main road connection between Helwan and Greater Cairo area passes through Maadi (south of Nile corniche road).

The total cumulative frequency of OCIs for the 10 listed links is 53.7%. In other words, by locating the charging stations at only 10 locations out of 53 possible ones, more than 50% of the whole intended area was covered. Therefore, this algorithm can be used to maximize the charging infrastructure coverage while optimizing the number of stations required. This can result in significant saving in the initial CAPEX for the investor.

While the proposed method was applied to Greater Cairo area, it can nonetheless be adapted to any metropolitan area. Furthermore, the algorithm could definitely be further elaborated to account for more real-life effects such as

- Road traffic congestion-related route choice
- Topographical profile to exploit EV regenerative braking
- Prioritizing locations based on power supply availability
- Exploiting distributed renewable energy generation sources for powering

REFERENCES

1. US Energy Information Administration, "International energy outlook 2017," 2017.
2. Global Fuel Economy Initiative (GFEI), "The global fuel economy initiative: Delivering sustainable development goal 7," 2018.
3. K. Ernest, "Promoting electric mobility in developing countries," in *Sustainable Transport in Egypt: Progress, Prospects and Partnerships*, Cairo, 2016.
4. Boston Consulting Group, "The comeback of the electric car? How real, how soon, and what must happen next," 2009.
5. Electric Vehicles Initiative (EVI), "*Global EV Outlook 2019, Scaling-up the Transition to Electric Mobility*," International Energy Agency (IEA), 2019.
6. R. Akbiyikli and D. Eaton, "*Comparison of PFI, BOT, BOO and BOOT Procurement for Infrastructure Projects*," Research Institute for Built and Human Environment, University of Salford, Salford, 2003.
7. X. Shi, J. Pan, H. Wang, and H. Cai, "Battery electric vehicles: What is the minimum range required?," *Elsevier Energy*, no. 166, pp. 352–358, 2019.
8. Q. Cui, Y. Weng, and C.-W. Tan, "Electric vehicle charging station placement method for urban areas," *arXiv:1808.09660v1*, August 2018.
9. E. Y. ElBanhawy and K. Nassar, "A movable charging unit for green mobility," in *International Archives of the Photogrammetry, Remote Sensing and Spatial Information Sciences*, vol. XL-4/W1, London, 2013.
10. Y. Xiong, J. Gan, B. An, C. Miao, and A. L. C. Bazzan, "Optimal electric vehicle charging station placement," in *Proceedings of the 24th International Joint Conference on Artificial Intelligence (IJCAI)*, 2015.
11. S. Wagner, M. Götzinger, and D. Neumann, "Optimal location of charging stations in smart cities: A point of interest based approach," in *34th International Conference on Information Systems*, Milan, 2013.
12. T. Franke, I. Neumann, F. Bühler, P. Cocron, and J. F. Krems, "Experiencing range in an electric vehicle - understanding psychological barriers," *Applied Psychology: An International Review*, vol. 61, no. 3, pp. 368–391.
13. Q. Yuan et al., "Investigation on range anxiety and safety buffer of battery electric vehicle drivers," *Journal of Advanced Transportation*, 2018.
14. International Electrotechnical Commission (IEC), "IEC 61851-1: Electric vehicle conductive charging system - Part 1: General requirements," 2017.
15. International Electrotechnical Commission (IEC), "IEC 62196-1: Plugs, socket-outlets, vehicle connectors and vehicle inlets - Conductive charging of electric vehicles - Part 1: General requirements," 2014.
16. Electric Power Research Institute (EPRI), "Electric vehicle supply equipment installed cost analysis," 2013.
17. Z. Sun, Y. Zhao, Y. He, and M. Li, "A novel methodology for charging station deployment," *IOP Conference Series: Earth and Environmental Science*, vol. 113, 2018.
18. T. Sweda and D. Klabjan, "An agent-based decision support system for electric vehicle charging infrastructure deployment," in *IEEE Vehicle Power and Propulsion Conference*, Chicago, 2011, pp. 1–5.
19. D. Hall and N. Lutsey, "Emerging best practices for electric vehicle charging infrastructure," *The International Council on Clean Transportation*, 2017.

14 Environmental Impact of the Recycling and Disposal of EV Batteries

Zeeshan Ahmad Arfeen
University Technology Malaysia
The Islamia University of Bahawalpur (IUB)

Rabia Hassan
Institute of Business Management Sindh

Mehreen Kausar Azam
Institute of Business Management Sindh
N.E.D University of Engineering & Technology

Md Pauzi Abdullah
University Technology Malaysia

CONTENTS

14.1 Introduction 314
    14.1.1 Battery Repurposing and Clearance for Sustainable Society 314
14.2 Delaying Recycling through Repurposing 316
    14.2.1 Repurposing 317
14.3 Economic Aspects 319
    14.3.1 Identifying Domestic Demand 319
    14.3.2 Identifying Industrial Demand 319
14.4 Standards for Reusing EV Batteries 320
14.5 Environmental Impacts of EV Batteries | EVBs 320
    14.5.1 Raw Material Manufacturing Effects 321
    14.5.2 Battery Manufacturing Effects 321
    14.5.3 Thermal Gas Emission 321
    14.5.4 Chemical Hazards 321
14.6 Battery Dismantling and Handling Health Hazards 322
    14.6.1 Lithium-Ion Battery Landfill 323
    14.6.2 Impact of Recycling on the Environment 323
    14.6.3 Recycling of EV Batteries 324

14.7	Environmental Aspects of Reuse	324
14.8	Environmental Aspects of Recycling	325
14.9	Recycling	326
	14.9.1 Recycling Methods	326
	14.9.2 Mechanical Procedure I MP	327
	14.9.3 Pyro Metallurgical Procedure I PM	327
	14.9.4 Hydrometallurgical Procedure I HP	327
	14.9.5 Direct Recycling Procedure I DRP	328
14.10	Best Practices of Lithium-Ion Battery Recycling	329
	14.10.1 Umicore Company	329
	14.10.2 Retrieve Technologies	330
	14.10.3 Onto Technology	330
14.11	Safety Indicators	330
14.12	Dismantling and Storage	330
	14.12.1 Reorganizing and Screening	330
14.13	Technological Initiatives	331
14.14	Conclusion	331
14.15	Recommendations and Future Directions	332
References		332

14.1 INTRODUCTION

Electric vehicles (EVs) have substantial advantages contrasted with internal combustion-based engine vehicles due to their eco-friendly nature. EVs are preferred as they are more economical to maintain and run as well as they add more value for life because of their negligible air contamination and less reliance on oil as well as since no poisonous substances emerge which can destroy the ozone layer [1]. Furthermore, electric engines and batteries add to the vitality of electric vehicle. Many studies observed that the materials utilized in batteries are perilous for Earth. The mining and preparing of items such as lithium, copper, and nickel require much exuberance, and this can lead to the emission of harmful substances. In the chapter, it is comprehensively discussed how the second-life batteries, which already completed their useful life, could be reconditioned. Furthermore, if they need to be disposed of what are the best ways of doing so to opt.

14.1.1 Battery Repurposing and Clearance for Sustainable Society

We live in a period of sustainability, which underpins the crucial endeavoring to improve the natural wellbeing and personal satisfaction for the network. As reusing the paper, cardboard boxes, and utilize containers of spaghetti sauce, however, what is demanded by EV battery after it has lost its dynamism? So analysts of the world are dealing with the territories where they can create ways on which these EV batteries which are disposed of can be reused. EV sales are skyrocketing. Figure 14.1 below shows the best five EV automotive organizations for the year 2019.

As EV sales increase, the number of batteries that will get access for a second life outside of the vehicle increases. EV batteries are a gauge to surpass what might be compared to about 3.4 million packs by 2025, contrasted and around 55,000 of every

Recycling and Disposal of EV Batteries 315

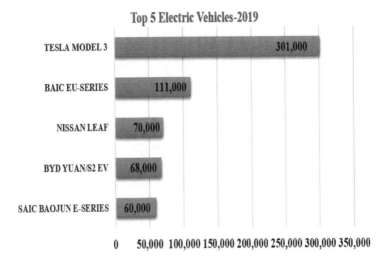

FIGURE 14.1 Sale of top 5 EV manufacturing companies in 2019 [2].

2018 as expressed in Bloomberg Businessweek. Does the inquiry rise that where will these batteries land? Various roads have been inquired about for the repurposing of the EV battery, and a portion of the major is examined in detail in this chapter. Figure 14.2 signifies that the battery's production volume increases with a fall of net cost yearly.

There are several manufacturers of EVs. A compact summary of the types of plug-in EVs is given in Table 14.1.

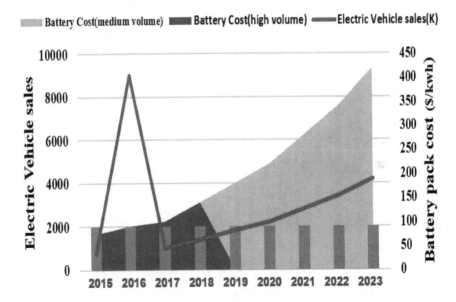

FIGURE 14.2 Sales of EV increases with falling battery cost [3].

TABLE 14.1
Types of Plug-in EVs [3–5]

Types of EVs	EV Models	Prominent Features
Plugin EV (gasoline + battery)	Toyota Prius, Ford Fusion, Audi A3 E-Tron; BMW 530Le	It is powered by dual-fuel similar to Hybrid EV with another option of plug-in charging with any electrical outlet. The energy stored in battery packs supplied power to an electric motor or by an internal combustion engine that runs on conventional fuel or regenerative braking. It employs both in the charge sustaining mode as well as a charge depleting mode of operation.
Hybrid, EV (gasoline + battery)	Honda Civic Hybrid, Ford Fusion Hybrid, Toyota Prius Hybrid, LEXUX NX	HEVs cannot be charged from outside outlets like the power grid. The battery bank charged by IC engine default built-in and by regenerative braking, i.e., by transforming its kinematic energy into chemical energy that deposited in the banks of the battery.
Extended range EV/PEV	Chevrolet, Cadillac	This is an example of a merger between PHEV and BEV, which has a more efficient system for fuel and fewer emissions of toxic nature
Battery, EV (Battery)	Nissan Leaf, BMW i-3, Mitsubishi I MiEV, Tesla Roadster Model S, Ford Focus Electric, Benz EQC	Zero-emission is achieved in this type by the merger of batteries, which can be recharged by an electrical outlet. It works in the charge depleting states of operation and requires enough power battery packs and high power. As the fixed cost is high which does not break even quickly because of short mileage, the charging duration also long makes this option less feasible.
Fuel cell EV, (FC +SC/battery)	Nissan Motor	It accumulates hydrogen as a fuel that is produced from natural gas. The energy, in this case hydrogen, provisions of a fuel cell within combustions with the oxygen from around produces a flow of electrons, which is due to electrolysis in the reverse direction, thus resulting in heat and water as byproducts of this chemical reaction

14.2 DELAYING RECYCLING THROUGH REPURPOSING

Three kinds of procedures are right now accessible for reusing EV batteries. In the first place, the batteries can experience a refining procedure, accessible on an enormous scale for different kinds of batteries, including lithium-particle and nickel-metal batteries. These batteries are encouraged into the smelter to recuperate significant metals. Remaining materials, for example, lithium, are lost to slag, which can be used for the manufacturing, not accessible for a wide range of batteries, including isolating parts through different physical and synthetic procedures, including hydrometallurgical advances, and afterward recouping any battery-grade materials directly. Finally, transitional procedures include extricating exclusively perilous battery segments toward the end of the battery life, which is

Recycling and Disposal of EV Batteries

TABLE 14.2
Pros and Cons of Reconditioned EV Batteries

Sr. No	Pros	Cons
1	Making their repurposing for a second and even third life much progressively significant to expand their monetary and ecological incentive before reusing.	The reusing lithium-particle batteries involve expense and potential waste.
2	The batteries can be used for stationary storage, which is monetary favorable, thus saving on the initial amount of the product. This battery property is another factor that will be making EVs popular.	New batteries are used for stationary storage at the moment which has a high density and has a bigger life than the batteries that have served their first life. At the moment, companies are not issuing warranty to batteries that are used.
3	The stability of the grid and the ability of the grid to integrate renewable energy are aided by the secondary storage.	The size and shape, as well as the performance of the EV batteries, differ. The batteries perform under different climatic conditions and stressors and for different car models.
4	As per the costing, this is the most economical system where the battery packs are of the uniform constitution.	When the battery packs are of the nonuniform constitution, the output has to be regulated by using software that is noneconomical. Research is being conducted to attain performance by incorporating old batteries with new ones, but it will take time and resources to reach the desired results.

useful in limiting the measure of dangerous substances that advance into the earth. Table 14.2 gives a very precise view of the merits and demerits, which can be attained by reusing the EV batteries [6].

Figure 14.3 sheds light on the EV batteries currently used and how they can be further utilized in second life.

14.2.1 Repurposing

Synthetic batteries in EVs give a quick and enormous force supply. Most flow module EVs – characterized as vehicles that attach into the lattice for a few or the entirety of their capacity – use lithium-particle batteries [7]. These batteries, though in various configurations, normally utilized inconvenient customer hardware, for example, phones and PCs. In contrast, their higher vitality per unit mass is comparative with other electrical vitality stockpiling frameworks. Nevertheless, the specific science of vehicle batteries regularly contrasts from buyer hardware, batteries, just as from one another relying upon the automaker. In general, batteries in vehicles have a higher all-out force limit and size.

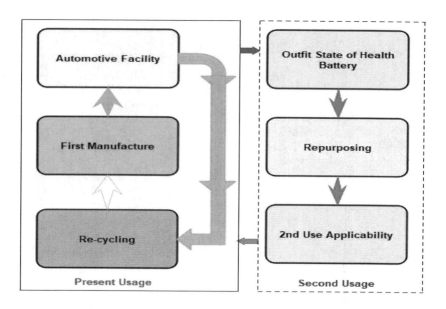

FIGURE 14.3 Possible vehicle battery life [2].

Automakers lean toward lithium-particle batteries since they convey predominant execution in both force and vitality thickness, permitting them to accomplish a high weight-to-execution proportion. Moreover, most segments of lithium-particle batteries can be recycled. Most of the industries are using LiBs because of their economic and ecological perspectives; re-using the LiB batteries which are no longer to be used in their prime application field is preferred [8]. Participants at the assembling noticed that expanded assembling and economies of scale have discounted costs and broadened their helpful life cycle, with an anticipated value diminishing around seven-to-eight percent for each year. Table 14.3 depicts the mechanical design of the battery for different brands of EVs.

Rather than reusing them promptly, the huge number of batteries that will be leaving EVs in the coming years could be repurposed, prompting a surge of reasonable batteries that can give vitality stockpiling administrations to clients, utilities, and matrix administrators. Since second-life batteries will hold critical limits, they might be appropriate for different clients and lattice applications, especially whenever amassed for mass vitality stockpiling [10,11].

TABLE 14.3
Common Zero-Emission Vehicle Battery Packs [9]

Battery Types	Cell Design	Company	Cells/Modules	Energy in kWh
Tesla model S	Cylindrical	Panasonic	7,104 /16	85-large pack
Nissan leaf	Pouch	NEC	192/48	24
Chevy volt	Prismatic	LG Chem	288/9	16.5

14.3 ECONOMIC ASPECTS

Electricity is usually generated by nuclear or hydroelectric means in developing countries, whereas developing and underdeveloped countries are still stuck with hydroelectric resources. A country's demand for electricity is never-ending and increases every day as population and industry are on the rise, and most of the appliances use electrical energy to operate. As we enter the digital age, the electricity demand has increased many folds. A nuclear supply can only be operated on a level load, as it is not feasible to turn down or increase the level easily. To cater to these demands, nonrenewable energy sources need to be used. During the peak energy situation, the produced energy could be stored somewhere else. Thus, at this point, reusing the EV batteries is not a choice but a necessity.

Reused batteries for EVs, HEVs, and PHEVs present a superb, financially perceptive alternative for vitality stockpiling applications. The organizations involved reusing lithium-particle batteries to form a circular economy are currently working with battery creators to embrace effectively disassembled item structures and are preparing them to take up the new green science forms for the green recuperation of all esteemed battery segments.

From a business stance, the risk is marginalized for investments by putting the resources into an enormous number of little applications than few huge applications following the age-old saying of "Don't put all your eggs in one basket". Smaller applications can incorporate various private customers and a wide assortment of business clients, including telecommunication organizations requiring 5–10 EV battery packs each, light business structures requiring 10–15 packs each, and nourishment dissemination focuses requiring 30–40 packs each. The batteries can be used in several applications. A few of them are discussed below.

14.3.1 IDENTIFYING DOMESTIC DEMAND

Energy storage systems (ESSs) are capturing the market by storm. It works by reusing the stationary Li-ion EV batteries as energy storage devices. It helps to counter the effects of supply and demand gaps. It can be used to power households and industries by improving the economy and decreases greenhouse gas (GHG) emissions.

Now ESS is an established system that is controlled centrally. The demand side used it for deployment at the microgrid side. The system is currently catering to the efficient usage of renewable energy resources; other resources such as power demand shortfalls and standby operator as the weather conditions change, and the usage of peak demand rate factors and climate changes. However, the efficiencies are not up to the standard, and the high cost of installations makes the system lagging and gives room for further research.

14.3.2 IDENTIFYING INDUSTRIAL DEMAND

The impact of solutions provided by ESS in the areas of the smart grid is quite visible. The utility grid problems can be stabilized by using ESSs. The ESS optimizes the quality of the grid as it maintains the power as constant. Renewable energy sources are used to balance the load curve, and the frequency is controlled. The capability of the transmitting lines can be enhanced, as well as the voltage fluctuations are mitigated by using them [4].

The industry needs systems that provide capacity with a high rating, and the electric wires that are used should be of high grade as well to handle the rise in voltage and current. This is provided by centralizing power generation systems.

14.4 STANDARDS FOR REUSING EV BATTERIES

The organization that is currently working for the EV is known as EVs and the Environment Informal Working Group (EVE IWG) which was formed in March 2012. This organization is currently working on environmental issues related to EVs [12].

A new guideline in China presently considers EV creators answerable for the recuperation of batteries, expecting them to set up reusing channels and administration outlets where used batteries can be gathered, put away, and moved to reuse organizations. Before the finish of Feb 2019, 393 carmakers, 44 rejected vehicle destroying ventures, 37-course use endeavors, and 42 reusing undertakings had just joined the new discernibility stage to follow starting point and proprietors of disposed of batteries [13].

Worldwide battery reusing prerequisites are either missing totally or where they exist; they vary a lot, as along with the profundity of inclusion. This presents another interesting area of research on a worldwide premise because of the unpredictable idea of practices and mentalities toward reusing EV batteries around the world. What is prescribed is that assets be assigned to assess the benefit of creating producing for recyclability necessities [12].

14.5 ENVIRONMENTAL IMPACTS OF EV BATTERIES | EVBs

The design of LiBs depends on the selection of active elements to be used for cathode, whereas these materials have a substantial impact on the health of *Homo sapiens* in the surrounding environment. These impacts are due to the design procedure of materials, the use of energy, and the way they are extracted from the surface of the earth. The elements which are needed for the manufacturing of LiBs contribute more toward metal degradation and GHGs. An Environmental Protection Agency (EPA) study found that the batteries which used aluminum have higher rates of contributing to ozone depletion [14]. Environmental risks arising from lithium-ion batteries must be addressed during execution of recycling and waste strategies. Lithium-ion batteries contain a high proportion of hazardous heavy metals that are harmful to the environment as well as for the human body [15]. In 2005, the used LiBs were gathered, and the result showed that from 4,000 tons of LiBs, 1,100 tons of heavy metals and around and above 200 tons of toxic materials were produced; toxic materials in lithium-ion batteries include nickel, copper, and organic and lead chemicals [16]. LiBs are categorized as class-nine dangerous materials. As LiBs are hazardous, especially the possibility of rapid discharge, in this case, the batteries generate avalanche currents, which cause overheating and risk of explosion. Therefore, lithium-ion batteries might be very hazardous when transported, particularly by air flights [17].

The energy demand for the manufacturing of LiBs has estimated at approximately 90 MJ kg^{-1} and 12.5 kg of carbon dioxide per kg. The main contribution toward the health and environmental effects starts with the development of raw materials accompanied by battery manufacturing, supply, uses, services, recycling, and waste management [18].

14.5.1 Raw Material Manufacturing Effects

Extraction, casting procedure, smelting, leaching, and filtering are the main steps used to convert the metal and its elements in the particular material used for the batteries. In the production of LiBs for EVs, the aluminum alone produces 2–3 kg of CO_2 equivalent/kg of battery. The manufacturing of electrode like lithium manganese oxide ($LiMn_2O_4$) generates about 800–1,000 kg of carbon dioxide equivalent/kg of battery. However, the manufacturing of electrolytes such as dimethyl carbonate and lithium hexafluoro phosphate ($LiPF_6$) generates 100–500 kg of CO_2 equivalent/kg of battery. The manufacturing of such raw materials has high threats to the environment [19].

14.5.2 Battery Manufacturing Effects

Hazardous air emissions, solid waste, and water pollution can generate damage to health and the environment during the manufacturing of LiBs. The automotive industry has surged the demand for LiB, with the dire negligence of GHG (GHG) emissions. It is reported that three categories of cathode materials are mostly used in the manufacturing of LiBs that are mainly responsible for GHG emission. It is reported in Ref. [20] that the manufacturing of 28 kWh battery, the cathode materials of cobalt oxide, lithium nickel manganese, lithium-ion manganese oxide, and lithium iron phosphate produced GHG emissions approximately 2,912 Kg.CO_2.eq, 2,705 Kg.CO_2.eq and 3,061 Kg.CO_2.eq, respectively.

14.5.3 Thermal Gas Emission

The major thermal run-away effects are gas and heat emissions, which are highly flammable. The structure of the batteries and cells typically involve protections to release the gas without generating a risk of cells or batteries exploding. Likewise, nonflammable materials like plastics are also avoided to prevent the additional participation of plastic combustion to the generated heat. Table 14.4 shows the emission of gases throughout the lithium-ion battery thermal lifecycle.

14.5.4 Chemical Hazards

Lithium-ion battery leaking is equal to hydrofluoric acid production. Since the electrolyte is fluid, there is a possibility that this fluid may be leak inside the battery and make contact with water and air. Two chemical reactions may cause the hydrofluoric acid production, the electrolyte ions of PF_6^- in water presence and that PF_6^- ion combustion. PF_6^- ion hydrolysis takes place in the existence of water in basic and acidic

TABLE 14.4
Lithium-Ion Battery Thermal Run-Away Emission of Gases [10]

Compound	Concentration (Percentage)
Carbon monoxide	40
Hydrogen	30
Carbon dioxide	20
Hydrocarbons	7
Hydrogen fluoride	Less than 3

(PH level between 1 and 12) medium, whereas hydrolysis production is not favorable. Through this reaction, hydrofluoric acid is produced. The hydrofluoric acid may cause serious chemical accidents when it interacts with the eyes or skin.

14.6 BATTERY DISMANTLING AND HANDLING HEALTH HAZARDS

The batteries of EVs are available in a wide variety in terms of chemistry and forms. To recycle batteries of an EV, the first important thing is sorting and classifying the chemistry of the battery through inspection, and after that forwarding the batteries for dismantling safely toward workbenches. Trained technicians dismantle the packs of batteries to the module or cell, circuitry, wiring, and pieces of assembly segregated from individual cells. Disassembling through the manual process is performed by manual and power equipment as cells and packs of the battery are constructed with complexity since it is arduous to devise a cost-effective end-of-life (EOL) process plan for EV batteries. Safety training is vital for the personnel who are responsible for the dismantling of these items, and the instructions for dismantling should be given by the manufacturers.

Lithium-ion battery disassembly is a complicated, costly, and lengthy process because of the extensive range of battery composition, components, and potential hazards due to high voltage and chemical products in cells of the battery. All of these elements must be taken into consideration while planning the dismantling process and appropriate workstations. A lithium-ion battery includes toxic chemicals and by-products, which are either destructive, or by the reaction with other materials, they produce toxic, heat-generating, or flammable chemicals. The details of the main components, material incompatibility, and their health effects are summarized in Table 14.5 below.

The component of the cathode has a significant risk of toxicity. Therefore, the probability of health effects is termed as low as the elements are solids, and a high temperature is used to melt them. If these elements are disclosed, they will involve the breaking of a battery with the chemical component separation or vaporization. In a battery of lithium-ion, the cathode and anode are covered by toxic substances such as the most hazardous chemicals like solid electrolyte interface layer, which is coated on anode and cathode, electrolyte salt, carbon anode, and electrolyte solvent.

Recycling and Disposal of EV Batteries

TABLE 14.5
Component Details of Lithium-Ion Battery [21]

Components	Examples	Effects on Health	Incompatibility	Noninflammable/ Flammable
Cathode (negative)	Lithium cobalt (LiCoO$_2$); Lithium manganate (LiMn$_2$O$_4$); Lithium phosphate (LiFePO$_4$)	Respiratory, skin, eye, and gastrointestinal irritant; possible carcinogens	Nil, but elude of extreme temperature and fire	Noninflammable
Anode (Positive)	Graphite (C)	In solid form no effect (evade dust)	Nil, but elude of extreme temperature and fire	Noninflammable
Salt electrolyte	Lithium hexafluoro phosphate (LiPF$_6$)	Causes respiratory, eye, skin and gastrointestinal	Agents of oxidizing; Water; strong acids	Noninflammable
Solvent electrolyte	EC; DMC; PC; DEC	Respiratory, skin, eye, and gastrointestinal irritant;	Agents of oxidizing; alkalis and acids;	Flammable
Separator	Polyethylene; polypropylene	Nil	Nil, but elude of extreme temperature and fire	Noncombustible

14.6.1 LITHIUM-ION BATTERY LANDFILL

The reuse of LiBs is done in fewer countries, while in other countries, LiBs were landfill or thrown away. Discarded batteries have eventually reached landfills and saturated, and this may also harm the environment [22]. GHG emissions from organic components like electrolyte solvent, plastic, and paper are produced during landfill processing. Besides, heavy metals can be leached to the atmosphere and can enter water or soil based on the landfill state. Landfilling possesses more danger to the environment in comparison to the recycling of the batteries.

14.6.2 IMPACT OF RECYCLING ON THE ENVIRONMENT

Growing waste required more land space for disposal, as these batteries contain harmful chemicals, so ultimately, these chemicals return to the environment. However, with the comparison to other batteries, the LiB contains small toxic materials, so recycling of batteries may have a less impact on the environment. Particularly by recycling lithium-ion batteries, the considerable reduction in energy consumption, GHG emission, and the use of natural resources are found in comparison to the manufacturing of new LiBs.

Additionally, by recycling, we can retrieve lithium and other materials. In particular, the use of lithium is on the rise due to the increasing need for LiBs, which are used in EVs. Around 25% of the total amount of lithium is used by battery manufacturing industries for EVs, and this number will surge exponentially shortly [23].

14.6.3 Recycling of EV Batteries

The demand for EVs is rising and will be increased in the coming years. It is estimated that post-vehicle battery pack application will be increased from 1.4 to 6.8 million by 2035 [24]. LiBs are a suitable choice for EVs; secondly, the disposal of LIBs is vital to conserve the environment and health of *Homo sapiens*. Regrettably, the LiBs' recycling percentage is only 3%, whereas its revival is negligible. It is estimated that if the rate of recycling will not increase, the requirement of lithium transcends by 2023 [25]. Figure 14.4 summarizes that after the manufacturing of lithium-ion battery, its life is approximately 8–10 years. When the battery completes its life, it should be sent toward service and retail centers where these batteries can be repaired. If the product has the end of life, then it should be sent toward the recycling process from where raw materials can be obtained to manufacturing new ones [26].

14.7 ENVIRONMENTAL ASPECTS OF REUSE

Assuming that the second life of batteries of EV may reduce the strain of producing a new battery or eventually reduce the usage of resources, to adequately evaluate the environmental aspects of using secondary batteries, the environmental impacts of EV batteries should be empirically measured over the entire life cycle. Contemplating

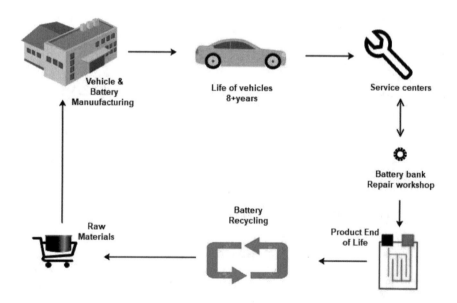

FIGURE 14.4 The lifecycle of a lithium-ion battery.

the reuse of nearing retirement of EV batteries, the complete environmental effect of the production, primary reuse, and recycling of EV batteries should be dealt with consideration in the analysis of the life of the batteries. Reusing an EV retired lithium-ion battery in a stationary system might reduce global warming potential (GWP) by 15% when compared to nonstationary use. It has been reported that LiBs' refurbishment may reduce GWP effects by 12%–46% compared with the similar battery functionality of Pb acid [27]. The production phase of the battery pack produced the majority of greenhouse gas emissions by approximately 40% when compared with both phases of usage and re-usage (31% and 26%), thus contributing to greenhouse gas emissions significantly [28].

14.8 ENVIRONMENTAL ASPECTS OF RECYCLING

When EV batteries are decommissioned, the overall impact on the environment must be evaluated while considering the processes of battery recycling, which include collection, transport, dismantling, and recycling. Transporting lithium-ion batteries to recyclers has a slight contribution to the environment as compared to manufacturing and usage phases. However, there is a need for adequate transport to minimize or exclude the environmental effects of the process. Research has shown that truck and rail transport is capable of reducing greenhouse gas emissions related to transportation by about 23%–45%. Besides, they combined geospatial models and life cycle analysis to recognize the anticipated infrastructure needs, categorize emissions, and material flows and evaluate human health risk from the lithium-ion battery recycling system in California [29]. These findings indicate that developing of facilities more than two for dismantling would probably reduce investment returns. Optimum recycling technologies and multimodal supply chain collection and transport will help in dealing with the economic and environmental challenges of LiB production.

Pyrometallurgical methods of recycling can minimize greenhouse gas emissions and primary energy utilization by 23% and 6%–56%, respectively, compared to virgin production [29]. Hydrometallurgy recycling provides enormous electrical energy savings and slight air pollution; furthermore, pyrometallurgical recycling has significant benefits concerning water utilization. The significant environmental challenges regarding the processes of recycling are waste material landfills, plastics incineration, and electricity consumption, specifically in the procedures of smelting, which releases a lot of energy. The hydrometallurgical processes' major impact on climate change is landfill, while plastics incineration is the major impact of pyrometallurgical processes. Mechanical and hydrometallurgical techniques may retrieve more materials consuming lesser energy than the pyrometallurgical technique.

The resources required for battery recycling is just 387.4 MJ for the manufacturing of 1 kg of cathode material, whereas the manufacturing of new battery production required 795.4 MJ of 1 kg of cathode material. The recycling of batteries may also recuperate 51% of the valuable resources used during the production of new batteries, while nuclear energy consumption and fossil fuel were decreased by 57.25% and 45.3%, respectively [30].

14.9 RECYCLING

Recycling of the EV batteries will be the major concern in upcoming decades, whereas in the manufacturing of EV batteries, different chemical materials are used, which makes the recycling process more complex. Lithium-ion battery packs have a complex cell, module, and structure due to cylindrical or prismatic cells associated with a series-parallel configuration such as wire bonding, welding, and mechanical appending [31].

14.9.1 Recycling Methods

In each cell of lithium-ion, batteries require a broad spectrum of materials, which results in a more complex recycling process, including

- Lithium, copper, manganese, aluminum, zinc, cobalt, and steel
- Components of plastic: polyethylene terephthalate and polypropylene
- Graphitic carbon
- Sulphuric acid electrolyte and solvent
- Fiberglass
- A management system of battery/coolant

The flow process of a lithium-ion battery recycling is depicted in Figure 14.5.

During recycling, all of the above materials must be separated from one another. The following are the various recycling methods/processes currently used for lithium-ion batteries.

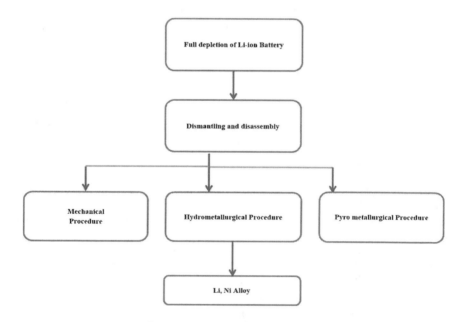

FIGURE 14.5 Lithium-ion battery-recycling procedures.

14.9.2 Mechanical Procedure | MP

A mechanical method is used to split the components of EV batteries where electrolyte securely separates to cut down the metals so that the process becomes easier. Batteries can be fed into equipment through a conveyer, and the components of the battery are shredded into small pieces and then moved toward recycling.

14.9.3 Pyro Metallurgical Procedure | PM

This is the process of extreme temperature, which includes two operations. Lithium-ion batteries firstly kindle in the smelter at which substances are smashed down, and other materials like separators and plastic are burned away. Then by carbon curtailment, new alloys are produced, and metal alloys are further extracted to retrieve materials [32]. During this process, costly metals like copper (Cu), nickel (Ni), and cobalt (Co) are retrieved with maximum competence. The electrolyte, plastic, and anode are oxidized and provide the energy for the procedure. The major advantages of this process are as follows: (1) a mature and simple process, (2) reduction in size and sorting out is not mandatory – a composition of LiB and nickel-metal hydride batteries, which can be recovered, and (3) production is comprised of basic building blocks, which might be used to synthesize materials of the cathode by various chemistries. On the other hand, the main drawbacks are as follows: (1) carbon dioxide formation and higher power consumption during the procedure of smelting, (2) alloy needs more refining due to which the cost of recycling will increase, (3) many products in lithium-ion batteries like graphite, plastic, and aluminium were not retrieved; the procedure recuperates cobalt and nickel from the material of cathode and copper from anode current collector, which accounts for only 30% w.r.t. lithium-ion batteries for electronics, (4) the business strategy might not work effectively for EV batteries, because of the lower concentration of cobalt. Furthermore, the aim of the industries is cobalt reduction or eventually to produce cobalt-free cathode materials. Recently, the pyro-metallurgical process exists extensively in the industry due to its naivety and higher efficiency. There is no standardized procedure for pyro-metallurgical recycling of the batteries, and every current method is peculiar to the concerned company. Given Figure 14.6 portrays the whole recycling processes.

14.9.4 Hydrometallurgical Procedure | HP

In the hydrometallurgical procedure, material retrieval is obtained through aqueous chemistry, by leaching method, and later, concentration and filtration are done. In lithium-ion batteries, solution ions are segregated using different technologies like solvent extraction, ion exchange, electrolysis, and chemical precipitation [33]. In this process, various steps are involved in recycling. The following steps are for dismantling, crushing, and sieving and after that, the leaching process takes place where metals in current EOL of lithium-ion batteries are dissolved by the leaching process and the latter leachate goes through by further process to isolate the metal ions. The last step is the separation of different solvents and production of the cathode from where one can develop a new battery. The major benefits of the hydrometallurgical

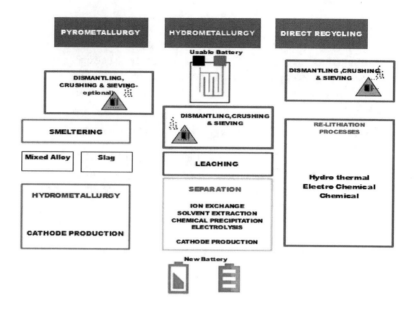

FIGURE 14.6 Recycling methods.

process are as follows: (1) generation of high purity materials, (2) most constituents of a LiB can be retrieved, (3) the operation of the hydrometallurgical process is at a lower temperature, (4) the emission of CO_2 is lesser in the comparison of a pyro-metallurgical process [34]. The major disadvantages of this process are as follows: (1) sorting is required, which needs more storage space due to which cost increases and process complexity arises, (2) difficulty in the segregation of elements like nickel, iron, cobalt, aluminum, and copper from the solution as they have similar properties, and this again results in higher costs, and (3) water waste treatment and associated costs that are associated with them.

14.9.5 Direct Recycling Procedure | DRP

DRP method of recovery is suggested for the direct processing and recovery of active lithium-ion battery material, while maintaining the structure of its origin [35]. In this procedure, battery components are separated with the use of methods such as magnetic separation, physical separation, and thermal process to evade active materials' chemical breakdown. Active materials are then refined, and defects are retrieved by the hydrothermal or re-lithiation process. Cathodes might be a combination of two or more active materials, and the segregation of these materials is difficult. The major advantages of this process involve the following: (1) relatively easy procedure, (2) after regeneration of active materials, they can be used directly, and (3) in the comparability of hydrometallurgy and pyro-metallurgy process, the recycling process generates fewer emissions and secondary pollutants [20]. The major disadvantages of this process are as follows: (1) involves preprocessing/comprehensive sorting for active chemical substances, (2) untested technology, which to date only exists at the laboratory scale, (3) significant susceptibility to changes in the data stream, and (4) this process is not flexible because the material is not purified, and therefore, the cycle might not be acceptable to the cathode chemistry.

14.10 BEST PRACTICES OF LITHIUM-ION BATTERY RECYCLING

The best practices for recycling LIBs in the industries are discussed below

14.10.1 UMICORE COMPANY

Umicore company exploits the pyro-metallurgical procedure for the recycling of LIBs by using the method of ultra-high temperature. This company is able to handle 7,000 metric tons/year, and the process of ultra-high temperature creates an alloy, copper (Cu), nickel (Ni), and cobalt (Co), and generates slag for construction materials.

However, the further steps of segregation or purification involved the technique of hydrometallurgy, which includes leaching, extraction of solvents, and precipitation for the production of new cathode substances like nickel-manganese-cobalt (NMC) and lithium cobalt oxide (LCO). The lithium had existed previously in the slag, which was mainly used for building [36], but Umicore recently illustrates that by doing the further process, the lithium-ion battery slag can be converted into Li-recovery flow sheets with external collaborators [37]. Figure 14.7 showed the systematic process.

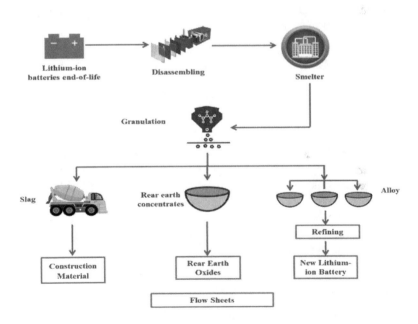

FIGURE 14.7 Umicore recycling process [38].

14.10.2 Retrieve Technologies

In North America, the company Retrieve Technologies utilizes a hydrometallurgical procedure to recycle lithium-ion batteries. They manually dismantled the lithium-ion battery pack to the cell level or module. Then, it is fed into a lithium brine flooded hammer mill for crushing to lower the battery reactivity, neutralization of electrolyte, and avoid the emissions of gas [39]. This company follows a three-stream process of materials that include plastic fluff, metal solids, and metal-enriched liquid. The metal-enriched liquid contains mainly lithium-ion, precipitated and purified, and solid metals might contain copper (Cu), aluminium (Al), and black mass. The solid metals are provided to downstream smelters who are interested in nickel (Ni) and cobalt (Co) contents. On the other hand, plastic may be disposed of or recycled. However, lithium is recovered as lithium carbonate [40,41].

14.10.3 Onto Technology

Onto Technology recycled lithium-ion batteries by using the direct recycling method where they obtain the electrode from lithium-ion batteries. After that cathode, the material is obtained by combining and detaching the electrode from the current collector in an aqueous alkaline solution. The devalued cathode content might revive and is possible to use in the development of cells by using hydrothermal and additional heat treatment.

14.11 SAFETY INDICATORS

Safety precautions are the most important aspect of lithium-ion batteries while dismantling, sorting, reorganizing, and recycling the retired batteries and must require proper management to avoid risks.

14.12 DISMANTLING AND STORAGE

The lithium-ion batteries must be reassigned for reuse, and after cessation, the module and battery pack must be dismantled with the elimination of the exterior circuit and removal of battery scarp and data sensors. The remaining ability, immensely high voltage, an exterior circuit, and reliability of connectivity need the dismantling of demolished batteries to be carried out in an open area with safety precautions. Disassembling processes might also prevent permanent impairment to the battery. To achieve nondestructive and secure dismantling, it is suggested to dismantle the module or pack level instead of cell level soft connection modules. Dismantling at the cell level is not viable both economically and technically. To minimize safety risks, transport and storage of demolished batteries should evade transportation of long-distance, long-term storage, fire sources, rain exposure, and other collisions.

14.12.1 Reorganizing and Screening

Decommissioned batteries cause several risks like flatulence, maligned casing, leakage, insulation failure, and short circuit. Batteries' screening out with long service life and good performance is a difficult task while maintaining the protection of decommissioned batteries. Traditional screening techniques are based primarily on cells that are singled out, comprising assessment and testing for performance and appearance through nano-CT to evaluate the three-dimensional structure qualitatively; to create an electrochemical model, and to evaluate the electro-chemical external parameters' performance. Some technologies have recently been developed, like contact-type ultrasonic, nondestructive screening technology that can quickly and correctly determine the cell's state of health and state of charge by correlating ultrasonic signals with a performance of electrochemical and nondestructive and noncontact screening methodology. Detection procedures are fairly successful for battery modules and might make visible module, hence lowering the risk of short circuits and "dead Li". Furthermore, at this point, research has not come forward with any technology or equipment that can detect and view battery modules.

The modules of batteries have to be rearranged after the screening, depending upon database compiling demand details for secondary usage, functional details, and test requirements comprising of the capacity, residual cycle life, internal resistance, and material system. The battery management system of decommissioned batteries seems more systematic when concerning about safety. Thus, new packs of the battery must consider the following factors while reorganization to minimize the safety risk of separate integration of decommissioned batteries: optimizing the module model connection to enhance flexibility; implementing intelligent technology of time-division; equipping an effective temperature monitoring system and current equalizers and the module protection through high-voltage and thermal monitoring.

14.13 TECHNOLOGICAL INITIATIVES

Research is simultaneously being done in three major avenues to curb the issue of used batteries in the current scenario.

Composition: Different compositions such as lithium-sulphur batteries and lithium-air batteries are being tested for performance and cost in comparison to LiBs.

Available resources: The reserves of lithium will become scarce as they are being consumed at a very high rate since its demand increases exponentially with the increase in EVs. Presently, we are expecting an exceed in the lithium cost if the battery recycling is not made efficient and economical.

Recycling methods: It has been observed that the methods of hydrometallurgical and mechanical procedures are giving better results in the recollection of materials, as well as it is a low energy-seeking procedure in comparison to the pyro-metallurgical procedure. The Lithium-ion Battery Recycling Initiative (LiBRi) is teaming with different companies that are working on ways to recollect lithium from the slag content. Research is underway to recover the maximum amount of lithium during the dismantling of the batteries

14.14 CONCLUSION

Researchers are experimenting with different combinations of chemicals to replace lithium and cobalt. The alternatives can be the materials that are found in abundance and are less toxic. Nevertheless, if new batteries are not efficient or more costly than lithium, they could wind up negatively affecting the earth severely in general. If recycling of EV batteries is taken as a business practice, key areas to be considered can be a system to collect expired batteries, storing them, making logistical arrangements, etc.

14.15 RECOMMENDATIONS AND FUTURE DIRECTIONS

Further research can be done on developing a supply chain model of closed loop for the batteries which can have properly marked and labeled pieces right at the manufacturing unit. This will help in the retrieval of these batteries after their end life. Also, sustainability can be applied to the whole manufacturing procedure for LiBs, which help in reusing the batteries to achieve zero carbon emissions.

REFERENCES

1. Z. A. Arfeen, A. B. Khairuddin, A. Munir, M. K. Azam, M. Faisal, and M. S. B. Arif, "En route of electric vehicles with the vehicle to grid technique in distribution networks–Status and technological review," *Energy Storage*.
2. R. Irle. (2019). *Global BEV & PHEV Sales for 2019*. Available: www. EV-volumes.com.
3. P. Slowik, N. Pavlenko, and N. Lutsey, "Assessment of next-generation electric vehicle technologies," *White paper, International Council on Clean Transportation (October 2016)*, 2016.
4. A. R. Bhatti and Z. Salam, "A rule-based energy management scheme for uninterrupted electric vehicles charging at constant price using photovoltaic-grid system," *Renewable Energy*, vol. 125, pp. 384–400, 2018.
5. X. Lü, Y. Wu, J. Lian, Y. Zhang, C. Chen, P. Wang, and L. Meng "Energy management of hybrid electric vehicles: A review of energy optimization of fuel cell hybrid power system based on genetic algorithm," *Energy Conversion and Management*, vol. 205, p. 112474, 2020.
6. (2019). *Better World Solutions*. Available: https://www.betterworldsolutions.eu/repurpose-ev-batteries/.
7. Z. A. Arfeen, M. P. Abdullah, R. Hassan, B. M. Othman, A. Siddique, A. U. Rehman, and U. U. Sheikh "Energy storage usages: Engineering reactions, economic-technological values for electric vehicles – a technological outlook" *International Transactions on Electrical Energy Systems*, 2020.
8. J. Falk, A. Nedjalkov, M. Angelmahr, and W. Schade, "Applying lithium-ion second life batteries for off-grid solar powered system—a socio-economic case study for rural development."
9. L. Canals Casals and B. Amante García, "Assessing electric vehicles battery second life remanufacture and management," *Journal of Green Engineering*, vol. 6, pp. 77–98, 2016.
10. k. Environmental, "*Research Study on Reuse and Recycling of Batteries Employed in Electric Vehicles*," energy API, 2019.
11. M. Pagliaro and F. Meneguzzo, "Lithium battery reusing and recycling: A circular economy insight," *Heliyon*, vol. 5, pp. e01866–e01866, 2019.
12. (2019). *Electric Vehicle Regulatory Reference Guide*. Available: https://www.unece.org/.
13. *Battery Recycling as a Business; 2019*. Available: https://batteryuniversity.com.
14. S. Amarakoon, J. Smith, and B. Segal, "Application of life-cycle assessment to nanoscale technology: Lithium-ion batteries for electric vehicles," 2013.

15. P. B. Tchounwou, C. G. Yedjou, A. K. Patlolla, and D. J. Sutton, "Heavy metal toxicity and the environment," in *Molecular, Clinical and Environmental Toxicology*, ed. Springer, 2012, pp. 133–164.
16. J. Ordoñez, E. Gago, and A. Girard, "Processes and technologies for the recycling and recovery of spent lithium-ion batteries," *Renewable and Sustainable Energy Reviews*, vol. 60, pp. 195–205, 2016.
17. (2019). *"Hazard class 9- hazard class labels explained,"* Available: [Online]. Available: https://www.general-data.com/about/blog/hazard-class-9-hazard-class-labels-explained
18. M. C. McManus, "Environmental consequences of the use of batteries in low carbon systems: The impact of battery production," *Applied Energy*, vol. 93, pp. 288–295, 2012.
19. P. Meshram, A. Mishra, and R. Sahu, "Environmental impact of spent lithium ION batteries and green recycling perspectives by organic acids–A review," *Chemosphere*, p. 125291, 2019.
20. H. Hao, Z. Mu, S. Jiang, Z. Liu, and F. Zhao, "GHG emissions from the production of lithium-ion batteries for electric vehicles in China," *Sustainability*, vol. 9, p. 504, 2017.
21. D. Stephens, P. Shawcross, G. Stout, E. Sullivan, J. Saunders, S. Risser, and J. Sayre "Lithium-ion battery safety issues for electric and plug-in hybrid vehicles," *National Highway Traffic Safety Administration, Washington, DC Report No. DOT HS*, vol. 812, p. 418, 2017.
22. J. Yang, F. Gu, and J. Guo, "Environmental feasibility of secondary use of electric vehicle lithium-ion batteries in communication base stations," *Resources, Conservation and Recycling*, vol. 156, p. 104713, 2020.
23. L. Gaines and P. Nelson, *"Lithium-ion batteries: Possible materials issues,"* in 13th international battery materials recycling seminar and exhibit, Broward County Convention Center, Fort Lauderdale, FL, 2009, p. 16.
24. S. Rohr, S. Müller, M. Baumann, M. Kerler, F. Ebert, D. Kaden, and M. Lienkamp "Quantifying uncertainties in reusing lithium-ion batteries from electric vehicles," *Procedia Manufacturing*, vol. 8, pp. 603–610, 2017.
25. A. Sonoc, J. Jeswiet, and V. K. Soo, "Opportunities to improve recycling of automotive lithium ion batteries," *Procedia cIRP*, vol. 29, pp. 752–757, 2015.
26. J. Weisman. (2019). *Tesla LIB and PV Recycling Programs.* Available: www2.calrecycle.ca.gov.
27. K. Richa, C. W. Babbitt, N. G. Nenadic, and G. Gaustad, "Environmental trade-offs across cascading lithium-ion battery life cycles," *The International Journal of Life Cycle Assessment*, vol. 22, pp. 66–81, 2017.
28. L. Ahmadi, S. B. Young, M. Fowler, R. A. Fraser, and M. A. Achachlouei, "A cascaded life cycle: Reuse of electric vehicle lithium-ion battery packs in energy storage systems," *The International Journal of Life Cycle Assessment*, vol. 22, pp. 111–124, 2017.
29. T. P. Hendrickson, O. Kavvada, N. Shah, R. Sathre, and C. D. Scown, "Life-cycle implications and supply chain logistics of electric vehicle battery recycling in California," *Environmental Research Letters*, vol. 10, p. 014011, 2015.
30. J. Dewulf, G. Van der Vorst, K. Denturck, H. Van Langenhove, W. Ghyoot, J. Tytgat, and K. Vandeputte, "Recycling rechargeable lithium ion batteries: Critical analysis of natural resource savings," *Resources, Conservation and Recycling*, vol. 54, pp. 229–234, 2010.
31. A. Das, D. Li, D. Williams, and D. Greenwood, "Joining technologies for automotive battery systems manufacturing," *World Electric Vehicle Journal*, vol. 9, p. 22, 2018.
32. L. Gaines, "The future of automotive lithium-ion battery recycling: Charting a sustainable course," *Sustainable Materials and Technologies*, vol. 1, pp. 2–7, 2014.

33. X. Li, Y. Mo, W. Qing, S. Shao, C. Y. Tang, and J. Li, "Membrane-based technologies for lithium recovery from water lithium resources: A review," *Journal of Membrane Science,* p. 117317, 2019.
34. R. E. Ciez and J. F. Whitacre, "Examining different recycling processes for lithium-ion batteries," *Nature Sustainability,* vol. 2, pp. 148–156, 2019/02/01 2019.
35. Y. Shi, G. Chen, F. Liu, X. Yue, and Z. Chen, "Resolving the compositional and structural defects of degraded $LiNi_xCo_y MnzO_2$ particles to directly regenerate high-performance lithium-ion battery cathodes," *ACS Energy Letters,* vol. 3, pp. 1683–1692, 2018.
36. L. Gaines, J. Sullivan, A. Burnham, and I. Belharouak, "Life-cycle analysis of production and recycling of lithium ion batteries," *Transportation Research Record,* vol. 2252, pp. 57–65, 2011.
37. (2020). *"Recycling Process,"* Available: [Online]. Available: https://csm.umicore.com/en/recycling/battery-recycling/our-recycling-process.
38. J. N. Kevin Popper, A. Lanfrankie, and F. Caro. (2019). *The (potential) value of labeling in the lithium ion battery supply chain.* Available: https://blogs.anderson.ucla.edu/.
39. B. Swain, "Recovery and recycling of lithium: A review," *Separation and Purification Technology,* vol. 172, pp. 388–403, 2017.
40. J. Heelan, E. Gratz, Z. Zheng, Q. Wang, M. Chen, D. Apelian, and Y. Wang "Current and prospective Li-ion battery recycling and recovery processes," *Jom,* vol. 68, pp. 2632–2638, 2016.
41. (2020). *"Lithium-Ion,"* Available: [Online]. Available: https://www.retrievtech.com/lithiumion.

15 Design and Operation of a Low-Cost Microgrid-Integrated EV for Developing Countries
A Case Study

Syed Muhammad Amrr
Indian Institute of Technology Delhi

Mahdi Shafaati Shemami
Aligarh Muslim University Aligarh

Hanan K. M. Irfan
Abul Kalam Azad University of Technology

M. S. Jamil Asghar
Aligarh Muslim University Aligarh

CONTENTS

15.1 Introduction	336
15.1.1 Central Power Station System	337
15.1.2 Distributed Generation System	337
15.2 The Design Scheme of Proposed Microgrid System	340
15.2.1 Modifications in the Proposed Grid-Connected PV System	341
15.2.2 Layout of the Proposed Control Strategy	341
15.3 Detailed Controller Design and Its Working	343
15.3.1 Mode Selector Controller	343
15.3.2 Source Selector Controller	346
15.4 Hardware Implementation of the Designed Controllers	347
15.4.1 The Experimental Setup and Results	347
15.5 Hardware in the Loop Testing of Proposed Strategy	351
15.5.1 Hardware in Loop Results	351
15.6 Conclusion	356
References	357

15.1 INTRODUCTION

The overall prosperity of any country can be measured by electrical energy production and its accessibility to its citizens. The socioeconomic development of a country cannot be improved without an adequate and uninterrupted source of energy. Conventional sources of energy viz., coal, gas, oil, etc. had been extensively exploited to meet the ever-growing demand for energy. The power demand was continuously increasing linearly in the past, and the same trend is predicted for the future. Relying on coal, gas, oil, etc. is increasingly becoming unpopular due to their adverse climatic and health impact so much so that 40% of CO_2 production is attributed to power and energy industries [1]. Further, with the increase in demand for power, the continuous supply of these nonrenewable resources is not possible due to their limited stock. As the awareness of the harmful effects of coal and fuel has been promulgated, scientists and researchers have shifted their focus toward cleaner sources of energy [2,3].

The last few decades have seen a significant jump in the process of energy harnessing and storage from renewable energy sources. As of today, several developed as well as developing countries have significantly shifted their focus toward the harnessing of "green energy", which is not only beneficial to the environment and population but is economical too. Out of the many available options of generating power, i.e., hydro energy, wind energy, geothermal energy, etc., it is solar energy that has attracted main attention and investment globally. Due to its global availability and ease of harnessing and storage, it has turned out to be quite popular in the underdeveloped countries where nonrenewable resources are limited [4]. Solar photovoltaic (PV) system is the most preferred means of tapping solar energy. Moreover, the wide commercialization of the PV system is made possible because of power electronic converters, which helps in energy conversion and grid integration of distributed generating system [5].

In today's world, electricity is no longer a luxury, and it is of utmost importance that even the most remote regions receive enough energy to power their homes. Unfortunately, in the developing and under-developed countries, the demand for power is more than the power generation. In countries like India, owing to factors like the population size, the demand for electric power is significantly higher than the power generating capacity of conventional power plants [6]. Thus, a major percentage, up to 70%, of power generation in India, is still done in thermal power plants, whereas renewable energy constitutes only 22% of the total generated power. The Indian government, like several other developing countries, has recognized the potential of power generation through solar PV and has included incentives to encourage investment in the same. The Indian government hopes to reach 175 GW of renewable energy power generation, which includes 100 GW from solar, 60 GW from wind, 10 GW from biomass, and 5 GW from small hydropower by 2022 with a huge investment of around US$100 billion [7].

Considering the ambitious goal of achieving 100 GW solar energy, the Government of India has introduced many subsidized schemes for the public to generate solar power on their own. The power generation from the solar PV systems can be broadly classified as [8] follows:

15.1.1 CENTRAL POWER STATION SYSTEM

A power station or a generating station is the conventional power generation unit from where power is transmitted to the grid only. No mechanism is used for energy storage. Moreover, power generation entirely depends upon whether condition (solar insolation), and no power is generated at night. From morning to noon and from noon to evening, the power generation does not remain the same, and it varies.

15.1.2 DISTRIBUTED GENERATION SYSTEM

Instead of one centralized location of power generation, smaller units of different types of generating plants are clubbed together, which are generally closer to the end consumers. In general, there is a great saving of transmission and distribution losses in the distributed generation system, which varies between 10% and 30%. In many under-developed countries, it is about 30%. Distributed generation systems can be further divided as follows:

a. *Stand-alone systems.* It is mostly used in remote areas that have no connections to a power grid. A stand-alone system focuses on meeting the energy demands of a small area like a colony or a village. It is built to store energy from the Sun in batteries during the hours of availability, which is then utilized in the day as well as at night. When solar energy or storage energy (battery) is not available, a diesel generator or gas generator is used. The stand-alone system could be very simple having solar PV modules, controller, and load, and there could be a complex system containing several power sources (solar, wind, small hydro, etc.), sophisticated controllers, and energy storage units for a typical load. The capacity of such a system is around 10 Wp–100 kWp.
b. *Grid-connected system.* In this case, to enhance the reliability of a solar system, the grid is also connected to this system. The generated power is supplied to the load as well as to charge the battery. When solar or stored energy is not sufficient to meet the load demand, power is received from the grid, which increases the reliability of such a system.
c. *Grid-tied system.* This system is similar to a grid-connected system. Instead of a conventional DC-to-AC inverter, it has a grid-tied inverter to synchronize with and to supply surplus power to the grid. Solar energy is supplied to load for charging of the battery, and the surplus power is supplied to the grid during the daytime. Bidirectional or smart meters are required for the net metering of energy. Such systems are expensive, and legal permission/agreement is also required for net-metering. During the daytime, surplus power is supplied to the grid, and at low insolation or during the night, the system receives power from the grid.

The proposed chapter is based on the *grid-connected PV system* or *grid assisted PV system*, where power is generated from the solar PV modules for the home loads only. The surplus power is not fed to the grid at any point in time. The system is

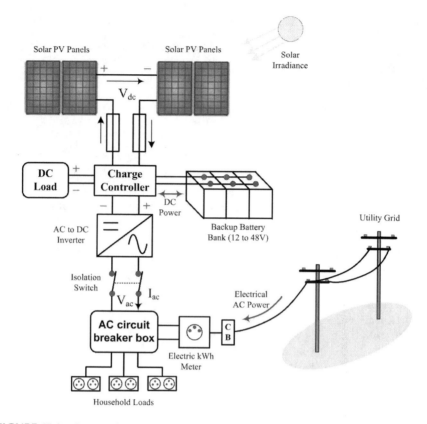

FIGURE 15.1 A general solar grid-connected system.

connected to the grid to receive power from it when solar or battery storage energy is not sufficient for the load. The basic diagram of the grid-connected PV system is shown in Figure 15.1.

These systems assist in reducing the stress on the utility grid during the daytime peak hours by isolating the home load from the grid to the PV system [9]. In developing countries, such a system (solar inverter) is commercially available in the market at a reasonable price (100 US$ for 1 kVA system). The batteries in these systems get charged from solar PV through the power conditioning unit (DC-DC converter), and the home load demands are fulfilled through the utility grid. However, in the absence of sunshine hour, the utility grid charges the batteries. During the blackout or load shedding hours (nonavailability of grid power), the home load demands are met from the batteries through the DC-to-AC inverter of the system. These commercial solar inverters have some drawbacks in their control schemes, which makes inefficient utilization of solar power. First, when the batteries are fully charged, the solar PV system is underutilized, and only the trickle current is supplied from the solar PV to the batteries. Because the battery cannot be charged further, stored energy can be used only during the load shedding hours. Second, even in the presence of sufficient solar power, the home loads receive power from the utility grid, and solar power is kept as a backup supply for the load-shedding period only. This results in the consumption of a costly tariff-based grid power instead of a freely available solar power.

The charging preference scheme of the commercially available solar grid inverter is based on the voltage level of the battery. Therefore, solar PV power charges the batteries when the voltage level of the battery is greater than a preset voltage value. Otherwise, the utility grid supplies power for charging the batteries. Moreover, in the evening or night (in the absence of solar irradiance) and if the battery is in a discharged condition, the utility grid will charge the batteries throughout the night. Consequently, in the next morning, the batteries become fully charged (under the assumption of no blackout or load shedding for longer hours at night). As a result of this, the solar PV power, which will be available in the morning, will be unutilized or underutilized since the batteries are already fully charged or partially charged in the night from the utility grid. This implies that although such inverters will ensure the reliable backup power supply, underutilization of solar PV power still exists in these control schemes. Thus, solar grid inverters require a new control algorithm to address the drawbacks of underutilization and inefficient harnessing of solar energy.

This chapter establishes a very simple, low-cost, and effective control strategy to illustrate a complete or total harnessing of solar energy, while reducing the stress on the utility grid and shaving the peak loads. The proposed microgrid is established with the use of a conventional (low cost) single-phase DC-to-AC inverter with the integration of a solar charge controller to operate as a grid-connected solar inverter. The microgrid control strategy is governed by the proposed low-cost controller circuits, which works without any microcontroller or digital signal processor (DSP).

The contributions of this chapter are as follows:

- The problems associated with commercially available solar grid-connected inverter are identified.
- The existing backup UPS is modified into a solar grid-connected inverter.
- Two simple analog controllers are developed without microprocessors for the proposed strategy for total harnessing and utilization of solar energy.
- Hardware lab testing of controllers and hardware in the loop testing of the proposed scheme has been successfully implemented for the household loads.

This chapter is organized as follows:

- *Section 2*: discussion about the solar grid-connected inverters and modification in the proposed control strategy.
- *Section 3*: the details and working principle of proposed controllers.
- *Section 4*: hardware lab testing of controllers.
- *Section 5*: closed-loop application of the proposed strategy with comparative analysis for a case study.
- *Section 6*: concluding remarks.

15.2 THE DESIGN SCHEME OF PROPOSED MICROGRID SYSTEM

For a poorly maintained power distribution system, blackout and load shedding for hours are a common practice to meet the power demands. Normally, people have some backup power supply for emergency loads, e.g., emergency lights and fans. The most common solution for the problem is the installation of conventional low-cost DC-to-AC inverters with a lead-acid battery as power storage for emergencies. An AC-to-DC charger circuit is also integrated with this inverter. Moreover, the aspect of energy requirement in rural and urban areas is significantly different. The demand in urban areas is for economic electric power, and in rural areas, it is reliable electric power. With the availability of solar power and the low-cost PV modules, a commercially available grid-connected inverter could solve these problems, i.e., offering both economical and reliable sources of energy. The commercial name of this inverter is the solar inverter. The grid-connected inverters have a conventional DC-to-AC inverter, an AC-to-DC charger circuit, and an additional DC-to-DC buck-boost converter, which charges the batteries from solar PV modules.

The proposed grid-connected PV system has the load classification in three different categories as shown in Figure 15.2. The high rating appliances are clubbed in a heavy load category, which is powered by the utility grid only. The heavy loads cannot be run during the load shedding hours by the proposed inverter scheme. The moderate and low rating appliances are categorized in normal loads, which derive the power from the proposed grid-connected PV system. The essential and emergency loads are mainly powered by the proposed system. However, in the worst scenario, when both grid and the proposed scheme gets exhausted, then a standby power supply will be used for the emergency loads.

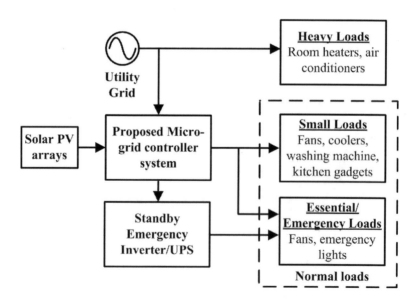

FIGURE 15.2 Block diagram of the proposed hierarchical control with load classification.

Design and Operation of a Low-Cost EV

15.2.1 Modifications in the Proposed Grid-Connected PV System

In a commercially available solar inverter, the batteries are more likely to be charged from the grid than from the solar PV modules, which causes the problem of underutilization and inefficient storage of solar energy. This problem can be resolved by integrating two controllers in the system. One at the supply side and another on the load side. Moreover, the existing backup inverter (which is normally available in every house in the developing countries) can be transformed into a solar grid-connected inverter by incorporating a solar MPPT charge controller with it. Moreover, the proposed controller circuits are an addendum to the system. Therefore, the customer need not buy a new inverter, and it can easily be implemented in the existing inverter. Thus, the new system is based on grid assistance, which is a combination of stand-alone and grid-connected systems. The proposed system will make sure that the proper and complete utilization of solar energy takes place whenever solar irradiance is available without drawing power from the grid.

Furthermore, the daytime peak load demand matches the maximum solar irradiance hours. Therefore, the total and effective utilization of PV power by letting off the household loads from the utility grid will also assist in reducing the stress on the grid. Thus, the proposed system also acts as a peak power plant and called home-to-grid (H2G) system [10].

15.2.2 Layout of the Proposed Control Strategy

The proposed grid-connected solar inverter system consists of a solar charge controller, a single-phase DC-to-AC inverter, and two proposed controller circuits as shown in Figure 15.3. There are two input power sources, i.e., utility grid and solar PV modules. The charging of batteries and the home load demands can be fulfilled by both input sources. However, charging from the solar PV modules is given the priority whenever solar irradiance is available. The solar MPPT charge controller

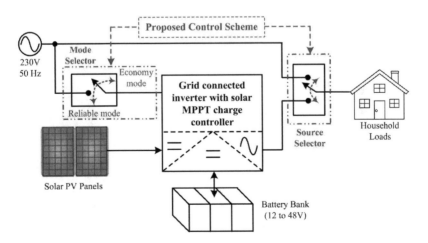

FIGURE 15.3 Proposed model of the solar inverter with two additional controllers.

optimizes the power output from the solar PV module, which is then stored in the battery bank. The battery bank of the electric vehicle can also be used as an energy storage element and thus can work as a vehicle-to-home (V2H) system [11,12].

On the other hand, the source of power supply to charge the battery is determined by the input controller called *mode selector*. This controller works on two modes: (1) *economy mode* and (2) *reliable mode*. In economy mode, the objective is to trap maximum solar power so that no or minimum energy gets unutilized from the PV modules. This scheme is more suitable for the urban localities where load shedding does not happen for a long duration. Whereas in rural areas, long hours of scheduled and unscheduled load shedding are very common, and therefore, reliable mode operation is more suitable. In reliable mode, batteries are charged either through solar PV module or grid depending upon their availability to fulfill the load demands during the long hours of load shedding. The block diagram of the mode selector is shown in Figure 15.4. The capacitor in the RC filter circuit 2 imitates the charging and discharging behavior of the battery voltage. The filter circuits are used to tune the charging and discharging rate of the capacitor to match the battery condition. Based on the capacitor voltage, the decision of selecting the economy and reliable mode is made with the help of level detectors and the SR flip-flop. The detail of its working is demonstrated in the next section.

The second controller is placed at the load side, and it is termed as *source selector*. This controller decides whether the household loads should meet their requirement through solar energy stored in the battery or utility grid. The controller diagram of the source selector is illustrated in Figure 15.5. This controller is set up with the help

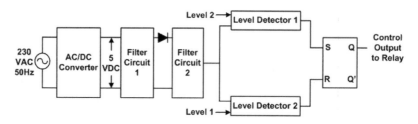

FIGURE 15.4 Block diagram of mode selector controller.

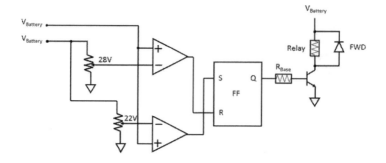

FIGURE 15.5 Circuit diagram of the source selector controller.

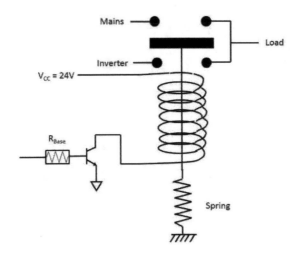

FIGURE 15.6 Schematic diagram of the read-type relay operation.

of two comparators with an adjustable predefined input reference voltage, SR flip-flop, MOSFET, and static relay to switch the load. The basic schematic of load and source connection using a read-type relay is demonstrated in Figure 15.6. So, once the output of the SR flip-flop is high (shown in Figure 15.5), the read-type relay will get actuated. The read-type relay is used for galvanic isolation of the load from either of the power supply, i.e., utility grid or solar inverter. The detailed working of the source selector is explained in the next section.

Using this controller, the consumer will be able to customize how they want to utilize the stored solar energy. Even if there is no load shedding, the consumer can choose to use free solar power instead of drawing power from the grid supply. This makes the entire system more reliable and economical and solves the problem of underutilization of solar energy.

15.3 DETAILED CONTROLLER DESIGN AND ITS WORKING

In this section, the working details of two analog controller circuits have been explained individually along with their flowcharts.

15.3.1 Mode Selector Controller

The circuit of a mode selector, as shown in Figure 15.7, has a rectifier, which is used as an indicator for grid supply availability and the load shedding. The output of LM7805 gives the logical output, i.e., logic 1 during the availability of grid supply and logic 0 for load shedding period. The RC filters, which have large time constants, adjust the charging and discharging time of capacitor C_2. The voltage V_{C2} is the reference voltage for two-level detectors (LD) LM324. The other input port of these LD has a predefined set and reset values that are related to the modes, i.e., reliability and economy.

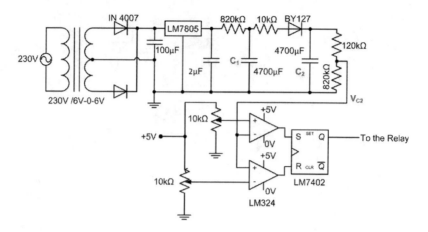

FIGURE 15.7 Detailed circuit diagram of mode selector controller circuit.

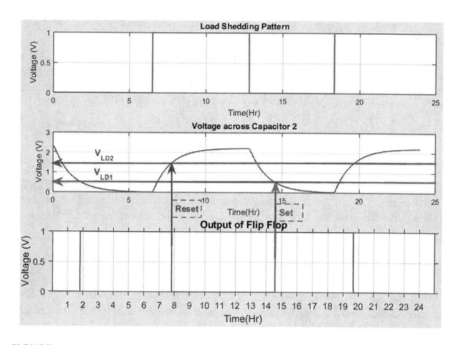

FIGURE 15.8 Control operation of mode selector controller.

During a prolonged power cut of grid supply, the voltage V_{C2} keeps on decreasing, and when V_{C2} reaches the predefined voltage V_{LD1} of LD1, the output of S-R flip-flop becomes set ($Q=1$) as shown in Figure 15.8. This means the reliable mode operation gets activated, and the battery is going to be charged from the grid whenever the mains supply gets restored. On the other hand, during the availability of grid supply for the long hours, V_{C2} keeps on rising, and when it crosses the V_{LD2} level, the output of LD2 will reset the flip-flop ($Q=0$). The relay will change the charging connection between the grid and battery to the solar PV module and battery, and now, the system

Design and Operation of a Low-Cost EV

TABLE 15.1
Operation of Reliable and Economy Modes Using Reference and LD Voltages

$V_{C2} \le V_{LD2}$ Output of LD2 Set	$V_{C2} > V_{LD1}$ Output of LD1 Reset	$Q_{FF}(t+1)$	Mode of Operation
0	0	No change	-
1	0	Charging by mains + PV	Reliable
0	1	Charging by PV only	Economy

will operate on the economy mode. Afterward, V_{C2} will remain at a high level, and SR flip-flop will remain at a low level. One thing to note is that both modes of operations can be tuned using V_{LD1} and V_{LD2} according to the needs of the customer and load shedding pattern of the area. The control operations of the mode selector can be summarized using Table 15.1. Moreover, the working of this controller is also described by the flowchart in Figure 15.9.

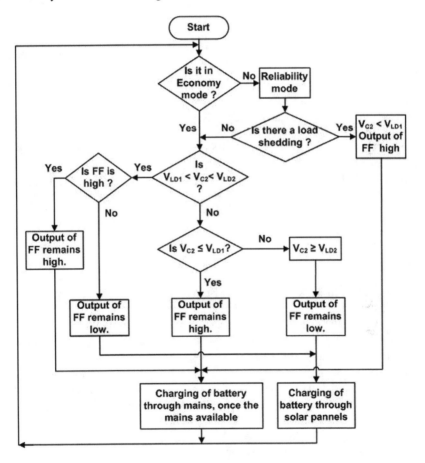

FIGURE 15.9 Flowchart of mode selector controller.

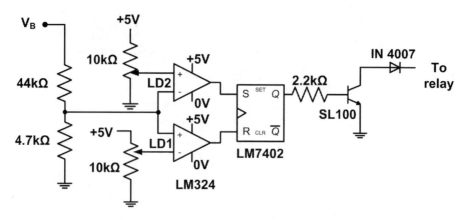

FIGURE 15.10 Detailed circuit diagram of source selector controller circuit.

15.3.2 SOURCE SELECTOR CONTROLLER

The second controller is attached to the load side to switch the load from inverter supply (battery) to mains supply and vice versa according to the battery voltage condition. The detailed circuit diagram source selector is shown in Figure 15.10.

The source selector controller also has two level detectors where the inputs are attenuated battery voltage and tunable voltage levels of set and reset as shown in Figure 15.10. The battery starts to get charged either from the PV module or from the mains depending upon the mode selector operation, i.e., reliable or economy modes. The working principle of this controller is similar to the mode selector. For instance, the output of LD1 is low and LD2 is high. Then, S will be 1 and R will be 0, and hence, the flip-flop output will get set, i.e., $Q=1$. The actuation of SR flip-flop will make load connected to the inverter supply. The output will remain under this state until battery voltage V_B becomes equal to V'_{LD1}. Then, $S=0$ and $R=1$, and flip-flop will become reset, i.e., $Q=0$. Once the output of flip-flop becomes reset, the load will get disconnected from the inverter and will get connected to the grid supply. Now, the load will meet the power requirement from the utility grid.

Under this scheme, the battery can exist in two conditions. In condition 1, the battery will get charged and utilized for home load power demand, simultaneously. This condition will exist during the daytime. The input of SR flip-flop will remain as $S=1$ and $R=0$. Thus, the output of SR flip-flop will remain high ($Q=1$), and the load will receive the power via inverter supply (battery).

Whereas in condition 2, the battery gets only discharged when solar power would not be there, and the mode selector controller could not have actuated yet to switch over to the mains. The battery voltage will start to decline (with Q being high) till $V_B = V'_{LD1}$. At this condition, $S=0$ and $R=1$, which will reset the flip-flop, and thus, the load will get connected to mains. The working of this scheme is also demonstrated through a flowchart as shown in Figure 15.11.

Design and Operation of a Low-Cost EV 347

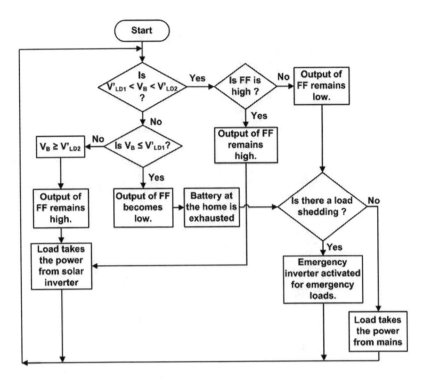

FIGURE 15.11 Flowchart of source selector controller.

15.4 HARDWARE IMPLEMENTATION OF THE DESIGNED CONTROLLERS

The hardware circuits of the proposed controllers are shown in Figure 15.12. The working of these circuits is first experimentally verified, and then, they are implemented for the hardware in the loop analysis, which is discussed in the next section. For brevity, the hardware analysis of the mode selector controller is only shown in this section. The performance of the controller is tested according to the load shedding condition, and its input-output results are recorded using a data logger as shown in Figure 15.13. The data logger has been used for the continuous monitoring of controller performance. The data has been stored at the sampling time of 20 seconds.

15.4.1 THE EXPERIMENTAL SETUP AND RESULTS

The experimental setup comprises of mode selector controller, data logger, 5 V DC power supply for saturation voltage (V_{CC}) of flip-flops, multimeter, relay, and 230 V AC power supply. The hardware testing of the mode selector controller along with the relay operation is shown in Figure 15.14. The relay operation using the controller has been accomplished with the proposed charging and discharging capacitor voltage-based control strategy.

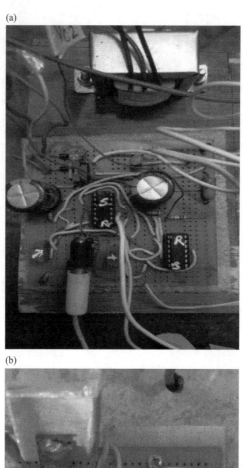

FIGURE 15.12 Hardware control circuit of (a) mode selector and (b) source selector.

Design and Operation of a Low-Cost EV

(a)

(b)

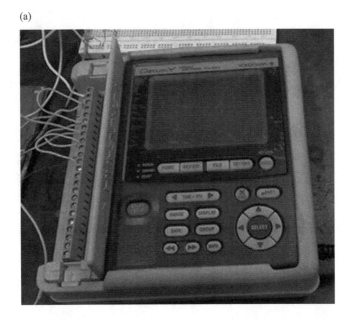
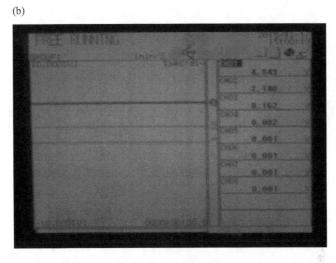

FIGURE 15.13 (a) Data logger and (b) screenshot of the running status.

The recorded performance of the mode selector is shown in Figure 15.15. The sky-blue line in Figure 15.15a represents the availability of grid power supply. In Figure 15.15b, the grid supply voltage level falls to zero, implying the occurrence of load shedding. The red line represents the voltage across the capacitor C_2 in the mode selector controller. The set and reset voltage levels of this experiment are $V_{LD1} = 0.5\,V$ and $V_{LD2} = 1.5\,V$, respectively.

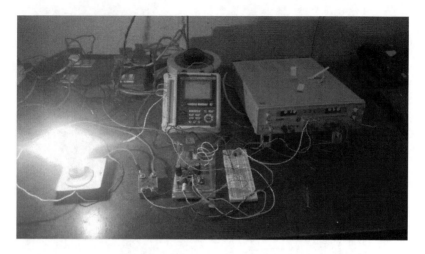

FIGURE 15.14 Hardware testing setup of mode selector controller in the lab.

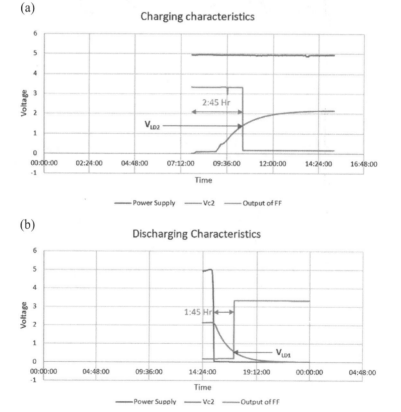

FIGURE 15.15 Hardware performance analysis of mode selector controller. during (a) charging phase, and (b) discharging phase.

Initially, the capacitor voltage V_{C2} is zero. But when the grid power supply comes in, the V_{C2} starts to rise, and after around 2.45 hours, it becomes equal to 1.5 V. Then, the output of S-R flip-flop becomes low ($Q=0$) as shown in Figure 15.15a. This implies, by the action of relay operation, that the battery charging will get disconnected from the mains and will now be charged only through the solar PV modules. On the other hand, during load shedding, C_2 is in the discharging condition. As V_{C2} decreases to 0.5 V, then the S-R output becomes high ($Q=1$) as shown in Figure 15.15b. The relay actuation will change the charging status, and now the battery will be charged through the mains, whenever available. Thus, the experimental results are found to be consistent with the proposed theoretical control strategy.

15.5 HARDWARE IN THE LOOP TESTING OF PROPOSED STRATEGY

In this section, the working of the proposed control scheme integrated with the hybrid solar grid-connected inverter system has been demonstrated. The system has been tested for the variable home loads for a period of 24 hours. These loads consist of five ceiling fans of 100 W, four tube lights of 40 W, and five CFL bulbs of 12 W.

The hardware in the loop setup of the proposed scheme is shown in Figure 15.16, where four rooftop solar PV panels of 250 W_P each (Figure 15.16a) with two lead-acid batteries of 12 V, 200 Ah for solar energy storage are used. Moreover, the comparison between the proposed hybrid inverter control system and commercially available solar inverter is also illustrated in this section. In the proposed hybrid inverter, a separate MPPT solar charge controller is used with a rating of 24 V and 40 A. The AC-to-DC and DC-to-DC converters in both the system are taken the same for better comparison. The complete experimental setup is shown in Figure 15.16b. The performance of a 1.44 kVA commercial solar grid-connected inverter and the proposed system is demonstrated in the next subsection.

15.5.1 Hardware in Loop Results

First, the experimental results of the commercial solar inverter are demonstrated on a typical day of the month of October from 6:00 to 18:00 hours, and the results are recorded in Table 15.2. The three important parameters, i.e., power demand, solar power, and grid supply are plotted from Table 15.2 in Figure 15.17 for better visualization. Moreover, the power demand requirement of home load met by different supply sources is also illustrated in Figure 15.18. It is evident from Figures 15.17, 15.18, and Table 15.2 that the solar PV modules come into usage only during the load shedding hours (10 am–1 pm), and the rest of the times, the mains supply is used to feed the power to the home loads. Similarly, when the mains power is restored, the battery is getting charged by the grid even when there is enough solar power available from the solar PV modules. Thus, this demonstrates that the commercially available solar inverter is ineffective in trapping and utilizing solar energy, as identified in the previous section.

On the other hand, the proposed system controls the effective charging of battery as well as the load allocation to ensure the complete utilization of solar energy.

(a)

(b)

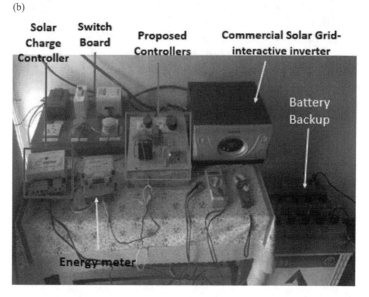

FIGURE 15.16 (a) 1 kWP rooftop PV panels and (b) hardware setup of the proposed system.

Under the economy mode during the sunshine hours, the charging of the battery is done by the solar PV modules, and the home load requirements are met by the stored energy using the DC-to-AC inverter. As the solar irradiance decreases to zero when the sun sets, the stored battery power feeds the power to the load. When the battery voltage V_B decreases to the value V'_{LDl} of the source selector, the flip-flop will get reset ($Q=0$), and the load will get connected to the mains supply. In this way, both the proper harnessing of solar irradiance (using mode selector) and the complete utilization of solar energy (using source selector) have been achieved. The results of the proposed scheme on a different day for a similar load demand pattern is recorded for 24 hours in Table 15.3. The power demand and supply from different sources under

Design and Operation of a Low-Cost EV

TABLE 15.2
Performance of Commercially Available Solar Grid-Connected Inverter

Time (Hr)	I (W/m²)	Temp (in) (°C)	PV Current (A)	Inverter I/ DC input (A)	PV Volt (V)	Battery Volt (V)	AC Load (A)	Status (Mains/PV)	Energy Meter Reading (kWh)	Solar Power Output (W)	Mains Power (W)	Power Demand (W)
6	2	25	0.02	7.3	26	28.8	1.62	M	0.4	0.52	400	372.6
7	32	25.2	0.05	6.9	31.1	28.6	1.77	M	0.45	1.555	450	407.1
8	290	28.6	0.03	6.1	32.5	28.5	1.56	M	0.4	0.975	400	358.8
9	522	34.8	0	5.1	32.8	28.5	0.95	M	0.3	0	300	218.5
9.8	866	40	0	5.4	32.6	28.5	0.9	M	0.2	0	200	207
10	866	40.2	9.79	1.25	27.6	26.8	0.9	PV	0	270.204	0	207
11	978	42.2	9.8	1.52	27.4	26.8	0.86	PV	0	268.52	0	197.8
12	1,024	42	5.86	5.23	29	28.6	0.96	PV	0	169.94	0	220.8
12.8	955	42	5.9	5.5	30.2	28.5	0.92	PV	0	178.18	0	211.6
13	948	42.2	0.03	1.15	31	26.4	0.81	M	0.2	0.93	200	186.3
14	920	44.1	0.04	1.03	30.9	26.4	0.83	M	0.2	1.236	200	190.9
15	796	43.2	0.03	0.9	31.3	26.4	1.22	M	0.3	0.939	300	280.6
16	670	41.9	0	0.47	31.6	26.5	1.52	M	0.4	0	400	349.6
17	160	34.8	0.03	0.72	31	26.5	1.98	M	0.5	0.93	500	455.4
18	5	31.7	0.04	0.71	25.3	26.5	1.03	M	0.2	1.012	200	236.9

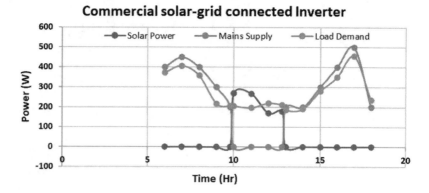

FIGURE 15.17 Performance analysis of commercially available solar PV inverter.

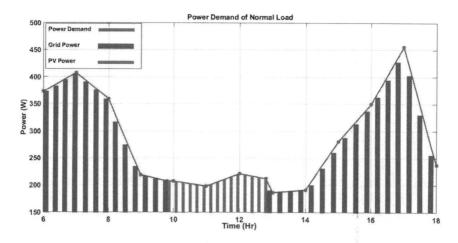

FIGURE 15.18 The commercial solar grid-connected inverter fulfils the home load power requirement on a typical day.

the proposed system is also illustrated in Figure 15.19. As can be seen in Table 15.3 and Figure 15.19, once enough solar irradiance is available to the system (at around 7 am), the battery gets charged through the solar PV module irrespective of the load shedding. The mains supply gets cut off from the home load at around 9 AM, and till 6:25 pm, solar energy feeds the power to the home load. It is important to note that solar irradiance becomes negligible after 5 pm, and even then, the stored solar energy continues to fulfill the power demand of the load.

The power demand pattern under the proposed scheme is depicted in Figure 15.20. It can be seen from Figure 15.20 that the PV power gets utilized as soon as the voltage level of the battery reaches the predefined set value, and when the battery voltage level comes down to the reset value, then the load gets connected to the grid supply. Thus, the proposed scheme ensures the complete utilization and harnessing of solar energy.

Design and Operation of a Low-Cost EV

TABLE 15.3
Performance of the Proposed Solar Grid-Connected Inverter

Time (Hr)	Ins (W/m²)	Temp (in) (°C)	PV Current (A)	Inverter I/DC input (A)	PV Voltage (V)	Battery Volt (V)	AC Load (A)	Status (Mains/PV)	Energy Meter Reading (kWh)	Solar Power Output (W)	Mains Power (W)	Power Demand (W)	battery backup (W)
0	0	23	0	0	0	24.3	2.84	M	0.7	0	700	653.2	0
1	0	23	0	0	0	24.3	2.84	M	0.7	0	700	653.2	0
2	0	23	0	0	0	24.3	2.84	M	0.7	0	700	653.2	0
3	0	23	0	0	0	24.3	2.84	M	0.7	0	700	653.2	0
4	0	23	0	0	0	24.3	2.84	M	0.7	0	700	653.2	0
5	0	24	0	0	0	24.3	2.84	M	0.7	0	700	653.2	0
6	1.8	26	0.03	0.5	24.8	24.3	2.0	M	0.6	0.7	600	460	12.15
7	130	25.3	0.85	0.72	24.5	24.4	2.45	M	0.5	20.8	500	563.5	17.57
8	353	28.9	1.73	0.71	24.8	24.8	2.11	M	0.6	42.9	600	485.3	17.6
9	670	33	12.42	9.55	26.2	26	0.98	PV	0.3	325.4	300	225.4	248.3
10	835	40.2	13.85	8.46	25.9	25.8	0.84	PV	0	358.7	0	193.2	218.2
11	1038	42.3	14.12	9.45	25.7	25.6	0.95	PV	0	362.8	0	218.5	241.9
12	1,064	42.8	13.76	8.7	25.7	25.7	0.86	PV	0	353.6	0	197.8	223.6
13	1,030	42.4	12.9	7.6	26	25.8	0.78	PV	0	335.4	0	179.4	196.08
14	928	42.1	12.5	7.9	25.9	25.8	0.83	PV	0	323.7	0	190.9	203.8
15	703	42.8	12.02	14.36	25.2	25.1	1.53	PV	0	302.9	0	351.9	360.4
16	535	41.5	7.85	18.46	24.1	24	2.0	PV	0	189.2	0	460	443.04
17	165	35.7	2.45	16.35	23.4	23.4	1.62	PV	0	57.3	0	372.6	382.6
18	3	31.9	0.06	36.67	21.6	21	3.46	PV	0	1.3	0	795.8	770
18.25	0	31	0	37.9	0	19.8	3.52	PV	0	0	0	809.6	750.4
19	0	29	0	0	0	23.2	3.35	M	0.78	0	780	770.5	0
20	0	28	0	0	0	23.2	3.28	M	0.8	0	800	754.4	0
21	0	26	0	0	0	23.2	3.34	M	0.8	0	800	768.2	0
22	0	24	0	0	0	23.2	2.75	M	0.7	0	700	632.5	0
23	0	23	0	0	0	23.2	2.1	M	0.75	0	750	713	0

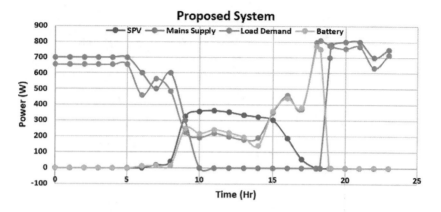

FIGURE 15.19 Performance of the proposed system for home load.

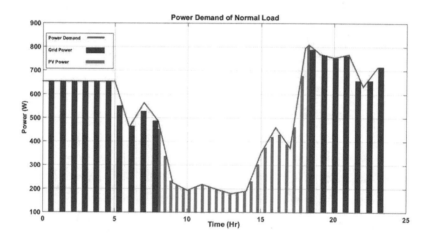

FIGURE 15.20 The proposed solar grid-connected inverter control scheme fulfils the home load power requirement on a different day.

15.6 CONCLUSION

The proposed work develops a solar grid-connected inverter using a conventional single-phase convention backup inverter along with the solar charge controller and two proposed controllers. The control strategy can be used in economical or reliable modes, depending upon the grid supply availability. The proposed technique enables the utilization of solar power even when there is no load shedding. The system also indirectly works as a home-to-grid mode and thus reduces the electricity bill by letting off the home load from the grid during the hours of high tariff rate (i.e., peak hours). The proposed control strategy illustrates that the modified grid-connected solar inverter is harnessing the solar energy more efficiently with better utilization of stored solar energy. Moreover, the control scheme is built with low cost and does not require any expensive equipment. Thus, the design can be implemented in existing home-based inverters; therefore, the customer does not have to invest a lot for the complete system setup. The experimental results of the proposed strategy and their comparison with the commercially available solar inverters illustrate that the developed scheme ensures the complete utilization of solar energy in contrast to the commercially available system. Furthermore, the proposed system has surpassed months of testing and has proved to be effective in its practical applicability. The future extension of this work could involve more intelligent control features with a greater number of input scenarios by using microcontrollers or DSPs.

REFERENCES

1. A. Y. Saber and G. Venayagamoorthy, "Plug-in vehicles and renewable energy sources for cost and emission reductions", *IEEE Trans. Ind. Electron.*, vol. 58, no. 4, pp. 1229–1238, 2011.
2. F. Ahmad and M. S. Alam, "Feasibility study, design and implementation of smart poly-generation microgrid at AMU", *Sustain. Cities Soc.*, vol. 35, pp. 309–322, 2017.
3. F. Ahmad and M. S. Alam, "Economic and ecological aspects for microgrids deployment in India", *Sustain. Cities Soc.*, vol. 37, pp. 407–419, 2018.
4. M. Nehrir, C. Wang, K. Strunz, H. Aki, R. Ramakumar, J. Bing, Z. Miao, and Z. Salameh, "A review of hybrid renewable/alternative energy systems for electric power generation: configurations, control, and applications", *IEEE Trans. Sustain. Ener.*, vol. 2, no. 4, pp. 392–403, 2011.
5. J. T. Bialasiewicz, "Renewable energy systems with photovoltaic power generators: operation and modeling", *IEEE Trans. Ind. Electron.*, vol. 55, no. 7, pp. 2752–2758, 2008.
6. Access to electricity (% of population) – India, 2017. https://data.worldbank.org/indicator/EG.ELC.ACCS.ZS?locations=IN (Accessed: April 27, 2020).
7. https://pib.gov.in/PressReleaseIframePage.aspx?PRID=1514215 (Accessed: April 27, 2020).
8. B. H. Khan, *"Non-Conventional Energy Sources"*, 2nd Edition, Tata McGraw Hill Education, New Delhi, 2009.
9. A. Jhunjhunwala, A. Lolla and P. Kaur, "Solar-DC microgrid for Indian homes: a transforming power scenario", *IEEE Electrific Magaz*, vol. 4, no. 2, pp. 10–19, 2016.
10. S. M. Amrr, M. S. Alam, M. S. J. Asghar and F. Ahmad, "Low cost residential microgrid system-based home to grid (H2G) back up power management", *Sustain Cities Soc*, vol. 36, pp. 204–214, 2018.
11. M. R. Khalid, M. S. Alam, A. Sarwar, and M. S. J. Asghar, "A comprehensive review on electric vehicles charging infrastructures and their impacts on power-quality of the utility grid", *eTransportation*, vol. 1, p. 100006, 2019.
12. M. S. Shemami, S. M. Amrr, M. S. Alam, M. S. J. Asghar, "Reliable and economy modes of operation for electric vehicle-to-home (V2H) system", *5th IEEE International Conference on Electrical, Electronics and Computer Engineering (UPCON)*, Gorakhpur, India, pp. 1–6, Nov. 2018.

Index

Note: **Bold** page numbers refer to tables; *italic* page numbers refer to figures.

AC Level 1 **36, 38, 44**
AC Level 2 **36, 38, 44,** 120, 240
AC Level 3 **36, 38, 44**
active power 41, 42, 51–53, **54,** 117, 124, 271
active shielding 285–287
activity diagram 254, 255, *255*
agent based modeling (ABM) 305
aggregated electric vehicle 34, 35, 42–45, *45,* 50, 51, 55, 59, 242
ancillary service 34, 35, 43, 45, 47, 53, 58, 59
artificial intelligence (AI) 16, 23, 25, 27, 61, 250–252, 254, 258, 260–264
artificial neural networks (ANN) 25, 151, **152,** 251, 258, 259, 263
autoregressive moving average (ARMA) 258, 262, 263

backup power supply 339, 340
battery capacity 1, 2, 43, **44, 152,** 218, *219, 220,* 222, 268, **308,** 309
battery cell
    assembling 185–187
    mining for the material 209
battery charger 34, 40, *41, 42,* 51, 221
battery degradation 34, 57
battery disposal *see* battery recycling techniques
battery lifecycle 321, *324*
battery management system 143–177, 185, 187, **224,** 331
battery packaging 185, 187
battery pack assembly 188
battery purposing *see* energy storage systems (ESS)
battery recycling techniques 192–195
battery types capacities **318**
bidding
    framework 49, 50
    process 250, 260, 263
bidirectional charger 40, 42, 51, 242
bidirectional power flow 35, 58
blockchain 23, 25, 26, 60

carbon dioxide (CO_2) 183, 185, 252, 321, **322,** 327
carbon monoxide (CO) **322**
challenge 7–29, 33–61, 73–92, 121, 145, 146, 215, 216, 235, 237, 242, 250, 268, 279, 285, 295, 325
charging interface **38**

charging mode 37, **39,** 46, 55, 226
charging standard 4, **40,** 223, **224,** 225, 229, 230
charging station 1, 4, 5, 40, 51, 119–139, 267–278, 295–311
cloud-based 13, 252, 263
cobalt (Co) 185, **190,** 194, 195, 221, 321, **323,** 326–331
cobalt oxide 185, **190,** 194, 221, 321, 329
communication delay **24,** 45, 47
communication infrastructure 57
communication standards 38, **40**
comparative algorithms 196–201
conductive wires 275, 281–282
control function 34, *47, 48, 50–53,* 52, **54,** 61, 228
control loop 40, 42, 48, 50, 52, 53, 61
copper (Cu) 185–187, 271, 282–285, 314, 320, 326–328, 330

data flow diagram (DFD) 254, 255, *256*
day-ahead-market 49, 249–264
DC/AC inverter 35
DC charging 37, 60, 223, **224,** 226, 228–229, 231, 242, 302, 303
DC/DC converter 35, *41,* 42
DC Level 1 **36, 38, 44**
DC Level 2 **36, 38, 44**
DC Level 3 **36, 38, 44**
deep learning 251
dijkstra algorithm 309
direct recycling procedure (DRP) 328–329
distributed energy resource 11, 37, 98, **124,** 242
distributed optimization 49, 52
distributed resource 10, 11, 37, 51
distribution grid 34, 45, 51, 56, 80
distribution system 1–5, 48, 51, 57, 98, 100, 121, 340
dynamic programming 195, 250, 251

electric vehicle (EV)
    batteries manufacture 184–185, 252
    fleet 33–61
electromagnetic field shielding 283–288
end-of-life (EOL) 184, 322, 324
energy management system 49, 209
energy service 49, 50
energy storage 12, 34, 60, 98, 119–139, 143, 144, 176–177, 181–209, 221, 242, 269, 299, 319, 337, 342, 351

359

energy storage systems (ESS) 122, 123, 129–131, 134, 135, 139, 143, 176–177, 181–209, 319
entity-relation (ER) diagram 254, 256–258, 342
environmental protection agency (EPA) 309, 320
expert system 260, 263

finite element analysis 272
flammable chemicals 322
flux concentrator 281–283
forecasting 24, 196, 198, 205, 251, 258, 262
frequency control 46
frequency regulation 43, 46, 47–49, 53–54, 58, 61
frequency response 43, 45–47, 54
frequency support 46, 52, 242

generation following 49–50, **54**, 55, 61
global warming potential (GWP) 184, 187, **188,** 325
graphite (C) 185, 190, **323,** 327
graph theory 305
greenhouse gasses (GHGs) 2, 73, 75, 83, 184, 198, 200, 214, 252, 279, 295, 319–321, 323, 325
grid service 42–53, 59, 61, 242

hosting capacity 3, 34, 48–52, **54**
hydrocarbons **322**
hydrofluoric acid 321, 322
hydrogen 81, 144, 316, **322**
hydrogen fluoride (HF) **322**
hydrometallurgical procedure (HP) 327–328, 330

integrated charging system 35

lithium cobalt (LiCoO$_2$) **190, 323**
lithium cobalt oxide (LCO) **190,** 329
lithium hexafluoro phosphate (LiPF$_6$) 186, 321, **323**
lithium-ion battery (li-battery) 145, 321–326, 328, 329–331
lithium-ion battery recycling initiative (LiBRi) 331
lithium-ion manganese oxide 321
lithium iron phosphate (LFP) 185, **190,** 221, 321
lithium manganate (LiMn$_2$O$_4$) **323**
lithium manganese oxide (LMO) 185, 186, **190,** 221, 321
lithium nickel manganese 221, 321
lithium phosphate (LiFePO$_4$) **190, 323**
load following 50, **54,** 55
load frequency control 46
load levelling 54, 55, 58, 197
load shedding 338–340, 342, 343, 345, 347, 349, 351, 354, 356

machine learning 124, 125, 249–264
mean absolute error 258
mechanical procedure (MP) 327, 331
microgrid 1–7, 97–116, 249–264, 267–278, 335–357

nickel (Ni) 144, **151, 152,** 185, 190, 194, 195, 221, 314, 316, 320, 321, 327–330
nickel manganese cobalt (NMC) 185, 186, **190,** 221, 329

passive shielding 283–285
placement problem 299–301, 305, 306
power capacity 49, 50, 191
power electronic converter 35, 45, 58, 336
power electronic interfaces 34, 35, 41, 51
power factor 37, 40, 42
power factor correction (PFC) 40, 42
power flow 35, 40, 41, 51, 58
power level 35, 43, 229
power peak sheaving 191–192
power quality 37, 51, **54,** 121, 124, **236,** 242
power smoothing 47–49, **54,** 55
pyro-metallurgical procedure (PM) 327, 329

reactive power 36, 41, 42, 51–53, **54,** 56, 57, 61, 242, 271, 280
reactive power support 51, 53, **54,** 56, 242
reactive shielding 287–288
renewable energy resource 14, 34, 45, 49, 59, 60, 98, 319
residential microgrid 10–12, 17, 164, 206

smart grids 7–29, 61, 80, 84, 91, 189, 192, 195, 205, 206, 209, 250, 251, 262, 263, 319
software-defined networking (SDN) 23, 25, 26
solar photovoltaic (SPV) system 231, 268–269, 336–342, 344, 351, 352, 354
solution 4, 7–29, 56, 57, 59, 74, 80, 99, 105, 110–112, 114, 124, 132–134, 139, 166, 168, 169, 171, 176, 182, 188–190, 193–195, 198, 206, 208, 209, 230, 231, 235, 237, 250, 254, 301, 327, 340
spatial allocation
spinning reserve 50–51, **54,** 55–56
supervisory charging scheme 339
support vector machine (SVM) 260 , 263

time series 103, 117, 258, 259

unidirectional charger 40–42, 51, 52
unified modeling language (UML) 254
urban mobility 234, 254

Index

vehicle interface **36**
vehicle owner 34, 43, 80
vehicle to grid (V2G) 34, 36–38, 40,
    45, 52–61, 98–102, 105, 106,
    108, 116, 206, 212, **224,**
    242, 251
vehicle to home (V2H) 34, 342

vehicle type **44**
V2G *see* vehicle to grid (V2G)
voltage profile 50–52, 56
voltage support 51–53, 56

wireless charging 60, 223, **224,** 229–230,
    267–278